Urban Growth and Environmental Issues in India

Alpana Kateja · Rashmi Jain
Editors

# Urban Growth and Environmental Issues in India

Springer

*Editors*
Alpana Kateja
Department of Economics
University of Rajasthan
Jaipur, Rajasthan, India

Rashmi Jain
Department of Sociology
University of Rajasthan
Jaipur, Rajasthan, India

ISBN 978-981-16-4272-2 ISBN 978-981-16-4273-9 (eBook)
https://doi.org/10.1007/978-981-16-4273-9



This Springer imprint is published by the registered company Springer Nature Singapore Pte Ltd.
The registered company address is: 152 Beach Road, #21-01/04 Gateway East, Singapore 189721, Singapore

# Preface

*The right to the city is far more than the individual liberty to access urban resources: it is a right to change ourselves by changing the city. It is, moreover, a common rather than an individual right since this transformation inevitably depends upon the exercise of a collective power to reshape the processes of urbanization. The freedom to make and remake our cities and ourselves is, I want to argue, one of the most precious yet most neglected of our human rights.*

—David Harvey

Man's desire to develop an ideal living space, whether politically or architecturally, has led him to evolve cities as the bastion of power and culture. Plato in *Republic*, says that any city however small, is in fact divided into two, one the city of the poor and the other of the rich. These are at war with one another. Urban growth and related aspects have drawn the attention of scholars since times immemorial. The cities which are envisaged as engines of growth have also perpetuated newer form of exploitation and inequality.

Urban growth is accompanied with concurrent increase in aspirations of the people. The phenomenal urban growth has opened vistas of opportunity for the disadvantaged section of the society. A large number of people move to urban areas to fulfil their aspirations and dreams of enhancing the quality of life. The present work is divided into three sections dealing with myriad issues of urban growth and governance; environmental challenges and waste management processes and energy needs.

The editors would like to express their sincere thanks to the contributors in this volume. Their scholarly inputs in the form of their contributions will surely enrich the understanding of urban issues. We are also grateful to Springer Nature in general and Satvinder Kaur and the reviewers in particular for their patience and cooperation extended during the period of publication of this academic enterprise.

Jaipur, India

Alpana Kateja
Rashmi Jain

# Contents

# Editors and Contributors

## About the Editors

**Alpana Kateja** is Professor of economics at the University of Rajasthan, Jaipur, and also former Principal, University Maharani College, Jaipur. She is a gold medalist in economics, and she has formerly been associated with Banaras Hindu University, Varanasi, and JNV University, Jodhpur. Her published works include *Health Outcomes in India, Water and Health, Water Resource Management*, and *Lifestyle and Health.*

**Rashmi Jain** is Head and Associate Professor in Department of Sociology at the University of Rajasthan, Jaipur. She has interest in the areas of development communication, gender studies, sociology of law, and leisure studies. She is Convener of Working Group on Law and Migration of International Sociological Association. Her published work includes *Women, Media and Violence, Women Empowerment: The Global Scenario*, and *Health and Domestic Violence*. Her latest book (ed.) *Global Leisure and the Struggle for a Better World* has been published by Palgrave Macmillan.

## Contributors

**Maniklal Adhikary** Department of Economics, Burdwan University, Burdwan, West Bengal, India

**Narma Bhattacharya** Dr. B. L. Kapur Memorial Hospital, New Delhi, India

**P. D. Charan** Department of Environmental Science, MGS University, Bikaner, Rajasthan, India

**Sunetra Ghatak** National Institute of Public Finance and Policy, New Delhi, India

**Chandrasekhar Hajra** Nistarini College, Purulia, West Bengal, India

**Rashmi Jain** Department of Sociology, University of Rajasthan, Jaipur, Rajasthan, India

**Amithy Jasrotia** Department of Sociology, University of Rajasthan, Jaipur, Rajasthan, India

**Seema Joshi** Department of Commerce, Kirori Mal College, University of Delhi, Delhi, India

**Yogesh C. Joshi** Faculty of Management, Sardar Patel University, Vallabh Vidyanagar, Anand, Gujarat, India

**Alka Kataria** Department of Environmental Science, IIS (Deemed to be University), Rajasthan, Jaipur, India

**Alpana Kateja** Department of Economics, University of Rajasthan, Jaipur, Rajasthan, India

**Garima Kaushik** Department of Environmental Sciences, School of Earth Sciences, Central University of Rajasthan, Bandar Sindri, Rajasthan, India

**J. P. Kesari** Department of Mechanical Engineering, Delhi Technological University, Delhi, India

**T. I. Khan** Indira Gandhi Centre for HEEPS, University of Rajasthan, Jaipur, Rajasthan, India

**Sonia Maheshwari** Aayojan School of Architecture, Jaipur, Rajasthan, India

**Chandrakanta Mall** Department of Chemistry, Banaras Hindu University, Varanasi, Uttar Pradesh, India

**B. K. Nagla** Department of Sociology, M. D. University, Rohtak, Haryana, India

**Sandeep Narula** IIHMR University, Jaipur, Rajasthan, India

**Ravinder Palakurthy** Mahatma Gandhi Institute for Combating Climate Change (Govt. of NCT of Delhi), Delhi, India

**M. K. Pandit** Department of Geology, University of Rajasthan, Jaipur, Rajasthan, India

**P. Parthasarathy** E-Parisaraa Private Limited, Bengaluru, Karnataka, India

**Geetanjali Pawar** WTERT - India, Shree Samarth Building, J. A. Raul Marg, Prabhadevi, Mumbai, India

**A. D. Sawant** University of Rajasthan, Mumbai University, Mumbai, India

**K. C. Sharma** Department of Environmental Science, Maharishi Dayanand University, Ajmer, Rajasthan, India

**Kritika Sharma** Department of Life Sciences, Hemchandracharya North Gujarat University, Patan, Gujarat, India

**Prem Prakash Solanki** Department of Chemistry, Banaras Hindu University, Varanasi, Uttar Pradesh, India

**Prama Esther Soloman** Indira Gandhi Centre for Human Ecology, Environmental and Population Studies, University of Rajasthan, Jaipur, Rajasthan, India

**Bhawana Vasudeva** Department of Architecture, The Maharaja Sayajirao University of Baroda, Vadodara, Gujarat, India

**Sujata Warrier** Department of Sociology, N. G. Acharya & D. K. Marathe College, University of Mumbai, Mumbai, India

**Manish Yadav** Department of Sociology, University of Rajasthan, Jaipur, Rajasthan, India

# Abbreviations

| | |
|---|---|
| AAP | Ambient air pollution |
| ABC | Amala Bharatham Campaign |
| ACCCRN | Asian Cities Climate Change Resilience Network |
| AEGR | Annual exponential growth rate |
| AMRUT | Atal Mission for Rejuvenation and Urban Transformation |
| API | Active Pharmaceutical Ingredients |
| BCM | Billion cubic meters |
| BOD | Biochemical oxygen demand |
| BOOT | Built Own Operate and Transfer |
| BU | Billion unit |
| CA | Cellular automata |
| CAGR | Compound annual growth rate |
| CAS | Chinese Academy of Sciences |
| CBD | Central Business District |
| CBHI | Central Bureau of Health Intelligence |
| CBM | Coal-bed methane |
| CEDI | Composite Environmental Degradation Index |
| CFC | Chlorofluorocarbon |
| CGF | Critical Gap Funds |
| CHC | Community Health Centre |
| CMR | Child mortality rate |
| CNG | Compressed Natural Gas |
| COD | Chemical oxygen demand |
| CPCB | Central Pollution Control Board |
| CSIR | Council of Scientific and Industrial Research of Government of India |
| CSO | Central Statistics Office |
| DACs | District advisory committees |
| DALYs | Disability-adjusted life years |
| DAY-NULM | Deendayal Antyodaya Yojana-National Urban Livelihoods Mission |
| dB | Decibel |
| DDT | Dichloro-Diphenyl-Trichloroethane |
| DNPWC | Department of National Parks and Wildlife Conservation |

| | |
|---|---|
| DPCC | Delhi Pollution Control Committee |
| DTU | Delhi Technological University |
| EC | Electrical conductivity |
| EEC | Earth Engineering Centre |
| EIA | Environmental impact assessment |
| EKC | Environmental Kuznets Curve |
| ETL | Epidemiological transition level |
| FAO | Food and Agriculture Organization of the United Nations |
| FDA | Food and Drug Administration |
| FRBC | Federal Reserve Bank of Chicago |
| FRBM | Federal Reserve Bank of Minneapolis |
| GCIM | Global Commission on International Migration |
| GDP | Gross domestic product |
| GDPC | Gross domestic product per capita |
| GEF | Global Environment Facility |
| GHG | Greenhouse gases |
| GIS | Geographical Information System |
| GM | Genetically modified |
| GNP | Gross National Product |
| GPs | Gram Panchayats |
| GRIHA | Green Rating for Integrated Habitat Assessment |
| GW | Giga watts |
| GWC | Global WTERT Council |
| HDB | Housing Development Board |
| HDI | Human Development Index |
| HRIDAY | Heritage City Development and Augmentation Yojana |
| IARC | International Agency for Research on Cancer |
| ICCA | Indian Council for Citizen Action |
| ICCBD | Intergovernmental-—Governmental Committee on the Convention on Biological Diversity |
| ICIMOD | The International Centre for Integrated Mountain Development |
| ICT | Information and Communication Technology |
| IEA | International Energy Agency |
| IGSSS | Indo-Global Social Service Society |
| IHME | Institute for Health Metrics and Evaluation |
| IIHS | Indian Institute for Human Settlements |
| IMR | Infant mortality rate |
| IOM | International Organization for Migration |
| IPCC | Intergovernmental Panel on Climate Change |
| IPR | Intellectual Property Rights |
| IPT | Intermediate Public Transit |
| ISWM | Integrated Solid Waste Management |
| IUCN | International Union for Conservation on Nature |
| JNNURM | Jawaharlal Nehru National Urban Renewal Mission |
| KCAP | Kanchenjunga Conservation Area Project |

| | |
|---|---|
| kJ/kg | Kilojoules per kilogram |
| LLE | Loss of life expectancy |
| LMV | Light motor vehicle |
| LULU | Locally unwanted land use |
| mb/d | Millions of barrels per day |
| MGI | McKinsey Global Institute |
| MINAS | Minimum National Acceptable Standards |
| MMR | Maternal mortality rate |
| MNRE | Ministry of New and Renewable Energy |
| MNREGA | Mahatma Gandhi National Rural Employment Guarantee Act |
| MoEF&CC | Ministry of Environment, Forest and Climate Change |
| MOHUA | Ministry of Housing and Urban Affairs |
| MPCE | Monthly per capita expenditure |
| MSW | Municipal solid waste |
| mtce | Million tons of coal equivalent |
| MUD | Ministry of Urban Development |
| MW | Megawatt |
| MWe | Megawatt electrical |
| NAAQs | National Ambient Air Quality Standards |
| NCDs | Non-communicable diseases |
| NCDWM | North Carolina Division of Waste Management |
| NCR | National Capital Region |
| NCT | National Capital Territory |
| NCU | National Commission on Urbanization |
| NDMC | North Delhi Municipal Corporation |
| NEERI | National Environmental Engineering Research Institute |
| NGO | Non-Governmental Organizations |
| NGT | National Green Tribunal |
| NHM | National Health Mission |
| NLCP | National Lake Conservation Plan |
| NMT | Non-motorized transport |
| NRuM | National Rurban Mission |
| NSSO | National Sample Survey Office |
| NUHHP | National Urban Housing and Habitat Policy |
| NUHM | National Urban Health Mission |
| NULM | National Urban Livelihoods Mission |
| NWCP | National Wetlands Conservation Programme |
| OECD | Organisation for Economic Cooperation and Development |
| OLS | Ordinary least squares |
| ORF | Observer Research Foundation |
| PAHs | Polycyclic Aromatic Hydrocarbons |
| PHC | Primary Health Centre |
| PHED | Public Health Engineering Department |
| PM | Particulate Matter |
| PMAY-U | Pradhan Mantri Awas Yojana (Urban) |

| | |
|---|---|
| PPE | Personal Protective Equipment |
| PPM | Parts per million |
| PPP | Public–private partnership |
| PRB | Population Reference Bureau |
| PT | Public transport |
| PV | Photovoltaic |
| RAY | Rajiv Awas Yojana |
| RBI | Reserve Bank of India |
| RDF | Refuse derived fuels |
| RES | Renewable energy sources |
| RO | Reverse osmosis |
| SAARC | South Asian Association for Regional Cooperation |
| SAGY | Sansad Adarsh Gram Yojana |
| SBM | Swachh Bharat Mission |
| SCM | Smart Cities Mission |
| SDGs | Sustainable development goals |
| SDP | State domestic product |
| SEV | State equalized value |
| SHGs | Self-Help Groups |
| SJSRY | Swarna Jayanti Shahri Rozgar Yojana |
| SMHI | Swedish Meteorological and Hydrological Institute |
| SNDP | State Net Domestic Product |
| SNDPPC | State Net Domestic Product Per Capita |
| SPCBs | State Pollution Control Boards |
| SPM | Suspended particles |
| SPMRM | Shyama Prasad Mukherji Rurban Mission |
| SPV | Special purpose vehicle |
| SRITNE | Srini Raju Centre for IT and the Networked Economy |
| SWM | Solid waste management |
| TDS | Total dissolved solids |
| TFC | Total final consumption |
| TPD | Tonnes per day |
| TPES | Total Primary Energy Supply |
| TPY | Tonnes per year |
| TSS | Total suspended solids |
| UA | Urban agglomeration |
| UCCR | Urban climate change resilience |
| ULB | Urban local bodies |
| ULG | Urban local governance |
| UN | United Nations |
| UNCCD | United Nations Convention to Combat Desertification |
| UNFCCC | United Nations Framework Convention on Climate Change |
| UNFPA | United Nations Fund for Population Activities |
| UNICEF | United Nations Children's Fund |
| UPHC | Urban Primary Health Centre |

| | |
|---|---|
| URIF | Urban Reform Incentive Fund |
| USEPA | United State Environmental Protection Agency |
| USGS | United States Geological Survey |
| UT | Union territories |
| WASH | Water, sanitation and hygiene |
| WCMC | UN Environment World Conservation Monitoring Centre |
| WEEE | Waste electrical and electronic equipment |
| WHO | World Health Organization |
| WtE | Waste to energy |
| WTERT | Waste to Energy and Research Technology |
| WUP | World Urbanization Prospects |
| WWF | World Wildlife Fund |
| WWTP | Waste Water Treatment Plant |
| YLDs | Years lived with disability |
| YLLs | Years of life lost |

# List of Figures

# List of Tables

# Chapter 1
# Introduction

**Alpana Kateja and Rashmi Jain**

*Towered cities please us then, and the busy hum of men...*
*(L'Allegro, John Milton)*

**Abstract** Urban centers are vital for a vibrant, dynamic economy as they are engines of growth—attracting talent and investment, reservoirs of skill and entrepreneurial energies and provide the much-needed fuel for sustainable development in any country. Urban growth is not merely urban-focused—it uplifts the condition of rural areas by providing avenues of employment, cost-effective access to basic services, thus promoting inclusive growth. Cities are hubs for socio-economic-cultural transformation leading to human development by way of increased commercial activity, productivity and scientific growth. They can be the source of solutions to the challenges being faced by the world today. If urbanization is well planned and well managed, it can become a powerful tool for the sustainable development of developing as well as developed countries. The way human settlements are planned, developed, financed, governed and managed has a direct bearing on sustainability and resilience well beyond urban boundaries. Rapid urbanization in tandem with poor planning has led to problems of growth in slum settlements, creaking infrastructure, poor sanitation and sewage maintenance, sub-standard living conditions and accelerated environment degradation.

Cities have, through the ages, fascinated poets and writers alike who have written odes to the industriousness of city dwellers and richly described the pathos of city dwelling. It is no wonder that *A Tale of Two Cities* is considered one of the bestselling novel of all times, in which Dickens poignantly writes about the hopes and tribulations faced by city folk trying to eke out a decent living. Other books like *City of Joy*, *Maximum City* and innumerable movies have attempted to capture the essence of the enigma that is a city.

A. Kateja (✉)
Department of Economics, University of Rajasthan, Jaipur, Rajasthan, India

R. Jain
Department of Sociology, University of Rajasthan, Jaipur, Rajasthan, India



A. Kateja and R. Jain (eds.), *Urban Growth and Environmental Issues in India*,
https://doi.org/10.1007/978-981-16-4273-9_1

How do cities come to be? At its very core, urban centres culminate as a result of complementary economic, social and spatial impetuses. As these tendencies gain critical mass, urban growth accelerates creating a virtuous cycle that sustains and elevates these agglomerations. Urban centres are vital for a vibrant, dynamic economy as they are engines of growth—attracting talent and investments, reservoirs of skills and entrepreneurial energies and provide the much needed fuel for sustainable development in any country. Urban growth is not merely urban-focused—it uplifts the conditions of rural areas by providing avenues of employment, cost-effective access to basic services, thus promoting inclusive growth.

## Urbanization in India and Globally: A Snapshot

Tracing the urbanization trajectory in India, we find that by the 1950s, the urban population constituted 17% of the total population growing to a figure of 26% in 1990, 34% in 2018 and expected to touch 40% in 2030. By 2050, more people will live in India's cities than in villages. According to United Nations estimates, urban dwellers in India numbered 461 million in 2018 concentrated in the megalopolises of Delhi, Bombay, Chennai, Bangalore and Kolkata. Delhi ranks as the second most populous city in the world (29 million) after Tokyo (37 million). India presently has five megacities, defined as agglomerations with more than 10 million inhabitants. This trend is expected to accelerate with the addition of two more megacities, Hyderabad and Ahmedabad, by 2030. By 2050, India is expected to add 416 million urban dwellers taking the figure to over 800 million living in its cities (United Nations, 2019).

India's urbanization trends are similar to global patterns of urbanization—in 1950, 70% of the world population was found in rural regions, with the rural percentage in 2018 at 45%. Globally, 4.2 billion people were settled in urban centres with the rural population at 3.4 billion. By 2050, United Nations estimate indicates that 68% of the world's population would be residing in urban areas (United Nations, 2019). It further reports that 90% of the urban growth will emanate from the developing world in the coming decades. Thirty-five percentage of the urban growth by 2050 in the world is expected to be driven by a few countries—India, China and Nigeria leading the pack in that regard with addition of 416 million, 255 million and 189 million. This is in sharp contrast to the projected decreases in some cities in the developed countries like Tokyo, Busan, etc.

Unsurprisingly, the major contribution to world GDP will also be from the developing world. McKinsey report points to the fact that 600 cities will generate more than 65% of the world GDP by 2025, of which 440 cities from the emerging world including India, China Brazil will contribute to 47% of the expected GDP growth between 2010 and 2025 (Dobbs et al., 2011). Specifically for India, it is projected that urban centres could generate 70% of net new jobs created by 2030, produce more than 70% of Indian GDP and drive a four times increase in per capita incomes across the nation (MGI, 2010).

## Indian Urban Growth: A Critical Look

What has been India's performance with respect to urbanization? Has the urban growth contributed to upliftment of a significant section of the population from poverty? What has been the record with respect to improvements in productivity and contribution of urban centres to GDP? Has this been achieved in a sustainable manner? Any review of the successes and failures of urban growth must consider few perspectives.

All the highly developed countries are highly urbanized (more than 70%) generating more than half of their GDP in urban-based economic activities. In contrast, majority of population in the poorest countries resides in rural areas such as Sri Lanka and Uganda with rural population of 82 and 85%, respectively. Global trends corroborate the fact that all countries made good strides in achieving greater prosperity with the increase in urban population. Rapidly growing countries like Kenya, Vietnam and Indonesia have registered an urban population growth rate of at least 3% per year. Historically, very few countries have reached income levels of US $10,000 per capita before reaching about 60% urbanization (Spence et al., 2009). It has been forecasted that there will be a trend of greater concentration of economic activity in urban areas. Urbanization will thus provide an opportunity to developing countries to transform their economies and join the ranks of developed countries.

Empirical evidence from both the developed and the developing world suggests that there is a disproportionate contribution of urban centers to a nation's GDP and growth. Lewisian theory (Lewis, 1954) of urbanization and productivity indicates that urban agglomerations are expected to act as a magnet, creating a 'pull' effect on unutilized/underutilized rural workforce enamored by the higher wage potential in cities as well as responding to the 'push' effect from lower wages in agriculture. This phenomenon helps boost the industrial activities that are clustered in cities which are found to make a disproportionate contribution to productivity. The moving of 'excess' labour from the agrarian sector is also found to have positive connotations for the country's economic yield. Recent studies have also indicated that network effects of people concentrated within a geographical area leads to cascading positive externalities that can potentially drive greater productivity outcomes. It is well-established that as humans concentrate spatially, there is virtuous cycle of knowledge exchange and idea generation, needed to propel innovations and knowledge-based activities that are also key contributors to a nation's productivity, GDP and growth. A study on urban areas in the USA has indicated that a 10% increase in urban spaces can have an impact of 1% rise in productivity (Melo et al., 2016). Furthermore, the report finds distinct correlation between size of the urban areas and the average worker productivity (measured by wages).

A study of the urban growth rate and GDP growth suggests that as urbanization took flight in India with the 1960s onwards, GDP growth rate has also seen an upward trend thus marking a positive correlation between city growth and income generation. India's urban centres account for a significant share of the GDP—urban area contribution to GDP has risen from 38% in 1970 to 62% in 2009 and projected to be 75% in 2030 (GoI, 2013). Other positive outcome from urban growth has been the development of a vibrant industrial sector that has attracted capital inflow, skill development of the workforce, improvements in wages and social sector indicators like literacy rates, enrolment rates, life expectancy and infant mortality rate among others.

Cities are hubs for socio-economic–cultural transformation leading to human development by way of increased commercial activity, productivity and scientific growth. They can be the source of solutions to the challenges being faced by the world today. If urbanization is well planned and well managed, it can become a powerful tool for the sustainable development of developing as well as developed countries. The way human settlements are planned, developed, financed, governed and managed has a direct bearing on sustainability and resilience well beyond urban boundaries. Lack of planning and investment in physical infrastructure results in crucial challenges like insufficient urban services, congestion, infrastructure deficit, widening inequalities and environmental degradation, all undermining the benefits of the process of urban growth.

A deeper study reveals a more nuanced story with different implications of the urbanization story including pejorative impacts on some dimensions of the quality of life of the urban population. Rapid urbanization in tandem with poor planning has led to problems of over population, substandard living conditions, growth in slum settlements, creaking infrastructure, poor sanitation and sewage maintenance and accelerated environmental degradation. These have found greater prominence in urbanization discourses as the true costs of rapid urbanization are being understood and an increasingly politically vocal urban poor population is using the election machinery to bring light to these issues that impact them.

Rapid urbanization and poor policy planning has culminated in an infrastructure crisis for urban dwellers. By 2030, demand for affordable housing will exceed supply by 4 times and demand for private transportation will be double than the supply. There will be glaring gaps and shortfalls in the case of public goods—demand for rail-based mass transit will be 2.5 times the supply; water demand will be double the supply; demand for sewage treatment will be 3.5 times the supply (MGI, 2010).

Proliferation of slum settlements is also an unintended consequence of the urban growth story. Slums are a ubiquitous feature of Indian cities with 65% of Indian towns reported to have slum settlements. One in every six urban Indian living in a slum dwelling (Census, 2011) reflects messy urbanization on the peripheries of major cities. Over 60% of the slum dwellers live close to unsanitary drains and about 40% do not have access to treated water. This portends a major public health crisis in urban India. Fifty-five percentage households in urban areas were also found to be lacking in closed drainage connectivity for wastewater outlet, and 23% households did not have access to improved latrine and bathroom facilities (Census, 2011).

## Call for Action: Sustainable Development Goal 11

The 2030 Agenda for Sustainable Development (United Nations, 2015) sets the central objectives that 'no one will be left behind' and that it will 'reach the farthest behind first'. The central theme of inclusiveness will be an increasingly critical terrain to achieve in the current urban growth scenario. It will be extremely challenging to achieve the integration of the desired social, economic and environmental goals. A pivotal urbanization initiative being promulgated at a global level is the Sustainable Development Goal 11 (of the Sustainable Development Goal framework laid out by the United Nations) exhorting member countries to work on 'Making cities and human settlements inclusive, safe, resilient and sustainable'. Considering rapid trends of urbanization globally, with 68% of the world's population expected to be city dweller by 2050, United Nations has encapsulated the need to work on creating sustainable cities in an SDG. Some key tenets that encompass SGD 11 include tracking the proportion of urban population living in slums, access to public transport and public spaces, air quality and waste management. Cities are also centres of economic activity, with 80% of global GDP coming from them, necessitating the need for urban policy discussions like SDG 11 that promote conducive environments for urban populations (United Nations, 2015).

Globally, some good progress has been registered on achieving improved access to water, transportation, green spaces and more robust disaster management frameworks. However, challenges still exist with meeting the targets on slum populations, air pollution, sanitation, solid waste management. Concerted policy efforts with involvement of all local stakeholders can help with accelerating to the targets. (UNESCAP, 2018). The COVID 19 pandemic has severely imperiled the progress towards all the SDGs. With less than 10 years to deliver the Goals by 2030, United Nations has launched a 'Decade of Action' to accomplish the promises for 5 P's namely people, planet, prosperity, peace and partnerships. Rapid and haphazard urbanization in India has resulted in woefully underinvested infrastructure. It is estimated that per capita investment for infrastructure in India is at $17 annually, in sharp contrast to China which allocates $117 in per capita investment annually (Dobbs & Sankhe, 2010). It is well documented that bolstering up infrastructure investment could spur enormous growth and opportunity for creating sustainable cities. NITI Aayog, an economic policy think tank of the Government of India has measured India's progress against few SDG 11 indicators. Goa, Himachal Pradesh, Gujarat, Sikkim and Chandigarh are the top ranking States/UTs in their performance against SDG targets; while Mizoram, Tripura, Manipur, Nagaland and Meghalaya are the laggards (Niti Aayog, 2019).

## Focus of This Volume: Urban Growth and Environmental Issues in India

Cities and human settlements face unparalleled threats from unsustainable consumption and production patterns, loss of biodiversity, pressure on ecosystems, pollution, natural and human-made disasters and climate change and its related risks, all culminating in undermining the efforts for sustainable and inclusive growth (Fig. 1.1).

The present volume shines a spotlight on the myriad ways in which urbanization is leading to increased environmental damage and what are some mitigating actions that can be adopted. Here again what is unmistakable is the dichotomy between policy and governance planning leading to negative implications for sustainable urban development. Some of the perspectives included are on urban governance, biodiversity issues, air pollution, renewable energy, waste management, etc. These themes are finding resonance as 21 out of 30 most polluted cities in the world are situated in India. Delhi and Mumbai ranked abysmally in the Economist Intelligence Unit (EIU) Livability Index at 110, 115, respectively (Ellis & Roberts, 2016) much behind Baku (Azerbaijan), Manila (Philippines) and Tunis (Tunisia). Globally, cities account for about 70% of global carbon emissions and over 60% of resource use while occupying just 3% of the Earth's land (United Nations, 2015).

Another fallout of rapid urban growth is the encroachment of green spaces due to expansion of residential and industrial enclaves reducing the green space cover. This has implications for sustainable urban development. The urbanization phenomenon is responsible for creating upward pressure on energy consumption and $CO_2$ emissions. A recent study finds that higher energy consumption in newly industrialized

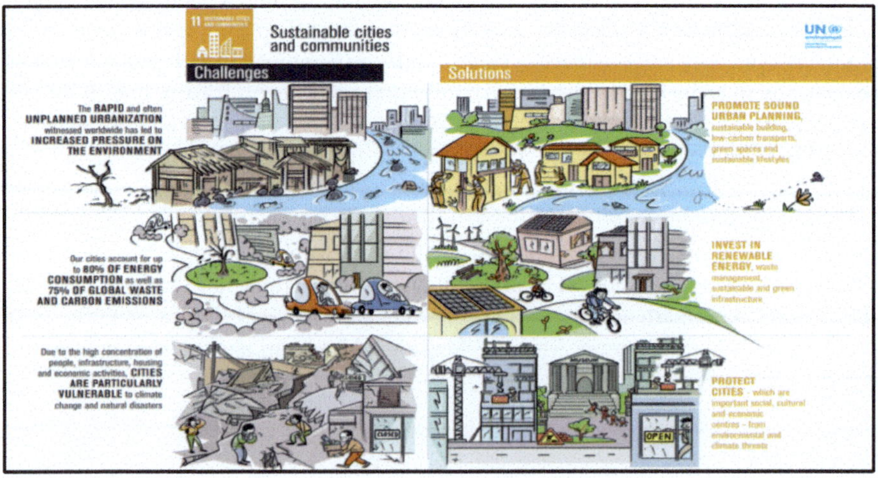


**Fig. 1.1** Sustainable cities and communities: challenges and solutions. *Source* https://www.unenvironment.org/explore-topics/sustainable-development-goals/why-do-sustainable-development-goals-matter/goal-11

countries, including India, gives rise to more $CO_2$ emissions, which will intensity environmental decay (Yazdi & Dariani, 2019). It is observed that urban areas that occupy less than 5% of the world's landmass, account for around 70% of both global energy consumption and greenhouse gas emission. Greenhouse gas emissions can be reduced by up to 1.5 billion CO2 annually by 2030, primarily through transformative change in transport systems in the world's 724 largest cities (OECD, 2020).

The recent COVID pandemic showed that the urban areas are also in frontline of a health disaster. The dense living quarters, high levels of pollution, increased the risk of spread of COVID-19. Many cities are also highly exposed to floods and storm surges. One in five people living in cities, or 613 million people, are exposed to a 100-year flood (Dottori, 2016). This exposure is heavily concentrated in a few cities. Of the top 20 cities with the highest population exposed to a 100-year flood, 3 are in India. The material damage and the loss of life in cities due to floods and storms will balloon if no action is taken. Reduction in air pollution, reducing the exposure to natural hazards, dealing with climate change issues and transition to low-carbon economy would be one of the biggest challenges in the coming times for cities.

With the rapid population growth experienced in Indian cities and resulting rise in waste generation per capita, safe and sustainable disposal of waste products is another pivotal public services challenge that needs to be urgently addressed. Solid waste disposal also has environmental implications since it has been found to be intensive in carbon-emission generation. It was estimated that 1.6 billion tonnes of carbon dioxide-equivalent greenhouse gas emissions were generated from solid waste management in 2016 which accounts for 5% of emissions globally. If the current waste management methodologies are followed without any improvements, emissions are estimated to increase to 2.6 billion tonnes of $CO_2$ equivalent by 2050 (Kaza et al., 2018). The need for efficient waste management is a crucial perspective as waste management is often the single highest item on municipal budgets with low-income countries expected to spend 20% of their municipal budgets.

A glance at India's urbanization trajectory illuminates the glaring need to pay attention to urban challenges of transport, housing, reduction of farm land, migrant population, waste disposal, biodiversity loss, etc. All these years, the successive governments have been focusing their attention on rural development since majority of the population still resides in villages. Though the urban growth is not high when seen in numbers, it has been making a significant impact on the lives of urban dwellers and poses a serious challenge if not attended to. India has shown a state of deep inertia about the urgency and scale of urban reforms. Since urban affairs are the responsibility of states, the issue of urban reforms becomes very complex. The 74th amendment and Jawaharlal Nehru Urban Renewal Mission (JNNURM) were India's first steps towards urban reforms. The state governments have demonstrated apathy towards urban issues and urban local self-government bodies are still trying to find a foothold in the governance structures. Lack of technical expertise, vision, political will and ability to leverage available funds to urban issues and also paying greater attention towards voter-driven issues has hampered the way states deal with urban challenges.

The urban population of India (2020) stands at 35% (worlometers.info/world-population/India) up by 4% since 2011. This translates into an addition of 91.1 million people in cities. As the urban population increases, demand for every key service will increase, resulting into increasing burden on already fragile urban infrastructure. The present book, divided into three sections dealing with urban growth and governance issues; environmental challenges; and waste management and energy issues, is an attempt to understand and analyse the issues plaguing urbanizing India. The three issues are intertwined, and there is a need to analyse them to be able to arrive at an understanding of the issues involved.

## Urban Growth and Governance Issues

Discussions on urban growth largely focus on the increasing size of the urban areas and the resultant governance issues. The burgeoning population in the cities has caused multitude challenges like overcrowding, need of affordable housing, availability of efficient transport system, overburdened public services, managing of waste, shrinking of green spaces and the like. Most of these challenges pertain to the way governance of urban areas is formulated. Due to the involvement of a large number of agencies and departments in providing for urban services, it has been found that there is a lack of coordination between the agencies resulting into poor delivery of the services. Also, the interests of the various stakeholders are not often aligned which makes it difficult to find common solutions.

Jain in her paper attempts to examine the urban transformation taking place both in terms of demographical and in terms of sociological dimensions. The urban growth heralded changes in the living patterns of the people, thereby transforming the way society functioned. The changes brought in by the urban transformation were not limited to the urban scape only. They also transformed the rural landscape. The author contends that the urban development policies and programmes should target to provision for adequate infrastructural support for economic development of the country, state or sub-region. The issues of social justice and equity should be inbuilt in the policies and programmes of urban development so that no one is left behind or left out and everyone is treated equally.

Vasudeva in her paper discusses the paradox of urban growth in India, the role of multiple stakeholders, the frame of physical planning and governance and changing dynamics of housing for the poor. The study attempts to explore formal or informal urban governance processes, underlying socio-economic or political forces, and changing socio-economic and spatial characteristics of shelter for the poor by focusing on the case of Vadodara. The research does not attempt to draw general conclusions; rather, it provides live shreds of evidence of particular situational dynamics. Understanding such urban growth dynamics and the study of other characteristics has the potential to improvise the current system of planning and governance of housing for the poor.

Joshi, using state-level data, makes a case for economic growth, industrialization and infrastructure being the three key determinants of urbanization rate in India. The author opines that as per capita income of a country increases there will be structural transformation from an agrarian economy to an industry service-based economy. Since industries and service activities are urban centric, the processes of industrialization and servicization of an economy contribute to the process of urbanization. The author also points out towards the positive relationship between economic growth and urbanization and negative between environment and urbanization.

Yadav examines the concept of smart cities and its impact on social relations. The cities are being transformed into smart cities through the use of state-of-the-art technological tools and new urban designs but not enough attention has been given to the resulting changes in social realm. Given the fact that humans have always sought to control nature to ensure comfortable living conditions through the development in techniques of means of production and mode of production, the use of technology has given rise to new social relations impacting the way citizens will interact in smart cities.

Urban growth cannot be seen as a separate phenomenon from rural growth. Evidence from all over the world shows that urban agglomeration and development of rural areas has developed side by side. Rurban areas have been defined as those areas which have urban characteristics of economy and lifestyle along with other rural characteristics. Joshi examines SPM Rurban Mission which aims at developing Rurban areas by providing for economic, social and physical infrastructure facilities. The author contends that Rurban area development is an increasingly acceptable and practical means for effective planning and managing rural areas either on the edge of urban areas or having potential of development of diversified economic activity.

Narula and Bhattacharya focus their attention on examining the public health infrastructure in urban areas. To reduce the burden of urbanization in the top tier cities, there is a need to ensure equity in developing policies and implementation for the rural and marginalized areas. Sufficient budget allocation, community partnerships, stronger infrastructures should be incorporated for a more sustainable growth.

Migration is considered to be an important factor in urban growth. The article by Nagla presents an overview of migration trends at a global and national level. The case of migration from rural areas to urban areas is not as high as envisaged. The first thought that comes to one's mind when one looks at the burgeoning cities is that people from rural areas have flocked to cities in large numbers but it is not so. Recently during the lockdown due to COVID pandemic, the issue of migrants leaving the cities presented a very dismal picture and called for a policy intervention.

## Environment Challenges

The growth of urban areas is associated with a concurrent threat to the environment. The most irreversible human impact on the global biosphere is rapid conversion of Earth's land surface for urban use. Forest cover and agricultural land are depleted to provide for the demands of urban life; trees are felled to make way for highways; dumpsites emerge due to increased amount of domestic and industrial waste; noise and air pollution levels rise; climate changes take place resulting into depletion of biodiversity. In rapidly urbanizing areas, agriculture intensifies on remaining undeveloped land and is likely to expand to new areas, putting pressure on land resources (Jiang et al., 2013). There is a great need to understand and analyse the environmental threats emerging due to fast-paced urban growth.

The second section on environmental challenges brings to the forefront some of the challenging issues related to rapid urban growth. Loss of highly productive farmland, increase in energy demand alters the climate, modifies hydrologic and biogeochemical cycles, fragments habitats and reduces biodiversity. Adhikary and Hajra look at the relationship between environmental damages and income growth using the Environment Kuznets Curve. Using empirical data from six selected SAARC countries, the authors used panel data over a period of 1980–2014. The Composite Environmental Degradation Index (CEDI) has been calculated on the basis of the index of air pollution, water pollution, destruction of soil quality, extraction of ores and metals, deforestation, depletion of natural resources and urbanization. The relationship between CEDI and per capita income at constant price of 2010 has been tested.

Increase in vehicular movement is a precondition of urbanization. As the urban areas grow in size, the need for public and private transport is accentuated. However, the increase in vehicular traffic results into pollution and low air quality and is a danger to public health. Kateja addresses the issue of vehicular pollution and threat to human health. The author looks at the trends and patterns of vehicular pollution in India and relates them with health outcomes of citizens in different states. She highlights the need for national sustainability effort for cleaning up air quality, reducing GHG emissions and improving environmental health.

Climate change is one of the biggest threats looming before human civilization. The advancement of human society has resulted into incessant exploitation of natural resources leading to global warming. The average global temperature has increased by 0.85 °C from 1880 to 2012. Pandit in his paper raises the issue of climate change and its consequences on the human population. According to him, the environmental degradation is not restricted to the source region of activity only but can affect far flung regions too. One such example is the incidence of largest ozone hole over the South Pole region. He makes a case for strengthening community participation to ensure sustainability.

Ghatak takes up the issue of forced migration due to climate-related vulnerabilities. She uses secondary data on migration of India to understand the social and economic impacts of forced migration on the overall development of the country. The paper also draws attention to the probable vulnerability of environment-induced migration and identifies the vulnerable classes associated with such migration. The author provides an analytical perspective on the overall picture of the forced migration, with special reference to the climate change and migration nexus in India.

Depletion of farm lands due to increased urbanization is one of the serious challenges for India. Maheshwari discusses the impact of climate change and usefulness of vertical farming as a suitable technique to meet the global food requirements with energy efficiency in indoor environments and suggestive waste management approaches for built environment. She argues that due to increasing urbanization trends, there is a loss of arable land leading to the challenge of having a consistent high-quality crop round the year. Vertical farming can be used as a multidimensional approach to sustain urban farming. This, though new in India, is well adopted in countries like Japan, Korea and Netherlands. The Vertical Farm System uses unconventional farming methods such as hydroponics and aeroponics to produce more yields at a faster rate.

Charan and Sharma highlight various aspects of freshwater ecology of semi-arid regions of Rajasthan and find out the restorative measures for conservation of the lakes and reservoirs of the state. The importance of ecosystem services provided by lakes and reservoirs is substantial. Due to unjustified and unsustainable use of lakes and their catchment areas, these valuable natural resources are facing a threat of extinctions.

## Waste Management and Energy Issues

Waste management is one of the humongous challenges of urban growth. Improper waste management leads to the quandary of insanitary living conditions which further increases the burden of diseases. Waste management techniques call for collection, separation, processing and disposal of waste to provide for a clean city. Sawant, Warrier and Panwar explore the relevance of socio-economic factors involved in the waste management. The authors argue that solid waste management continues to be a humongous challenge for urban planners. Management of Municipal Solid Waste (MSW) is a matter of great concern for the various government regulatory bodies and the pollution control departments. The efficient disposal of MSW will also reduce the disease burden of the country.

Palakurty and Kesari in their chapter raise the issue of solid waste management (SWM), the available treatment options and those practised in India along with some case studies of solid waste management projects in Delhi. Despite efforts by the urban local bodies (ULBs), the problem of collection and segregation of various types of waste generated has been a major obstacle in achieving desired results.

Environment knows no boundary and thus has the ability to affect any life form without any regard to caste, colour, creed, race, economic status or religion. Jasrotia and Soloman take up the issue of management of Municipal Solid Waste (MSW) and attempt to examine the impact of dumping sites. They also attempt to investigate as to which sections of the society are the worst sufferers of the developmental processes in the developing and the developed nations. Factors like socio-economic differences, environmental racism and environmental injustice play a vital role in determining the affected population in and around the municipal waste management process in both the developed and the developing nations.

Sharma and Kaushik opine that due to increase in demand for medication, the number of pharmaceutical industries is also increasing worldwide. The increase in pharmaceutical production brings forth the challenge of disposal of pharmaceutical waste. It has been observed that pharmaceutical waste has become an emerging problem in last few years. Even a very small concentration (microgram to nanogram per litre) causes very harmful impact on the environment. The authors provide a review of global pharmaceutical market, pharmaceutical waste generated, treatment and management of pharmaceutical waste.

The explosion of electronic gadgets in our lives, the need to upgrade the gadgets and the increased competition between manufacturers has been responsible for generation of a vast amount of e-waste. Parthasarthy articulates concerns on the vast amount of electronic waste generated by proliferation of technology. Electronic waste (e-waste) is one of the most rapidly growing solid waste problems of the world. E-waste contains hundreds of valuable resources which can be recycled, however some of which are hazardous and create serious environmental and health impacts by unscientific processing. The author discusses the current practices, both formal and informal, being followed in India for recycling of generated e-waste.

All economic activities in a modern economy depend on energy to flourish and sustain. A well-established energy system fuels all sectors from agriculture to high technology-intensive IT sector, from manufacturing to medicine, thereby making no development possible without fuelling the engine of growth. Between 2000 and 2019, around 750 million people in India gained access to electricity. In the current century, India is also experiencing a major shift to higher shares of variable renewable energy. Renewable energy sources such as wind, water, solar, biomass and geothermal energy are going to be the future of urban world as they are inexhaustible and clean. Renewable energy currently constitutes only 18% of the Indian energy mix.

India is the fourth largest consumer of energy. With the increase in energy demands of the country, there is a need to look at renewable energy sources. India's geophysical conditions are best suited for harnessing renewable sources of energy like solar power and wind power. Mall and Solanki examine the progress of India in improving access to energy and developing its indigenous sources of energy. Energy demand is set to rise 2–3 times over the coming decades; hence, India faces unique challenges of vast magnitude. The authors opine that, the way these challenges are met will have major ramifications for sectors, like water, food, urban planning and transport. Increasing awareness and advancing technology for exploiting renewable energy is must for sustainable energy prospects in India.

Increase in urban population raises the demand for energy. Developing countries like India are dependent on fossil fuels to meet the energy needs of its population. The use of fossil fuels to meet the energy requirements is also detrimental to the environment. Kataria and Khan attempt to review the state of energy sources and the dire need of shifting from non-renewable sources to renewable sources of energy. Fossil fuels, in particular coal, dominate India's electricity sector. India produced about three fourths of all electricity from coal in 2017–18. With abundant natural resources, conducive geographical and climatic conditions, India has one of the highest potentials for harnessing renewable energy. These conditions support the promotion of renewable energy technologies like solar, wind, biomass and small hydro.

The rapid urbanization also has the potential to improve the well-being of people by providing them access to a better life style, free from the shackles of traditional thought process. Cities are engines of growth, accounting for more than 80% of global domestic product. They are also home to younger people and provide ample opportunities for raising the living standards of its residents. The urban growth can be used to reap the demographic dividend in India provided we are able to provide for effective and citizen-centric governance. The world needs inclusive and sustainable urbanization as a milestone in the path towards socio-economic development, as recognized by SDG 11-sustainable cities and communities. Future human development prospects will largely hinge on how well these growing cities are managed (UNDP, 2017).

## References


Census. (2011). Census of India, Office of the Registrar General and census commissioner India.

Dobbs, R., & Sankhe S. (2010). Comparing urbanization in China and India, McKinsey Quarterly, https://www.mckinsey.com/featured-insights/urbanization/comparing-urbanization-in-china-and-India. Accessed on May 29, 2020.

Dobbs, R., Smit, S., Remes, J., Manyika, J., Roxburgh, C., & Restrepo, A. (2011). *Urban world: Mapping the economic power of cities*. McKinsey Global Institute.

Dottori, F., Salamon, P., Bianchi, A., Alfieri, L., Hirpa, F. A., & Feyen, L. (2016). Development and evaluation of a framework for global flood hazard mapping. *Advances in Water Resources, 94*, 87–102. https://doi.org/10.1016/j.advwatres.2016.05.002.

Ellis, P., & Roberts, M. (2016). *Leveraging urbanization in South Asia: Managing spatial transformation for prosperity and livability*. South Asia Development Matters. Washington, DC: World Bank. https://doi.org/10.1596/978-1-4648-0662-9. License: Creative Commons Attribution CC BY 3.0 IGO.

GOI. (2013). State of Slums in India—A Statistical Compendium 2013. Ministry of Housing and Urban Affairs, Government of India. http://nbo.nic.in/pdf/Slums_in_India_Compendium_English_Version.pdf. Accessed on May 17, 2020.

Jiang, L., Deng, X., & Seto, K. C. (2013). The impact of urban expansion on agricultural land use intensity in China. *Land Use Policy, 35*, 33–39.

Kaza, S., Yao, L., Bhada-Tata, P., & Van Woerden, F. (2018). *What a waste 2.0: A global snapshot of solid waste management to 2050*. Urban Development Series. Washington, DC: World Bank.

Lewis, A. (1954). Economic development with unlimited supplies of labour. *The Manchester School, 22*(2). https://doi.org/10.1111/j.1467-9957.1954.tb00021.x.

Melo, P., Graham, D.J., Levinson, D., & Aarabi, S. (2016). Agglomeration, accessibility and productivity: Evidence for large metropolitan areas in the US. Urban studies, https://doi.org/10.1177/0042098015624. Accessed on June 15, 2020.

MGI. (2010). India's urban awakening: Building inclusive cities, sustaining economic growth. McKinsey Global Institute.

Niti Aayog. (2019). SDG India Index and Dashboard 2019–20, https://niti.gov.in/sites/default/files/SDG-India-Index-2.0_27-Dec.zip. Accessed on June 17, 2020.

OECD/European Commission. (2020). Cities in the World: A New Perspective on Urbanization, OECD Urban Studies, OECD Publishing, Paris, https://doi.org/10.1787/d0efcbda-en. Accessed on December 23, 2020.

Spence, M., Annez, P. C., & Buckley, R. M. (2009). *Urbanization and growth (edited)*. The International Bank for Reconstruction and Development/The World Bank

UNDP. (2017). Rapid urbanization: Opportunities and challenges to improve the well-being of societies. Available at http://www.hdr.undp.org/en/content/rapid-urbanization-opportunities-and-challenges-improve-well-being-societies. Accessed on June 20, 2020.

UNESCAP. (2018). Sustainable Cities and Communities, https://www.unescap.org/sites/default/files/SDG%2011%20Goal%20Profile%20Final%20Edit%20260218.pdf. Accessed on June 15, 2020.

United Nations. (2015). Sustainable Development Goals, https://www.un.org/sustainabledevelopment/cities/. Accessed on June 3, 2020.

United Nations. (2019). World Urbanization Prospects 2018, Department of Economic and Social Affairs, Population Division, Highlights (ST/ESA/SER.A/421).

Yazdi, S.K., & Dariani, A.G. (2019). $CO_2$ emissions, urbanization and economic growth: evidence from Asian countries. *Economic Research Ekonomskalstraživanja, 32*(1), 510–530. https://doi.org/10.1080/1331677X.2018.1556107. https://www.tandfonline.com/doi/pdf/10.1080/1331677X.2018.1556107.

# Part I
# Urban Growth and Governance Issues

# Chapter 2
# The Urban Transformation: Issues and Challenges

**Rashmi Jain**

**Abstract** Today, half of the world population resides in urban areas. The countries of the world are becoming increasingly urbanized. Increasing urbanization can be seen as a global phenomenon that manifests differently across regions and development levels. It is forecasted that Africa and Asia will urbanize faster than other regions over the coming decades. When this happens, the landscape of human settlement will undergo a massive transformation. This will have significant implications in different parts of the world. As per the records of UN World Urbanization Prospects: The 2018 Revision, with 55% of the world's population residing in urban areas in 2018, the number of people living in urban areas is more than rural areas globally. In 1950, a little less than one-third of population was urban, while by 2050, 68% of the world's population is projected to be urban (UN WUP, 2019). Reports cite that 35% of the world's urban population growth between 2018 and 2050 will be in India, China and Nigeria. With mass urbanization taking place, it is imperative that urban growth is managed properly and efficiently. This is more important in low-income and lower-middle-income countries where the most rapid urbanization is expected between now and 2050 (UN WUP, 2018). The key aim of policies and programmes aimed at urban development should be to provide for sufficient socio-cultural, spatial and infrastructural support for development in the state or sub-region or country. Social justice and equity needs to be inbuilt in the policies and programmes of urban development so that no one is left behind and everyone is treated equally.

**Keywords** Urban transformation · Urbanism · Urbanity · Urban development programmes and policies

## Introduction

Urban transformation is a multidimensional concept involving several processes of transformation in the spatial, social, economic, environmental, cultural and governance arena. Often seen as a one-dimensional concept associated with the process of

R. Jain (✉)
Department of Sociology, University of Rajasthan, Jaipur, Rajasthan, India

 
A. Kateja and R. Jain (eds.), *Urban Growth and Environmental Issues in India*,
https://doi.org/10.1007/978-981-16-4273-9_2

improving the physical environment of urban areas, it is a complex process dealing with issues of unhealthy and therefore unsustainable urban areas in order to make them healthy and livable, focusing on their basic characteristics such as social, economic and physical space (Mehdipoura & Nia, 2013; Roberts, 2008). Merriam Webster dictionary defines transformation 'as an act, process, or instance of transforming or being transformed' Cambridge dictionary defines it 'as a complete change in the appearance or character of something or someone, especially so that the thing or person is improved'. Urban transformation, therefore, is not just about the increase in the number of cities or urban population, but it is concerned with both the changes that are brought about in the lives of the urban population and how the cities grow. Urbanization has been seen as a process of moving to cities, shifting of people from agriculture to other forms of occupations like trade, manufacturing, industry, service and management, and the consequent changes of behaviour patterns. For Hawley (1981), urbanization is the process of expansion in the entire system of interrelationships by which a population maintains itself in its habitat. A major element of urbanization is an increase in the size of towns and cities leading to growth of urban population. The rate of increase in the size of towns and cities impacts the growth of the urban areas. One of the dramatic consequences of the industrial revolution was agglomeration experienced in urban spaces (Clerici & Mironowicz, 2009). Experiencing the inhuman conditions of the working class and cramped and unhealthy conditions in major cities of Europe influenced many thinkers and planners. This accelerated the surfacing of the idea of transformation.

Whenever any discussion takes place on urban growth, it takes two dimensions into account: demographic and sociological. When urbanization is seen from the demographic point of view, the spotlight is on the size and density of population, the nature of work of adult population, the physical growth of cities and the like, but when the sociological lens is applied to the study of urbanization, the focus is on understanding the heterogeneity, impersonality, interdependence and quality of life in urban areas. Urban life had been attracting the attention of early classical sociologists who explored the dichotomy found in the society. Morgan (1877) conceptualized the distinction between savagery, barbarism and civilization and Maine (1931) differentiated the society based on status and contract. Further on, Tonnies (1957) distinguished between rural (*gemeinschaft*) and urban (*gesellschaft*) on the basis of social relationships; Durkheim (1964) talked of organic and mechanical solidarity found in the society; Weber (1958) and Simmel (1950) emphasized on dense living conditions, rapidity of change and impersonal interaction in urban settings.

Different approaches to study the urban transformation emerged, foremost being Chicago school which focused on the study of urban cultures. Burgess's (1925) concentric zone theory suggested that urban expansion causes a modern city to evolve in concentric zones, each characterized by typical land use. The different zones defined by Burgess are the central business district, zone of transition, working class residences, middle-class residences and commuter zones. He felt that zone development resulted as a result of competitive processes, i.e. competition for best site in the city. Hoyt (1939) argued that cities do not develop in concentric circles, but in sectors, which are characterized by different economic activities. The entire city is seen as a circle, and its neighbourhoods are seen as sectors radiating out from the core. Multiple Nuclei theory professes that cities do not have a single centre, but have multiple growth points or nuclei where growth takes place. Harris and Ullman (1945) argued that often a city starts with a single central business district (CBD), but over time, the activities scatter and get modified and attract people from surrounding areas and act as smaller nuclei in themselves. With time, these small nuclei gain importance and grow in size and begin influencing the growth of activities around them. Graham and Marvi's global cities (1996) in post-industrial time were spaces where malls, border cities, sub-centres and suburbs were located around the Global Command Centre CBD, and thus breaking the classic scheme of the compact city. Soja (2000) talked about six geographies that are part of the actual urban space, namely Flexcities (post-Ford industrial metropoli), Simcities (simulated cities and the theme park), Fractal Cities (dual city for well-living and poor living), Cosmopolis (global multicultural city), Exopolis (urban sprawl) and Fortress Cities (private city).

The issues like breaking up of elements of rural life, transformation in the lives of people due to industrialization, emergence of new forms of social interaction and cohesion and the like generated the interest of social scientists to examine the urban spaces. According to Castells (2002), urban sociology was built around the central theme of social integration in an urban society where the traditional institutions crumbled under the load of population growth, accelerated economic development, greater social mobility, and social struggles. Urban centres provided the sociologists with an opportunity to understand the ways in which integration took place when large number of people came in from rural areas to urban areas.

In order to understand the dynamics of urban growth in contemporary times, it is important to trace the development of urban spaces from a historical point of view. Urban spaces, although emerged in the centre of all discourses post-industrialization, but their presence can be seen during the ancient Indus valley civilization, Egyptian civilization, Greek civilization, etc. The urban centres in the ancient world emerged with the shift from exploiting livelihood (food gathering and hunting) to production livelihood (agriculture). The second division of labour corresponded to separation of crafts and their concentration in certain settlements. The emergence of the cities was the results of third social division of labour, i.e. the separation of the exchange of products from crafts. Pre-industrial cities were predominantly seats of political/administrative and military power, centres of trade, commerce and craftsmanship. At the start of industrial revolution, only 3% the world population was living in cities, and today, about 55% of the world population is living in the cities. Industrial development with thrust on mechanized work, mass production transformed the socio-economic landscape. It provided opportunity to the people to work in factories which resulted in mass movement of population from rural areas to urban areas. The opportunity to move to cities heralded not only economic gains but also loosened traditional barriers and restrictions of rural life. In the early stages of industrial growth, large-scale migration of rural people to urban centres took place leading to large-scale problems of overcrowding, poverty, lack of proper housing, development of slums, spread of diseases, moral degradation, crime, deviance, change in the social character of urban centres, to name a few.

## Urbanization, Urbanism and Urbanity

Urbanization happens as a result of numerous change agents in the society, but the most important elements are the redistribution of population from one region to another and a continual differentiation of the society, both in rural and in urban components (D'Souza, 1984). There are different reasons cited by the scholars regarding the redistribution of population. The process of urbanization has been used to understand and analyse the physical growth of the cities and the resultant complexity in the society. On the other hand, urbanism makes a distinction between the urban life and rural life on the basis of socio-politico-economic factors. The process of urbanization and the ensuing urban growth has attracted the interest across various disciplines. While geographers have focused their attention largely on the ways in which cities and towns are constructed, governed and experienced, architects lay emphasis on the built form and its relation with urban space; the discipline of anthropology examines the relationship between urban life and the reproduction of community, kinship networks and social mobility. While urban planners handle the problems of influence of various factors, natural, transport accessibility, economic, social, etc. (Bott, 1957; Gluckman, 1971; Lomnitz, 1977; Mitchell, 1969; Stack, 1974, 1996; Werbner, 1984), sociologists have shown interest in the problems of transformation of socio-cultural space of a modern city (Baum, 1999; Sachs-Jeantet, 1994).

Wirth (1938) mentions four characteristics of urban system, 'heterogeneity of population, specialization of function, anonymity and impersonality and standardization of behavior'. He argued that urbanism or the condition of living in a city was becoming the way of modern life. Transiency, superficiality, anonymity and individualism were the characteristics of urbanity. The urban areas are characterized by deep conflicts between norms and values; increased social differentiation; greater social mobility; higher levels of education and income; predominance of material possessions and individualism; impersonality of relationships and decline in intimate communication and increase in formal modes of social control, increase in anonymity, homogenization of behaviour patterns, etc.

More recent theorist such as Harvey (1985) stressed that urbanism is not a self-generating process; rather, it is part of larger political and economic processes and changes. He opined that the urban environment is built, destroyed and rebuilt to allow for a more efficient circulation of capital. Like Harvey, Castells too stresses on the idea that spatial form of a city is related to the larger process of the society. He further adds the dimension of the struggles and conflicts of various groups who make up the cities to it. The cities also represent a move towards higher degree of civilization. The online etymology dictionary attributes the use of the term urbanity to 1530s from French *urbanite* and from Latin *urbanitatem* meaning city life; refinement, city fashion or manners, elegance, courtesy. It means not only what is linked to city life but also a quality that is polished in manner, courteous or sophisticated. Even though the cities are confronted with many challenges, especially environmental and social, they also represent a region where business; arts, political activism, civil society aspirations, social entrepreneurship and networks of individuals intertwine, metamorphosing into a vibrant social fabric. The Latin word *urbanitas* carries connotations of refinement and elegance, the opposite of *rusticus*, associated with countryside. In the post-industrial, global society, interest in urbanism grew out of the belief that human habitat is problematic. The studies have very often focused on the need to find newer ways of looking at the urban space issues and also work towards addressing, how humans are supposed to create sustainable urban places.

## Global Urban Growth Trends

More than 50% of the world population lives now in urban areas. Increasing urbanization can be seen as a global phenomenon that manifests differently across regions and development levels. The developed nations already have a large part of their population residing in urban areas, and most of the urban growth now shall be concentrated in Africa and Asia. These trends will alter the scenario of human settlement and significantly impact the living conditions, environment and development in different parts of the world. As per the records of World Urbanization Prospects: The 2018 Revision, globally, with 55% of the world's population residing in urban areas in 2018, the number of people living in urban areas is more than rural areas. In 1950, a little less than one-third of population was urban while 68% of the world's population is projected to be urban by 2050 (UN WUP, 2019). The World Urbanization Prospects 2019 projects that nine countries will make up more than half the projected population growth between now and 2050 and India is one of them.

The world's urban population increased from 0.8 billion in 1950s to 4.2 billion in 2018. The global urban population is expected to grow approximately 1.69% per year between 2018 and 2030 and 1.28% per year between 2030 and 2050. Table 2.1 shows population data for world and development groups.

The increase in population started with humans settling down at one place and the subsequent shift to agricultural form of production. It is only in the early nineteenth century that the world population reached its first billion. Nevertheless, it is the twentieth century where population growth really took off and over the past hundred years the world population has more than tripled itself. It is expected that by 2100 nearly 85% of the population would be living in cities and urban population would increase from 1 billion in 1950 to 9 billion in 2100.

In 1990, there were 21 large cities and 10 megacities; and by 2018, both categories of cities had more than doubled (Fig. 2.1), with 48 large cities and 33 megacities. Majority of the megacities are now located in the developing countries, and this trend will continue as several large cities in Asia, Latin America and Africa are projected to become megacities by 2030. Table 2.2 shows the top ten megacities of the world in 2019. One of the important themes of urban growth has been the emergence of many large and mega cities, especially in low- and middle-income regions of the world.

The world city report suggests that the fastest growing urban centres are going to be the small and medium cities. It is predicted that the rapidly rising populations in

**Table 2.1** Total, urban and rural populations and their average annual rates of change, for the world and development groups, selected years and periods, 1950–2050

| Development group | Population (billions) | | | | | | Average annual rate of change (per cent) | | | | |
|---|---|---|---|---|---|---|---|---|---|---|---|
| | 1950 | 1970 | 1990 | 2018 | 2030 | 2050 | 1950–1970 | 1970–1990 | 1990–2018 | 2018–2030 | 2030–2050 |
| *Total population* | | | | | | | | | | | |
| World | 2.54 | 3.70 | 5.33 | 7.63 | 8.55 | 9.77 | 1.89 | 1.83 | 1.28 | 0.95 | 0.67 |
| More developed regions | 0.81 | 1.01 | 1.15 | 1.26 | 1.29 | 1.30 | 1.07 | 0.64 | 0.34 | 0.17 | 0.03 |
| Less developed regions | 1.72 | 2.69 | 4.18 | 6.37 | 7.26 | 8.47 | 2.23 | 2.21 | 1.50 | 1.09 | 0.77 |
| *Urban population* | | | | | | | | | | | |
| World | 0.75 | 1.35 | 2.29 | 4.22 | 5.17 | 6.68 | 2.95 | 2.63 | 2.18 | 1.69 | 1.28 |
| More developed regions | 0.45 | 0.67 | 0.83 | 0.99 | 1.05 | 1.12 | 2.06 | 1.04 | 0.64 | 0.46 | 0.34 |
| Less developed regions | 0.30 | 0.68 | 1.46 | 3.23 | 4.12 | 5.56 | 4.02 | 3.82 | 2.83 | 2.03 | 1.50 |
| *Rural population* | | | | | | | | | | | |
| World | 1.79 | 2.35 | 3.04 | 3.41 | 3.38 | 3.09 | 1.37 | 1.30 | 0.41 | -0.07 | -0.45 |
| More developed regions | 0.37 | 0.33 | 0.32 | 0.27 | 0.24 | 0.17 | -0.48 | -0.27 | -0.58 | -0.95 | -1.61 |
| Less developed regions | 1.42 | 2.01 | 2.72 | 3.14 | 3.14 | 2.92 | 1.75 | 1.52 | 0.51 | 0.00 | -0.37 |

*Source* UN World urbanization Prospects: The 2018 Revision

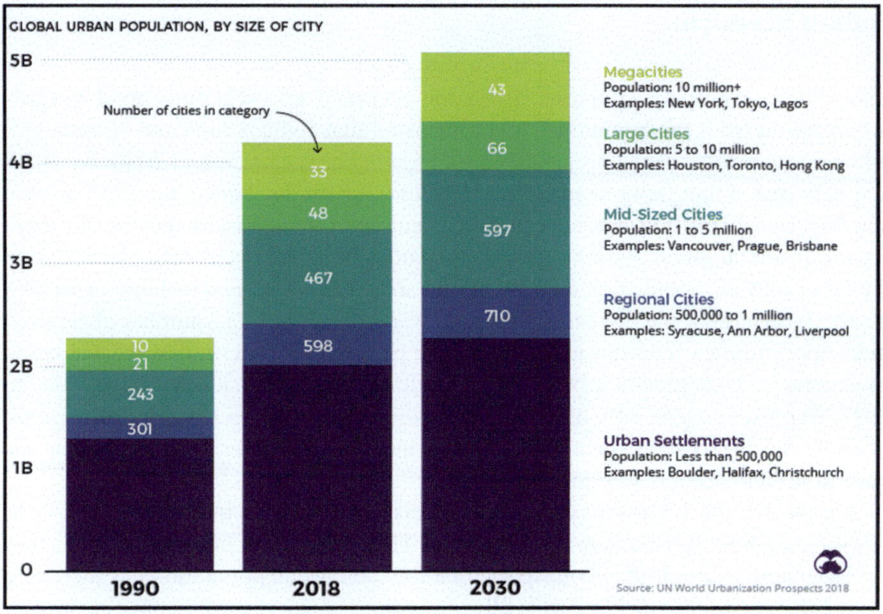


**Fig. 2.1** Global urban population by size of city. *Source* UN World Urbanization Prospects 2018

**Table 2.2** Top ten megacities of the world, 2019

| Rank | Urban area | Country | Population estimate |
|---|---|---|---|
| 1 | Tokyo-Yokoshama | Japan | 37,435,191 |
| 2 | Delhi | India | 29,399,141 |
| 3 | Shanghai | China | 26,317,104 |
| 4 | Sao Paulo | Brazil | 21,846,507 |
| 5 | Mexico city | Mexico | 21,671,908 |
| 6 | Cairo | Egypt | 20,484,965 |
| 7 | Dhaka | Bangladesh | 20,283,552 |
| 8 | Mumbai | India | 20,185,064 |
| 9 | Beijing | China | 20,035,455 |
| 10 | Osaka | Japan | 19,222,665 |

*Source* https://worldpopulationreview.com/world-cities/

megacities and major cities will be significant contributors in urban growth and the number of mid-sized cities will also swell drastically by 2030. Thus, the urban areas will become more influential economic hubs in the process (Ghosh, 2019).

## Indian Scenario

Like other developing countries, India too recorded an escalating trend towards urbanization from the beginning. It is forecasted that India, China and Nigeria will witness over a third of the projected urban growth between now and 2050. By 2050, it is anticipated that India would add 416 million urban dwellers, China 255 million and Nigeria 189 million. The census data reveals continuous rise in the rate of urbanization in India, albeit slow by global standards. This has been particularly during the latter half of twenty-first century. Some areas have emerged as large urbanized centres with large-scale industrial and trading activities offering a number of employment opportunities. This has led to shifting of population from rural to urban areas at a faster rate. The degree of urbanization in India increased considerably from 2007 to 2017. At present, about 34% of India's population lives in urban areas and an increase of about 3% is recorded since 2011 Census, indicating a remarkable increase in the pace of urbanization in our country.

A look at Table 2.3 reveals that India's level of urbanization increased from 17.6% (1951) to 27.8% (2001) and to 31.1% (2011). Though the level of urbanization is low, yet the magnitude is high when we see that the total urban population has increased from 286.1 m in 2001 to 377.2 m in 2011 in India. This increase is happening mostly as a result of reclassification of rural settlements into urban.

**Table 2.3** Comparing indices of urbanization: 2001 and 2011

| Indices | 2011 | 2001 | 1951 |
|---|---|---|---|
| Urban population (million) | 377.2 | 286.1 | 62.44 |
| Number of cities and towns | 7935 | 5161 | 2924 |
| (a) Statutory towns | 4041 | 3799 | |
| (b) Census towns | 3894 | 1362 | |
| (c) Metropolitan cities (+1 million) | 53 | 35 | 5 |
| Annual exponential growth rate (census decade) per cent | 2.76 | 2.74 | 2.3 |
| Per cent of urban to total population | 31.16 | 27.81 | 17.6 |
| (a) Per cent of population in cities with > 100,000 population | 70.24 | 68.62 | 41.8 |
| (b) Per cent of population in towns with (<100,000 population) | 29.76 | 31.38 | 58.2 |
| (c) Per cent of population in metropolitan cities (+1 million) | 42.62 | 37.82 | 11.75 |
| Urban population increase between 2001 and 2011 attributed to | | | |
| (a) Natural increase | 44.0[a] | 59.24 | |
| (b) Rural–urban migration | 21.12 | | |
| (c) Reclassification of rural settlements into urban | 56.0[a] | 9.73 | |
| (d) Boundary changes | 9.71 | | |

*Source* Census 2011, 2001, 1951, Registrar General and Census Commissioner, India
[a]Estimated interstate variation

**Table 2.4** Number of urban units

| No. of urban units | 2001 | 2011 | Increase |
|---|---|---|---|
| Towns | 5161 | 7935 | 2774 |
| Statutory Towns[a] | 3799 | 4041 | 242 |
| Census Towns[b] | 1362 | 3894 | 2532 |

*Source* Census of India, 2011
[a]All places with a municipality, corporation, cantonment board or notified town area committee, etc., are known as Statutory Town
[b]All other places which satisfy the following criteria are known as Census Towns: A minimum population of 5000; at least 75% of the male main workers engaged in non-agricultural pursuits; a density of population of at least 400 per $km^2$

Table 2.4 shows the census data for the number of urban towns from 2001 to 2011. It can be seen that a net increase of 2774 towns took place in 2011 (2532 census towns and 242 statutory towns) This unprecedented increase in the number of census towns implies that 'the increased level of urbanization in the country is not a result of acceleration in the growth rate of small and medium towns but because of an increase in the number of census towns' (Kundu, 2014).

India Habitat III National Report, 2016 outlines four major characteristics of India's urbanization. First and foremost, India ranks low in global urban standing with a meagre 31% urban growth. Secondly, India has witnessed modest population growth. Reversing the declining trend seen in Annual Exponential Growth Rate (AEGR) in the census decades of 1981–91 and 1991–2001, the decade of 2001–11 showed increased AEGR at 2.76% (Table 2.3). Thirdly, in the recent decades, the pattern of India's urbanization has undergone a significant shift. It is characterized by increase in number of metropolitan cities and an unparalleled increase in the number and population of census towns. The number of metropolitan cities has risen from 35 in 2001 to 52 in 2011 and accounts for 42.6% of the country's total urban population. Fourthly, India's urbanization is driven largely by high fertility rates and reclassification of rural settlements into urban. Rural–urban migration constitutes an estimated 20–22% of urban population growth (India Habitat III National Report, 2016). Urbanization process in India is not mainly 'migration-led' but a product of demographic explosion due to natural increase (Datta, 2006).

The number of million plus cities in India has gone up from 5 in 1951 to 53 in 2011. The corresponding increase in the population has been phenomenal from 11.75 million to 160.71 million. About 31% of urban population lives in the top 20 cities, and about 51% of country's urban population lives in top 100 cities. There are 468 Class I cities, and 264.9 m people live in these cities, constituting 70% of total population. There are three mega cities in India having over 10 million people (Greater Mumbai—18.4 m, Delhi—16.3 m and Kolkata—14.1 m).

A glance at Table 2.5 shows that the number of Class I cities in India has gone up from 76 in 1951 to 468 in 2011. As per 2011 census, 70% of the urban population is living in these Class I cities.

**Table 2.5** Trend of urbanization in India 1951–2011

| Census year | Urban population (in million) | Per cent of urban population | Annual exponential urban growth rate (%) | No. of Class I cities | Per cent of population in Class I cities |
|---|---|---|---|---|---|
| 1951 | 62.4 | 17.3 | | 76 | 44.6 |
| 1961 | 78.9 | 18.0 | 3.47 | 102 | 51.4 |
| 1971 | 109.1 | 19.9 | 2.34 | 148 | 57.2 |
| 1981 | 159.5 | 23.3 | 3.79 | 218 | 60.3 |
| 1991 | 217.5 | 25.7 | 3.09 | 300 | 65.2 |
| 2001 | 286.1 | 27.9 | 2.75 | 393 | 68.6 |
| 2011 | 377.1 | 31.2 | 2.76 | 468 | 70 |

*Source* Sabyasachi and Shupinder (2018)

Table 2.6 shows that the top three states have 34.6% of country's urban population residing in them. This is an indication of the top-heavy urbanization happening in the country.

The above discussion is merely to point out that the focus of majority of social scientists has been on examining the demographic data and not on the issues which lead to the development of urban world. While demographers have limited themselves to the census data and its analysis, the attention of policymakers has been drawn towards the challenges of urban transformation due to the large number involved. Before 1960, urban studies were not a common feature in India. Though sociologists like Mukherjee and Singh (1956) and Ghurye (1953) wrote on urban life, this domain was largely engaged by the geographers. The sociological literature on urbanization in India is largely confined to the study of urban life in contrast to rural life and the impact of the process of urbanization and urbanism on the lives of people. For

**Table 2.6** Urban population in states[b]

| Top three states | Absolute population (in million) | Share[a] |
|---|---|---|
| Maharashtra | 50.8 m | 13.5% |
| Uttar Pradesh | 44.4 | 11.8% |
| Tamil Nadu | 34.9 | 9.3% |
| *Bottom three states* | | |
| Sikkim | 0.15 | Negligible |
| Arunachal Pradesh | 0.31 | 0.1% |
| Mizoram | 0.56 | 0.1% |

*Source* Census of India 2011—provisional population tables—India

[a]Share to total population in the country

[b]Excludes UTs

Rao (1970), urbanization was a complex multifaceted process comprising of demographic, comparative, traditional and sociological elements. The empirical studies were largely concentrated on studying and analysing the role of caste in urban areas (Beteille, 1966; D'Souza, 1974; Ghurye, 1962; Gore et al., 1970; Kolenda, 1984; Lynch, 1969; Rao, 1974; Srinivas, 1962); impact of urban life on family structure and kinship relations (Desai, 1964; Kapadia, 1959; Ross, 1962; Singer & Cohen, 1968); urbanization and status of women (Cormack, 1961; D'Souza, 1963; Gore, 1968); urban politics (Kothari, 1970; Lynch, 1969, 1974; Rudloph, 1961); urbanization and rural life (Majumdar, 1958; Rao, 1974; Srinivas, 1962).

The last decade has witnessed a paradigm shift in India's outlook on urbanization. The urban planning is now in the forefront of developmental policymaking. The view that cities are central to the country's economic growth and development is gaining wider acceptance. This view is further strengthened by the increasing contribution of the urban sector to India's GDP. However, this whole discourse becomes meaningless till the dynamics of the urban transformation is understood from the angle of what pushes people to urban areas or what does the city provide to its dwellers.

The urban areas have always been pulling people from rural areas whether it is due to proximity to the economic or political centre or due to the availability of better opportunities. As suggested by India Habitat Report 2016, around one-fourth of the urban population growth is due to rural–urban migration. The major push factors associated with rural–urban migration are unemployment, lower wages, low agriculture productivity, crop failure, poor living conditions, poor health and inadequate educational facilities, poverty, lack of opportunities in the rural areas. The factors that pull people to urban areas are the availability of jobs, higher wages, quality of life, better education and health facilities, higher standard of living, emancipation from the fetters of traditions, egalitarianism and social freedom. In the case of India, it has been observed that people migrate to cities more as a result of rural push rather than urban pull. It is also observed that urbanization as a process puts strain on infrastructure, physical environment and social services, and it can make urban living difficult, especially for low-income groups. This is particularly evident in the cities of less developed regions (Chamie, 2019). The increase in urban population has led to unmet basic needs of food, water, housing, electricity, employment, education, health care, transportation, security, telecommunications, sanitation and waste management, to name a few. The unprecedented urban growth has thrown many challenges before planners, administrators and voluntary organizations.

In India, most of the rural migration is found to be towards Class I cities. People moved from rural areas without having ever lived in an intermittent city making them more vulnerable to the perils of urban life. Big cities provide cheap transportation, skilled manpower, continuous supply of power and water and other facilities attracting movement of people to them. One of the entrenched features of urbanization in India has been the unplanned and haphazard growth of informal settlements like slums, resulting into urban turbulence, contestations, disenfranchisement, dispossession, deprivation and marginalization of the urban poor. Davis and Golden (1954) describe the case of India as one of 'over-urbanization wherein urban misery and rural poverty exist side by side with the result that city can hardly be called dynamic' and where inefficient, unproductive informal sector (Kundu & Basu, 1998) becomes more and more perceptible. The urban areas have always been places where distinction based on wealth, power and prestige are highly visible and important. They are also reflected in the way urban policies are framed and implemented.

## Programmes and Policies for Urban Transformation

The reaction to urban growth in India has been to respond to challenges thrown by the urban transformation rather than planning the growth of urban areas. The policy initiatives have been targeted at solving the problems caused by the emerging urban needs. They failed to envisage the future growth of urban areas. The Five-Year Plans also reflect an urban bias. The development planning in India has largely concentrated on rural development, and it presumed that the urban areas will develop spontaneously. Ahluwalia (2017) observes, 'political economy of development in India has remained dominantly concerned with the development of rural areas implicitly assuming that urban areas can take care of themselves'. The first two Five-Year Plans majorly focused on building basic infrastructure, gaining self-sufficiency in food grains, development of heavy industries and improving the living standard of people. Some thought on urban issues was given in the third Five-Year Plan and master plans were made for some cities. It is from fourth Five-Year Plan that the issue of urban growth entered into policymaking and concern started being raised on the need for decongestion in urban areas. Indian economy witnessed many changes post-1980s, the foremost being the beginning of economic liberalization and also marking the end of Nehruvian socialism in the country. The plans thereafter focused on urban housing needs, slums, improvement of urban services, urban poor and infrastructure development. With regard to urban self-government, only two references are found in the Constitution of India. The first Entry 5 of Second List of 7th Schedule which falls in the state list talks about the constitution and power of urban local self-governing bodies and second Schedule 12 talks of powers and responsibilities of municipalities.

Despite the increasing trend of urbanization, there is no well-defined national urbanization policy in India. Prior to 1985, there was no separate ministry for dealing with urban affairs in the country. The government appointed various task forces, committees, commissions to examine the challenges of urban growth from time to time. A National Commission on Urbanization (NCU) was set up in 1988 by the government to review and analyse the urbanization process and formulating policies for integrated urban development. The committee recommended, among others, restructuring of Ministry of Urban Development and making it a nodal ministry to deal with urbanization, creating an Indian Council for Citizen Action (ICCA) to encourage citizens through organized voluntary effort; to start urban community development department. Thereafter, the 65th amendment bill was tabled, but it got defeated in Rajya Sabha. Urban local governance is a state subject. Yet since 1999 with the resolution of the 11th Finance Commission, the central government has shown increasing interest in the affairs of the city. Subsequently, the 74th constitutional amendment strengthened the local self-government bodies in cities and towns. It recommended major structural changes in local governance to ensure functioning of municipalities as democratic units and greater participation of people at grass roots in the decision making. One major reform was the devolution of municipal functions to urban local bodies (ULBs), but even till today, it has not been done in all states. Only 11 states have transferred all its functions to ULGs so far. Unless the ULBs have financial autonomy, they will not be able to perform their duties in an efficient manner. 'The central government funds must be matched the state government funds almost to an equal extent, which is often not accorded enough importance, and even the ULG's are expected to generate a certain amount of funds' says report published by IGSSS (2019). Most of the central funding is conditional. The provision of urban infrastructure has been the responsibility of the state. It is very daunting for the government to meet the demands for jobs, shelter, water, roads, transport, housing and other urban infrastructure. However, the government has been working in this direction, and there are now many programmes targeting urban developmental issues. There has been a great change in the urban policymaking and initiatives looking at the various issues in play. The shift from the centralized era of planned development to the decentralized, even localized era of participatory development is a major transformative process in India's urbanization. Now, it is also being backed by various design and technological interventions (Bhole, 2019). Table 2.7 below maps out the various urban initiatives taken in India.

A document of the Ministry of Housing and Urban Affairs entitled Transforming Urban landscape—2014–19 (2019) summarizes its approach as:

1. Having an inclusive approach focusing on empowerment in providing livelihoods and basic amenities; periodic review and state-level monitoring of missions to ensure adherence to timelines of projects;
2. Adoption of Public–Private Partnership (PPP) in various missions to give impetus to investment and create avenues for private sector participation;
3. Incentivizing and encouraging urban local bodies (ULBs) to become self-sufficient and mobilize resources for attaining viability through instruments like municipal bonds, etc.;
4. Emphasizing on improving urban governance for better service delivery, transparency and accountability.

All the urban initiatives are grandiose in design, but there are lacunae in implementation. The first major urban development programme was JNNURM in 2005. JNNURM was conceived as an umbrella programme with a project-based approach. It marked the start of large-scale central assistance in the urban sector. The mission sought to achieve this by integrating the development of infrastructure services and accelerating the flow of investment into urban infrastructure, through planned development, redevelopment and renewal of cities, inner-city areas, peri-urban areas, outgrowths and urban corridors; and through universal service delivery for the urban poor (PWC, 2016). However after 2014–15, the government rolled out many mission-based initiatives to focus on explicit goals and objectives in the specified areas. The

**Table 2.7** Urban initiatives in India

| Year | Initiative | Rationale |
|---|---|---|
| 1992 | The 74th Constitutional Amendment Act | Empower municipalities functionally, financially and politically |
| 2002 | Urban Reform Incentive Fund, (URIF) | Eliminate systemic weaknesses so as to strengthen municipal finance and functioning |
| 2005 | Jawaharlal Nehru National Urban Renewal Mission (JNNURM) | Eliminate structural and systemic weaknesses to create investment climate and to improve local governance and finance |
| 2007 | National Urban Housing and Habitat Policy (NUHHP)7 | Affordable housing for all with emphasis on vulnerable sections of society |
| 2012 | Rajiv Awas Yojana (RAY) | Working towards slum-free cities via property titling and tenure security |
| 2013 | National Urban Livelihoods Mission | Reduce poverty and vulnerability of urban poor households by providing them with access to employment opportunities |
| 2015 | Smart City Mission-development of 100 smart cities | Enhance the quality of urban life and provide a clean and suitable environment |
| 2015 | Housing for All (PMAY- U) | Address the housing requirements of the urban poor including slum dwellers |
| 2014 | Swachh Bharat Mission (SCM) | Access to sanitation<br>And Scientific management of solid waste |
| 2015 | Heritage City Development and Augmentation Yojana (HRIDAY) | Support core infrastructure aimed at revitalization of areas close to heritage sites |
| 2015 | AMRUT—Rejuvenation and transformation of 500 cities (>100,000 population) | Create infrastructure that has a direct link to the provision of better services to the people |
| 2016 | National RURBAN Mission(SPMRM) | Accelerate rural development with urban services |

*Source* India Habitat III Report, 2016

MOHUA has developed a three pronged strategy to tackle the challenges of urban transformation. Firstly, the Deendayal Antyodaya Yojana-National Urban Livelihoods Mission (DAY-NULM), Pradhan Mantri Awas Yojana-Urban (PMAY-U) and Swachh Bharat Mission-Urban (SBM-U) initiatives engage with the challenges of poverty alleviation, affordable housing and sanitation. Secondly, AMRUT looks after provisioning for basic infrastructure like water supply, sewerage, greening, etc., covering over 60% of the urban population. Thirdly, 100 cities are being developed under SCM to upgrade the living conditions and infrastructure. The main focus of this programme is area-based development of cities through increased use of digital technology. Much of the focus on urban renewal and transformation has been brought on by the forces of privatization and the international agencies. There has been a lot of criticism of the new initiatives, especially SCM. The critiques profess that most of the funds in these programmes lie underutilized due to lack of coordination, inadequate technical understanding and political biases. Another charge that is levied on the urban initiatives is that they rely too much on Western models which are not appropriate for a country like India facing social-economic inequalities. Prasad (2019) says, 'I believe that India must nurture its urbanization in indigenous ways, dealing with challenges by devising solutions that integrate a broader range of local factors. We must tap into lessons from ancient cities of India, while also drawing upon modern urban best practices that can be adapted in uniquely Indian ways. Understanding key emerging trends in India's urbanization is imperative to forge a new global framework of sustainable development'.

UN World Urbanization Prospects (2018) states, 'As the world continues to urbanize, sustainable development depends increasingly on the successful management of urban growth, especially in low-income and lower-middle-income countries where the most rapid urbanization is expected between now and 2050'. In the developing countries, rural development is still a challenge and now urban development is also being added to it. There is a proliferation of Rurban communities in India. Rurban communities are the rural socio-geographic spaces where styles of life and the standard of living have changed so much that they resemble those in urban localities (Parsons, 1949, p. 435). This has become possible due to increased interaction between rural and urban and development of a rural–urban continuum. There is a need to improve the lives of rural and urban population through integrated approach. Though urbanization has the potential to make cities more affluent and countries more developed, many cities are grossly ill-equipped for the multidimensional challenges associated with urbanization (UN Habitat WCR, 2016, p. 20).

Goal 11 of Sustainable Development Goals is also directed towards making cities and human settlements inclusive, safe, resilient and sustainable. Urban transformation is closely related to the three dimensions, namely economic, social and environmental of sustainable development. World Urbanization Prospects Report, 2018 states, 'Well-managed urbanization, conversant with the nuances of population trends over the long run, can help in maximizing the benefits of conglomeration while minimizing environmental degradation and other potential adverse impacts of a growing number of city dwellers'. The key aim of urban development policies and programmes should be to provide work as a facilitator for economic development in the country. There should be growth in all sectors like agriculture, extractive industries, manufacturing industries or the tertiary sector. Social justice and equity needs to be inbuilt in the policies and programmes of urban development so that no one is left behind and everyone is treated equally.

## References

Ahluwalia, I. J. (2017). Urban governance in India. *Journal of Urban Affairs*. Routledge. https://www.smartnet.niua.org/sites/default/files/resources/urban_governance_in_india.pdf. Accessed January 21, 2019.

Baum, S. (1999). Social Transformations in the Global City: Singapore. *Urban Studies,* No 36, http://usj.sagepub.com/content/36/7/1095.
Beteille, A. (1966). *Caste, changing patterns of stratification in a Tanjore Village.* Oxford University Press.
Bhole, M. (2019). India's urban transformation: A review of urban policies and challenges. *International Journal of Humanities and Social Sciences, 8*(3).
Bott, E. (1957). *Family and social network: Roles, norms, and external relationships in ordinary urban families.* Tavistock.
Burgess, E. (1925). The growth of the city. In R. Park & E. Burgess (Ed.), *The city.* Chicago University Press.
Cambridge Dictionary, https://dictionary.cambridge.org/dictionary/english/transformation. Accessed on November 12, 2019.
Castells, M. (2002). Urban sociology in the twenty-first century, concluding text. In: I. Susser's (Ed.), *The castells reader on cities and social theory.* Blackwell Publishers. Cities Magazine.
Chamie, J. (2019). The changing distribution of world population. http://www.ipsnews.net/2019/12/changing-distribution-world-population/. Accessed on December 30, 2019.
Clerici, A., & Mironowicz, I. (2009). Are landmarks essential to the city—Its development? In M. Schrenk, V. V. Popovich, D. Engelke, & P. Elisei (Eds.), *RealCorp 2009: Cities 3.0—Smart, sustainable, integrative, proceedings.*
Cormack, M. (1961). *She who rides a peacock: Indian students and social change—A research analysis.* Asia Publishing House.
Datta, P. (2006). *Urbanization in India.* https://pdfs.semanticscholar.org/9bd6/b2d2d78458680a48ce7f44d9f78524cd2bbb.pdf. Accessed on July 19, 2019.
Davis, K., & Golden, H. H. (1954). Urbanization and development in pre-Industrial Areas. *Economic Development and Cultural Change, 3*(1).
Desai, I. P. (1964). *Some aspects of family in Mahuva.* Asia Publishing House.
D'Souza, V. S. (1963). A case study of working mothers. *Social Action, 13*(11), 640–646.
D'Souza, V. S. (1974). Urban studies. In M. N. Srinivas, M. S. A. Rao, & A. M. Shaw (Eds.), *A survey of research in sociology and anthropology* (Vol. I). Popular Prakashan.
D'Souza, V. S. (1984). Urbanization as a perspective for social and historical analysis in developing societies. *Sociological Bulletin, 33*(1/2), 91–98. Retrieved from www.jstor.org/stable/23620098.
Durkheim, E. (1964). *The division of labour in society.* The Free Press.
Ghosh, I. (2019). https://www.weforum.org/agenda/2019/09/mapped-the-dramatic-global-rise-of-urbanization-1950-2020/
Ghurye, G. S. (1953). Cities of India. *Sociological Bulletin, March.*
Ghurye, G. S. (1962). *Cities and civilization.* Popular Prakashan.
Gluckman, M. (1971). Preface. In E. Bott (Ed.), *Family and social network* (2nd ed., pp. xiii–xxx). New Press.
Gore, M. S. (1968). *Urbanization and family change in India.* Popular Prakashan.
Gore, M. S., Desai, I. P., & Chitnis, S. (1970). *Field studies in the sociology of education: All India Report.* National Council of Educational Research and Training.
Graham, S., & Marvin, S. (1996). *Telecommunications and the city. Electronic spaces, urban places.* Routledge.
Harris, C. D., & Ulman, E. (1945). The nature of cities. *Annals of the American Academy of Political Science, 242,* 7–17.
Harvey, D. (1985). *The urbanization of capital.* John Hopkins University.
Hawley, A. H. (1981). *Urban society: An ecological approach* (2nd ed.). Wiley.
Hoyt, H. (1939). *The structure and growth of residential neighborhoods in American Cities.* Federal Housing Administration.
India Habitat III National Report. (2016). Ministry of Housing and Urban Poverty Alleviation, Government of India.

Indo-Global Social Service Society (IGSSS). (2019). *Looking Back at 25 Years: A Review of 74th Constitutional Amendment Act*—A National Level Roundtable. Available at https://igsss.org/wp-content/uploads/2020/01/74th-Constitutional-Amendment-Act.pdf.

Kapadia, K. M. (1959). *Marriage and family in India*. OUP.

Kolenda, P. (1984). *Caste in contemporary India—Beyond organic solidarity*. Waveland Press Inc.

Kothari, R. (1970). *Politics in India*. Orient Blackswan.

Kundu, A., & Basu, S. (1998). Informal manufacturing sector in urban areas—An analysis of recent trends. *Manpower Journal, 34*(1).

Kundu, D. (2014). Urban development pogrammes in India: A critique of JNNURM. *Social Change, 44*(4), 615–632.

Lomnitz, L. A. (1977). *Networks and marginality: Life in a Mexican Shantytown*. Academic Press.

Looking Back at 25 Years: A Review of 74th Constitutional Amendment Act—A National Level Roundtable. (2019). Indo-Global Social Service Society (IGSSS), New Delhi

Lynch, O. M. (1969). *The politics of untouchability-social mobility*. Columbia University Press.

Lynch, O. M. (1974). Political mobilization and ethnicity among Adi-Dravidas in Bombay Slum. *Economic and Political Weekly, 9*(39).

Maine, H. S. (1931). *Ancient Law its connection with the early history of society and its relation to modern ideas*. OUP.

Majumdar, D. N. (1958). *Caste and communication in an Indian village*. Asia Publishing House.

Mehdipoura, A., & Nia, H. R. (2013). Industrialization and city change; The concept and historical evolution of urban regeneration. *International Journal of Sciences: Basic and Applied Research, 12*(1), 176–181.

Merriam Webster.com Dictionary, http://www.merriam-webstre.com/dictionary/transformation. Accessed on July 12, 2018.

Ministry of Housing and Urban Affairs. (2019). *Transforming Urban landscape-2014–19*. Available at http://mohua.gov.in/upload/5c987f9e0fcecUTBook25March20191compressedcompressedcompressedmin11.pdf.

Mitchell, J. C. (1969). *Social networks in urban situations*. Manchester University Press.

Morgan, L. H. (1877). *Ancient society*. Macmillan and Co.

Mukherjee, R. K., & Singh, B. (1956). *Social profiles of a metropolis—Social and economic profiles of Lucknow*. Asia Publishing House.

Parsons, T. (1949). *Essays in sociological theory* (revised). The Free Press.

Prasad, S. (2019). World Economic Forum Annual meeting, https://www.weforum.org/agenda/2019/01/india-urbanization-why-the-world-should-watch/. Accessed on December 24, 2019.

PWC. (2016). Urban development in India: A special focus. *Public Finance Newsletter, Issue 11*, https://www.pwc.in/assets/pdfs/pf/archives/public-finance-quarterly-issue-xi.pdf. Accessed on December 24, 2019.

Rao, M. S. A. (1970). *Urbanization and social change: A study of a rural community on a metropolitan fringe*. Orient Longman.

Rao, M. S. A. (1974). *Urban Sociology in India*. Orient Longman.

Registrar General and Census Commissioner, India, Census reports, 1951, 2001 & 2011.

Roberts, P. (2008). The evolution, definition and purpose of urban regeneration. In P. Roberts & H. Sykes (Eds.), *Urban regeneration: A handbook* (pp. 9–36). SAGE Publications Ltd. https://doi.org/10.4135/9781446219980.

Ross, A. D. (1962). *Hindu family in its urban setting*. University of Toronto Press.

Rudloph, L. I. (1961). Urban life and populist radicalism: Dravidian politics in Madras. *The Journal of Asian Studies, 20*(3).

Sabyasachi, T., & Shupinder, K. (2018). Do negative externalities have any impact on populations agglomerations? Evidence from Urban India. *Theoretical and Empirical Researches in Urban Management, 13*(3), 5–24. Retrieved from www.jstor.org/stable/26472533.

Sachs-Jeantet, C. (1994). *Managing social transformations in cities. A challenge to social sciences*. http://www.unesco.org/most/sachsen.htm.

Simmel, G. (1950) In K. H. Wolff (Ed.), *The sociology of Georg Simmel*. Free Press.

Singer, M., & Cohen, B. S. (Eds.). (1968). *Structure and change in Indian Society*. Aldine.
Soja, E. (2000). *Postmetropolis. Critical studies of cities and regions*. Blackwell.
Srinivas, M. N. (1962). *Caste in India and other essays*. Asia Publishing House.
Stack, C. (1974). *All our kin: Strategies for survival in a black community*. Harper and Row.
Stack, C. (1996). *Call to home: African Americans reclaim the Rural South*. Basic Books.
Tönnies, F. (1957). *Community & society (Gemeinschaft und Gesellschaft)*. Transaction Publishers.
United Nations, (2019). *World urbanization prospects: The 2018 revision*. Online Edition. Department of Economic and Social Affairs, Population Division.
United Nations (2018). *World urbanization prospects: The 2018 revision, Key Facts*. Department of Economic and Social Affairs, Population Division.
UN Habitat. (2016). *World cities report 2016: Urbanization and development: Emerging futures*.
Weber, M. (1958). *The city*. Free Press.
Werbner, R. P. (1984). The Manchester School in South-Central Africa. *Annual Review of Anthropology, 13*(1), 157–185. https://doi.org/10.1146/annrev.an.13.100184.001105
WHO, Global Health Observatory (GHO) data available at https://www.who.int/gho/urban_health/situation_trends/urban_population_growth_text/en/.
Wirth, L. (1938). Urbanism as a way of life. *The American Journal of Sociology, 44*(1), 1–24. The University of Chicago Press, http://www.jstor.org/stable/2768119. Accessed on June 13, 2019.

# Chapter 3
# Shelter for the Poor: Conundrum, Visions and Methods of Urban Growth

**Bhawana Vasudeva**

**Abstract** In the last two decades, unlike the global North, the countries of the global South including India underwent a varied and swift urban transformation. During liberalization, economic change has become the most dominant aspect of explaining the phenomenon of urban growth in India. On the one hand, it symbolizes the narrative of India's ingress in the liberal world of global finance capitalism, in which both urban and the rural regions positioned transformed in the overhaul of technology and capital investment. On the other hand, the rapid urban growth resulted in lopsided Indian cities which are besieged by inadequate infrastructure and augmented land prices. In most Indian cities, the peri-urban region witnessed the development of plush gated housing societies and the growth of slums or development of Economically Weaker Section (EWS) housing with minimal basic facilities. In the growing scholarship on urban expansion in India, there are many potent accounts of the growth of slums, slum redevelopment and implementation of EWS housing schemes that have drawn our attention to the capricious and incongruous nature of urban policies and planning concerning shelter for poor. In the context of the above, this study discusses the paradox of urban growth in India, the role of multiple stakeholders, the frame of physical planning and governance and changing dynamics of housing for the poor. The study attempts to explore formal or informal urban governance processes, underlying socio-economic or political forces and changing socio-economic and spatial characteristics of shelter for the poor by focusing on the case of Vadodara. The research does not attempt to draw general conclusions rather it provides live shreds of evidence of particular situational dynamics. Understanding such urban growth dynamics and the study of other characteristics has the potential to improvise the current system of planning and governance of housing for the poor.

**Keywords** Urban growth · Housing · Peri-urban development · Shelter for poor · Socio-spatial structure

---

B. Vasudeva (✉)
Department of Architecture, The Maharaja Sayajirao University of Baroda, Vadodara, Gujarat, India

 
A. Kateja and R. Jain (eds.), *Urban Growth and Environmental Issues in India*,
https://doi.org/10.1007/978-981-16-4273-9_3

## Introduction

The cities are believed to be the growth engines of a country (Jacobs, 1961, 1969; Soja, 1989). In the last century, explosive urban growth and spatial expansion can be attributed to three divergent yet interconnected revolutionary forces, namely industrialization, transportation system and development of information technology. Though, during liberalization, economic change has become the most dominant aspect in explaining urban growth. Subsequently, the urban space was conceptualized as a social product, and it was observed that the pattern of spatial growth represents the internal reorganization of socio-economic and political processes in a particular time and society (Harvey, 1985).

Societies in global North or developed nations survived through the urbanization process which lasted more than 150 years. It had experienced a relatively smooth progression from rural to urban society. However, in the last two decades, unlike the global North, the countries of global South or South and East Asian countries including India experienced a diverse and rapid transformation.

Since economic liberalization, the expansion of urban activities has been rapidly transforming the peripheral regions of many Indian cities. It is observed that the changes in land use and the built environment of the urban periphery are not a natural process. It is considered to be a result of the new dynamics of capitalism which is happening under the prevailing impact of the global and local political economy of the world. It is believed that the contemporary urban expansions are shaped by the interlinked (socio-economic conditions, political–economic forces and formal and informal processes of urban governance) dynamics which include power, interest, choice and constraints of stakeholders (Banerjee-Guha, 2008, 2002; Harvey, 1978; Roy, 2005, 2015; Shaw, 2005; Sridharan, 2011; Webster, 2002; Yadav, 2011).

More recently, during the liberalization period, the phenomenon of urban growth is advancing in fast-growing economic regions of the world which led to the intense socio-spatial transformation in the periphery of many cities. This change has created several opportunities and also challenges for various groups of stakeholders in the developing countries.

As a result of rapid real estate urban growth, associated income has facilitated the improved living standards of cities of the global south, while on the flip side, urban areas have grown beyond their carrying capacity and created newer complexity for all stakeholders (Webster, 2002).

In many developing countries, the rapid urban growth is straining the capacity of the shelter delivery system of cities. As a result, slums or illegal urban settlements on public land grew in over a period of time in constant and irregular comportment. A slum is a ramshackle area of a city characterized by poor quality housing, poverty and deficient in tenure security. Dharavi is Asia's largest slum located on prime property in the central area of Mumbai (Jain, 2007).

Now, slum formation is observed as an integral part of urban growth and as a general phenomenon throughout the world which is a manifestation of the overall socio-economic policies and planning in the urban areas (Behera, 2007; Census of India, 2011).

In the past, Government of India has tried and implemented various implicit and explicit policies to improve the shortage of housing for the poor. Recently, the "Smart City Mission (SCM)", "Housing for All by 2022" Mission were initiated to develop basic amenities, construct houses for the Economically Weaker Section (EWS) of the society, and slum rehabilitation. However, it was observed that these policies and program implementation were not consistent with their objectives.

To reduce environmental threats and have sustainable, systematic and orderly development of urban areas, the cities need to be managed by using innovative planning approaches, new strategies and policy instruments and coordinated efforts of all stakeholders. In this context, through this study, an effort has been made to analyse and understand the changing socio-economic and political–economic dynamics of all stakeholders and its linkages with the issues and characteristics of shelter for the poor. The present study offers us a necessary base for policymaking and planning for providing housing and basic infrastructure facilities for the low-income group of people.

## Global Perspective on Processes of Urban Growth

Urbanization and urban spatial growth have been a worldwide phenomenon observed in various geographies of the world. Urban growth is considered as an index of change from traditional rural economies to a modern industrial one and is a process of concentration of population from expansive village settlements to the urban region (Davis, 1965). Currently, the globe is more urban than rural and the UN estimates that the world's urban population will increase from 54% in 2014 to 66% by the year 2050. At present 82% of the North American region is urbanized, and Asia remains the least urbanized region in which 40–48% population lives in the urban region. However, Asia is projected to become the most urbanized region in the future (refer to Fig. 3.1) (UN-Habitat, 2015). It is evident from the various studies that at present, the world is essentially urban where the temporally and spatial characteristics of urban growth differ from nation to nation (Castell, 1977; Gallion, 1986; Harvey, 1985).

Historically, the urban growth is considered an irreversible and influential phenomenon that impacted socio-economic conditions, built and natural environment and has brought immense global and local changes to all sections of the society including people living below the poverty line.

The Industrial Revolution, transportation and recent telecommunication revolution have made a multiplying effect on the pattern of urban growth. As a result, urban expansion has taken place in post-industrial cities of both the global north as well as fast-developing global south particularly the cities of Southeast Asian countries. Though, the phenomenon of urban growth became significant at the end of the nineteenth century (Castell, 1977; Gallion, 1986; Harvey, 1985).

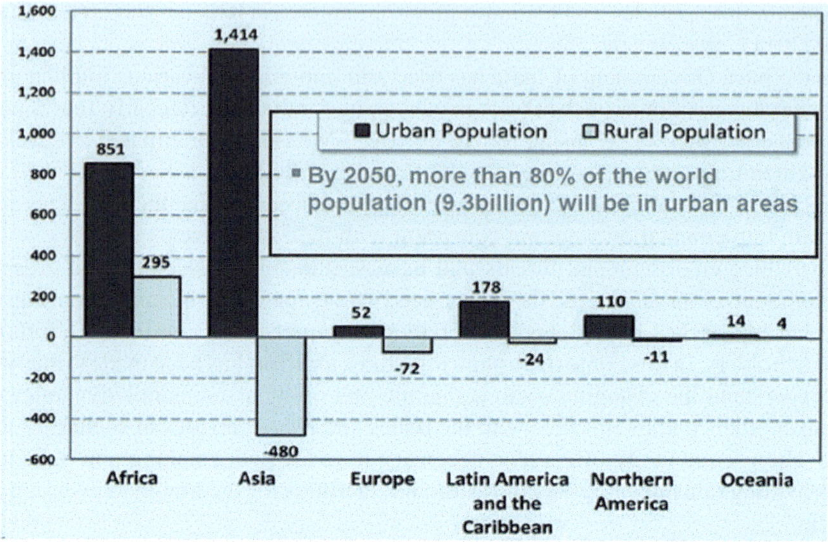


**Fig. 3.1** World urban population growth. *Source* New challenges of providing urban jobs, housing, energy and infrastructure seen by the UN. http://www.undp.org.rs. *Note* Changes in an urban and rural population in the major area between 2011 and 2050 in millions, urban growth estimation conducted by United Nations

It is observed that at different times, diverse yet interconnected forces, factors and processes were responsible for the rapid and volatile growth of urban regions of the world. Various studies demonstrate that 'urban growth' is identified differently and shows diverse characteristics and impact on various stakeholders, economies and social structure.

Primarily during the twentieth century, the process of urbanization in the global North has manifested in the form of suburban development, to be precise residential growth outside the boundary of the city. Great Britain and the United States of America are the best examples that have experienced the industrialization and growth of suburbs of the middle class much earlier than the countries of the global South.

The late twentieth century witnessed the restructuring of the global economy and the rise of information technology. The phenomenon is termed as the post-Fordist period and is associated with the deindustrialization of European and North American cities.

During this period, the manufacturing companies started outsourcing cheap labour from the global South. The investors moved towards industries, associated with design, architecture and entertainment. The economies of urban areas were revitalized and were inclined to benefit middle and upper-middle wealthier class urban residents. Financial markets, corporate service offices, banks firms and the headquarters of multinational companies were the important places for these transactions, which were located in urban regions (Sassen, 2006).

The post-Fordist urban centres, the image of the fragmented city has replaced that of a modern, functionally integrated city. Eventually, the development of shopping malls, office and commercial complexes, expensive business districts, full-service hotels, resorts and banquet halls and private residential communities and Economic Weaker Section (EWS) households and slum dwellers are all stakeholders of the new urban landscape.

## Urban Growth and Housing Patterns

Before industrialization in India, the surrounding land uses of cities were dominated by village settlements and agricultural farms. During the industrialization period, most of the industries and industrial residential townships were developed in the peripheral region of the cities.

However, in the post-industrial era, the growth and development of urban peripheries experienced a very specific form of residential and semi-residential communities that created social, cultural and economic changes within adjacent regions. The growth of urban settlements in India during 1970–80 was a manifestation of the post-independence economic policies (five-year plans) and ideological development processes which were mostly regulated by the State (Ramchandran, 1989). However, in a contemporary pattern of urban growth, the capitalist economy and liberalization of the market have become more important driving forces (Webster, 2002). In the post-liberalization period in India, the involvement of the state, private developers, local landowners and the choice of movement of an upper and middle-income group of stakeholders to the peri-urban region as well as the state-regulated physical planning decision-makers together have created a new arena, which seems to have both advantages and disadvantages.

According to Webster (2002), the general pattern of urban growth around urban areas in South Asia does not show the conventional clustered prototype of the village settlement. Individual stakeholders have built bungalows, country houses and residential complexes in a spread-out manner along the road, railway or canals.

Instead of migration, natural growth is an important factor that drove the urban growth and transformation in South Asia. The phenomenon of globalization is a new factor that has been attributed to increasing the speed and integration of Southeast Asia into the world economy.

The study conducted by Webster (2002) illustrates that the upper and middle-class residential development is one of the drivers of urban growth in South Asia.

The rapid urban expansions in India have been achieved with the help of specific logic of conversion of farmland available in the peripheral region, to various non-farm land uses. The elites and strong private groups indulge in informal real estate development practices by using the tools of formal spatial planning (Roy, 2005). According to Banerjee-Guha (2008), the newer forms of exploitation and respective social realities came to surface, and these were both motivations and opportunities for furthering the urban growth processes.

## Urban Growth and Governance Pattern: Shelter for the Poor

In the cities of the developing world, more jobs are created in the informal sector as the formal sector failed to provide jobs to the youth or migrant workers in the formal sector. The informal economy and urbanization had a very tangible impact on the housing environment and quality of life of the poor. The UN-HABITAT's audit on slum explains that in 2001 three out of ten residents were living in slums, substandard housing, often negotiating and modifying physical and environmental standards. Earnings vulnerability mass poverty, urban informal employment, urban land scarcity and lack of affordable housing options in the core urban areas led to the growth of slums (Behera, 2007; Jain, 2007).

Poor people build their substandard housing in the cheaper peripheral region of many cities. The cost of building a house in a periphery was cheaper, however, that increases their cost of transportation to work and other important employment opportunities available in the core city areas. The shelter for the poor also contributed to local environmental degradation mainly because of the lack of basic infrastructure available to them. It has taken a toll on their health and negatively impacted physical and natural resources existing in the vicinity. It is observed that poor quality of housing, non-availability of basic amenities and facilities in the peripheral area led to conflicts, urban violence and criminal activities (Mohanty & Mohanty, 2005).

The concentration of population in the urban growth region has led to several predicaments such as an increase in traffic congestion, encroachment and growth of slums, water and air pollution. Hence, it was recognized that urban infrastructure needs to be organized to take a greater load (UN-Habitat, 2007).

Consequently, in the early twenty-first century, the urban growth of developing economies was conceptualized with a governance overtone which discussed the issues of management of physical planning, infrastructure provision, financial provision, land monitoring system and role of various stakeholders involved in the process including slum dwellers (Allen, 2003). In this context, researchers recognized that the current urban growth processes are very complex and indistinguishable because of vagueness in the physical planning and governance structure of such regions and reflect in the form of various land use and built environment conflicts, conflicts among stakeholders (Sridharan, 2011).

The urban growth in South Asian countries has positively contributed to the local and regional economic development. However, foreign direct investment (FDI) or global capital has brought certain changes in spatial or built environment characteristics of urban growth. It has created segregation between the middle and lower-income groups of stakeholders.

Later after liberalization in India, rapidly growing middle class and upper-middle class choose to settle in the periphery of the cities. Economic transformation, the increasing social complexity among the rapidly growing middle class, socio-spatial segregation, multiple formal and informal stakeholders together pose unique challenges for urban planning and governance system (UN-Habitat, 2015).

## Objectives of the Study

The main theoretical frame of this study focuses on an examination of the role of stakeholders in interlinked urban growth processes. The present study aims at (a) Understanding the process, characteristics and role of all stakeholders in land use and built environment changes that have occurred due to the urban growth of Vadodara. (b) It also intends to critically analyse the processes of urban growth and the provision of shelter for the poor by understanding the nature of conflicts and exploring the relationship of changes in land use and built environment characteristics with socio-economic and demographic characteristics of all stakeholders.

## The Context

This exercise is an offshoot of a larger study of peri-urbanization. Vadodara is the main urban centre around which peripheral region is considered for this study which comprises the role of stakeholders in urban expansion. The choice of locating the fieldwork for this research in the periphery of Vadodara is, if anything, deliberate. Vadodara ranks sixteenth out of thirty-five urban agglomerations and is one of the fastest-growing metropolitan cities in India. This city has a very old history of human settlement regardless of numerous stages of ups and downs in its growth and development. The city has witnessed maximum planned development efforts, socio-economic reforms and adequate modern urban infrastructure development activities in nearly seventy-five years of span before independence in the year 1947.

In this study, the peri-urban region and villages selected for the case study are located in the Vadodara sub-district (Fig. 3.2).

The area of the Vadodara sub-district (urban) is 582 sq. km, population is 1,822,532, the area of Vadodara sub-district (rural) is 404.7 sq. km, and the population is 186,902.

As per the 2001 census report, the Vadodara sub-district is one of the most urbanized sub-districts of the state and has 87.49% urban population with an average population density of 6978 people per km$^2$. The decadal urban population growth (2001–2011) is 22.2%, while decadal rural population growth is −14%.

It can be concluded that the Vadodara sub-district continues to get urbanized at a faster pace. Negative rural population growth and increasing urban population growth indicate changing characteristics of the peri-urban region due to urban growth. Vadodara district is considered a socially, economically and culturally advanced district of Gujarat state and is known for its illustrious educational institutes (Census, 2011).

In the early nineties, the peripheral areas of Vadodara witnessed the economical weaker section (EWS) housing and slums. After a decade, dispersed tenements, luxurious duplexes, weekend homes and farmhouses sprang up in the middle of many agricultural fields in west and south-western, the north-eastern and eastern periphery

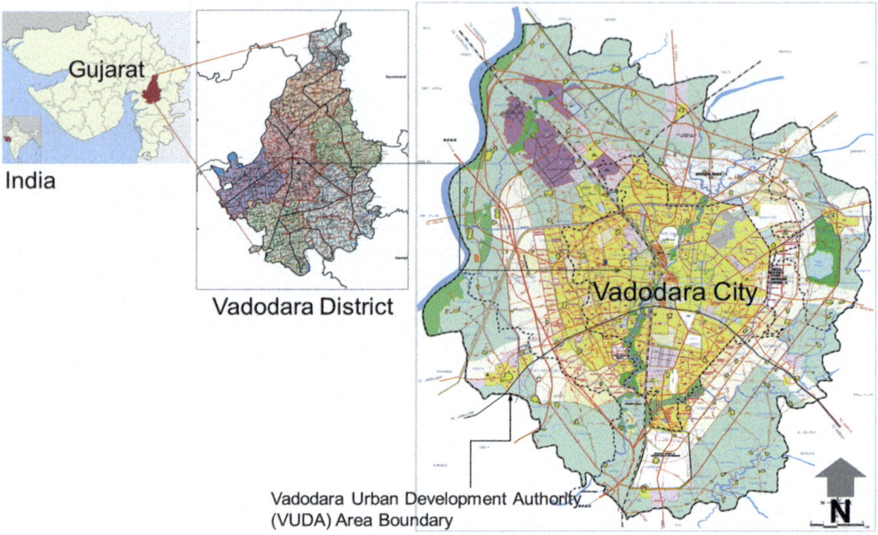


**Fig. 3.2** Vadodara and surrounding area—Location of research case study area, shown in VUDA boundary map. *Source* Vadodara district map sourced from www.mapsofIndia.com, Vadodara city and villages in VUDA boundary map sourced from VUDA office. *Note* The above image shows the Vadodara city and surrounding area. Most of the case study sites are located inside the Vadodara Urban Development Authority (VUDA) limit. The VUDA limit is located in the Vadodara sub-district and part of the Vadodara district in Gujarat

of Vadodara which includes a surrounding area of villages, namely Bhayli, Atladra, Samiyala, Talsat, Sevasi, Sherkhi, Ampad, Mahapura, Sindhrot, Vemali, Sayajipura, Sikandarpura and Kapurai, respectively.

## Data Sources and Methodology

The main conceptual frame of the research comprises the following three main areas and two sub-areas as delimitation for this research.

- The paradox of urban growth and linkages with the housing for the poor.
- Multiple stakeholders and their roles.
- The frame of physical planning and governance.
- Conflicts and crimes.

The research approach is based on the assumption that the recent spatial changes in urban growth of smaller cities of India are not a natural process. The economic and political forces (accumulation of capital in the real estate market) have been strongly influencing the way different stakeholders including the poor interact in a local arena of urban growth. It is observed that interests or apathy, knowledge or ignorance, weaknesses or strengths of stakeholders and consequent interrelated processes have been altering the built environment characteristics, economic structure and fragile ecosystem of the peripheral region of a city (Sridharan, 2011; Yadav, 2011).

This research draws on the findings of case studies that involve deskwork, fieldwork and interviews and methods of studying with the qualitative approach of data collection. The data collection method of the research is a mixed method consisting of document surveys, on-site observations, questionnaire surveys, in-depth interviews and key informant interviews conducted with stakeholders. Mixed methods are utilized to gain complimentary views about the phenomenon of urban growth and the role of stakeholders in built environment transformation. The mixed-method approach is used to obtain data from all five groups of stakeholders and includes face-to-face in-depth interviews, a questionnaire survey and key informant interviews and document reviews. An empirical study collected quantitative data from a total of 371 stakeholders from the category of peri-urban households and adjoining village residents, using the survey questionnaire as the primary data collection instrument.

The objective of the household survey, in-depth and key informant interviews was to explore the linkages or relationship between factors of residential and semi-residential, EWS housing development, slum formation and conflicts between stakeholders due to change in built environment and role of stakeholders in urban growth. The SPSS was used to perform various statistical analyses (chi-square test) to find linkages or association between variables. Qualitative analysis, content analysis, descriptive analysis, factor analysis, narrative analysis and thematic analysis methods are employed to find linkages or relation between two or more phenomena of change in the built environment.

## Urban Growth Processes and Shelter for the Poor: Characteristics and Issues

It is often observed that in developing countries urban growth is augmented by the rural–urban migrants. These migrant workers generally engaged in temporary jobs or work as contract labourers who comprised of a sizeable slum population in the cities and towns. Vadodara is the third-largest city in Gujarat and the 16th largest city in India. The city is the administrative headquarters of the Vadodara district and is located on the banks of the Vishwamitri River. The population of the city as per the 2011 census is 18.2 million. At present, 23% of workers were employed in the public sector, while 50% worked in the unorganized sector.

Peripheral areas surrounding Vadodara have become strategic zones where the expansion of urban growth and rural–urban transformation has been taking place. Considerable spatial transformation and land-use changes have occurred particularly concerning the agricultural land beyond the city corporation boundaries. Similar trends have also been observed in other Indian cities (Dikshit, 2011).

To understand the urban growth pattern in general, land-use change and built environment characteristics in particular and to reveal pertinent predicaments, it seems essential to view the changes in Vadodara (Gujarat) from a broad political and economic perspective that comprises the capital logic, economic and urban development policies, political rule, institutions, instruments and urban governance system.

Vadodara has an old history of human settlement and is known for socio-economic reforms and urban infrastructure development that was carried out during the period of ruler Maharaja Sayajirao III. Along with village landowners and residents, the other stakeholders, namely central government (economic and industrialization policies), Gujarat state (formed in 1961), VMC and VUDA, were involved in the process of urban development (DP, 2031).

The rapid industrialization, high population growth and extension of city limits during the year 1951–91 and later low population growth have influenced the land use and built environment of Vadodara. In the last three decades, the population and a population density of the Vadodara peri-urban area have grown by two and a half time. The area of the VUDA region was revised in 2011 DP to cover approximately seven times more area than in the last two decades.

The broad pattern of land use and built environment of Vadodara was similar to any other city which grows from a historic core and mainly consists of a residential area, roads, railways, industrial development, restricted areas, river, lakes, open spaces, gardens and commercial areas.

The thrust areas of the second five-year plan (the year 1956–1961), the discovery of oil and gas in the vicinity of Vadodara and the state's attitude towards industrial development have facilitated the development of public sectors shaped the urban growth of Vadodara.

Nine industrial units received less expensive gas fuel to run operations. The state government also offered concessional rates and exemption from certain duties to industries. The essential infrastructure such as land, water and power was made available easily by Gujarat Industrial Development Corporation (GIDC) along with the finances that were made available by Gujarat Industrial Investment Corporation (GIIC) (Cirvante, 1971).

The oil, petrochemical and chemical public sectors, namely Gujarat State Fertilizer and Chemical Limited (GSFC), Gujarat Refinery (GR) and Indian Petrochemical Corporation Limited (IPCL) now Reliance Industries Limited (RIL), Gas Authority of India Limited (GAIL) and Oil and Natural Gas Corporation Limited (ONGC) were established in the north-western and southern peri-urban region of Vadodara. Above industries and industrial residential townships were located near villages, namely Gorva, Undera, Nandesari, Bajwa and Makarpura. After the establishment of industries, the population growth of Vadodara increased in every consecutive decade with maximum increase took place between years 1961–1971. During this period the land use and the built environment of surrounding villages had changed (Rajyagor & Tripathy, 1979; Rangaswami, 2011). The VMC (earlier named as VMSS) area increased from 24.63 to 108.22 km$^2$.

As the urban development progressed in Vadodara, the north periphery has been occupied by public sector giants and big private industries, and the eastern periphery is developed due to proximity to the old city core. Initially, the southern and western peripheries of Vadodara were less populated as compared to northern and eastern peripheries.

A large part of poor people is employed as blue-collar workers in various industrial units developed in the peripheral area of Vadodara. City also witnessed a heavy population migrated from the rural areas which resulted in slums. These slums were located on the underprivileged sites such as river banks, near nallah, low lying areas of the core city or in peripheral areas near surrounding villages. They are typically a cluster of temporary shelters with rundown and infirm structures having common or no toilet facilities, suffering from lack of basic infrastructure. These shortages make the living conditions extremely sub-optimal, unhygienic, and result in a usually higher outbreak of various diseases among the slum dwellers.

In the early nineties, the urban growth in Vadodara witnessed the dispersed tenements, duplexes, weekend homes, farmhouses that sprang up in the middle of many agricultural fields in the west and south-western. EWS housing and slums were located in the north-western, north-eastern and eastern periphery of Vadodara which includes a surrounding area of villages, namely Bhayli, Atladra, Samiyala, Talsat, Sevasi, Sherkhi, Ampad, Mahapura, Sindhrot, Vemali, Sayajipura, Sikandarpura and Kapurai, respectively. Apart from residential and semi-residential pockets, EWS housing, slums, there are diverse land uses such as 40-m wide road in the western part, private school and universities in the south-western region, Narmada canal near Vemali in the northern region, transportnagar in the north-eastern region, and religious institutions such as ashrams, temples have developed in the eastern periphery of Vadodara. In recent years, many small and big industries have developed along highway NH 8. Land uses such as educational institutes, hotels, restaurants and other service industries are fast developing.

The formation of the new Gujarat state in the year 1960 brought many changes in the urban growth of the city. Many people and industries shifted base to Vadodara and settled in the peripheral areas. The VMC had prepared the first DP of Vadodara in 1970, and subsequently, many TPSs were also developed.

The Gujarat Town Planning and Urban Development Act 1976 (GTPUDA) and formation of VUDA in 1980 changed the land use and built environment scenario of Vadodara. The VUDA was made responsible for making a plan for 103 peripheral villages. In 1984, VUDA had prepared the second DP of Vadodara in which the city was decentralized in terms of functions and infrastructure.

At some point in the late twentieth century, a new type of urban growth emerged which had different characteristics as compared to previous patterns of development. Peri-urban development started to play an important role in the process of urbanization expansion, and now, peri-urban development is no longer just a substandard low-cost housing (EWS housing, shanty houses, slums) locality (Fig. 3.3).

Recent urban growth is planned to develop into self-sustainable settlements. The peri-urban settlements take the form of a clustered development which includes a combination of duplex, multi-storied buildings and bungalows which are different

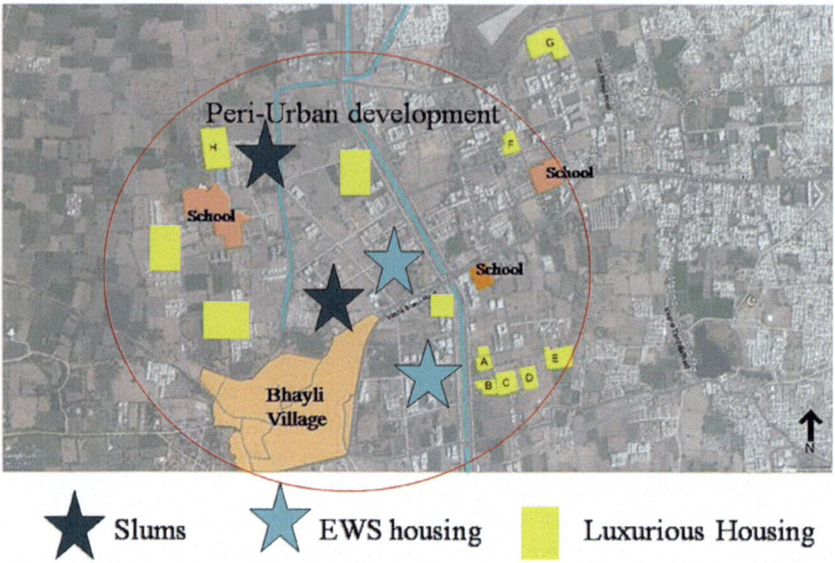


**Fig. 3.3** Urban growth, slums, economical weaker section (EWS) housing and luxurious residential area. *Source* Base map procured from Google Earth and compiled by the author. *Note* The above image shows the location of luxurious residential case study sites, slum and EWS housing, and Bhayli village located in the peripheral region of Vadodara city

from the former run of the mill built-up expansion or large industrial townships developed during mid-seventies and early eighties. The new form of development includes farmhouses, weekend homes and plotted development with infrastructure. The quality of amenities and facilities has spread extensively closer to the villages and the housing for the poor, situated in the vicinity.

**Case Study: Economical Weaker Section (EWS) Housing—The Built Environment of Sainath Co-Operative Housing Society, Bhayli**

Sainath co-operative housing society, economical weaker section (EWS) housing developed during the year 2012–13 near Bhayli village which is located in the western part of Vadodara. This residential complex comprises a total of 300 flats in four floor high buildings. There are a total of five linear buildings in an approximate area of 5500 m$^2$ and arranged on both sides of the internal roads. The common open spaces are located on the backside of the plot and used for various purposes such as marriage functions, get-togethers and celebration of festivals. This residential campus was built under JNNURM Awas Yojana and comprises two varieties of flats in which the number of bedrooms varies. The campus was projected as an affordable and safe living environment for a lower-income group of people with basic facilities such as bore well for water supply, electricity and soak pit (Fig. 3.4).

This society mainly occupied by a lower-income group of people with an average age of 30–40 years living in a joint family. The majority of them belong to the

**Fig. 3.4** Slums, economical weaker section (EWS) housing and luxurious residential area. *Source* The Sainath co-operative housing society, luxurious housing area, map procured from Google Earth, and photographs taken by the author. *Note* The above photograph shows various details that comprise information concerning views and details of Sainath co-operative housing society which was built in the year 2012–13 at the Bhayli peri-urban area of Vadodara. The remaining pictures show the slums and luxurious housing developed in the vicinity

uneducated or primary level educated category. They are engaged as blue-collar industrial workers, vegetable sellers, auto-rickshaw drivers, plumbers, electricians or as domestic help in nearby residential societies. Some of them are small-time entrepreneurs and running their small businesses. This society consists of a particular group of people who belong to a specific state of India and relocated to the core city after living in the slums. Subsequently, they shifted from core city to the peri-urban area after the construction of Sainath co-operative housing society under the slum rehabilitation program.

## Results and Discussion

Despite significant progress in economic growth and poverty alleviation programs, India's large percentage of the population is living below the poverty line. Poverty is multidimensional. It has a direct impact on the quality of life of people. Poor people lack access to minimum shelter and basic infrastructure facilities. With the beginning of the post-liberalization period (twenty-first century), the urban growth of Indian cities witnessed the influx of new projects including luxurious residential development. Since Gujarat has a high degree of urbanization, the characteristics of urban growth and shelter for the poor require probing. Vadodara is considered one of the most promising cities in the country as it is one of the best-connected cities in India by all three modes of transport. The city is an education and industrial hub and also known as the cultural capital of Gujarat (BCC, n.d.). The city offers ample employment opportunities for all stakeholders including the poor. Vadodara is considered one of the safest cities where crime has consistently remained low (Roy, 2007).

Under the umbrella of Smart City Mission, the city was envisioning slum-free, and it was the key component of the mission. Hence, 186 out of 397 identified slum pockets removed under the slum rehabilitation program. New 18,000 houses constructed to rehabilitate slum dwellers. Wellness and de-addiction centre was also planned in the newly developed facilities for the poor (MOUD GOI, 2016).

During the process of urban development and provision of housing for the poor in Vadodara, the Vadodara Municipal Corporation (VMC), Vadodara Urban Development Authority (VUDA) played an important role by absorbing more number of villages.

In the context of urban growth and shelter for the poor, the following are the main characteristics and issues that were observed.

## *Urban Growth, Land-Use Externalities, Conflicts and Shelter for the Poor*

The study reveals that processes of conflicts also play an important role in shaping the urban growth and shelter for the poor. Once the process of change begins, the individual choice, interest, power and investment capacity of different groups of stakeholders particularly the private sector played an important role to give basic form to city growth. The upper-middle class of stakeholders seeks affordable as well as luxurious residences, weekend homes and farmhouses, mostly for investment purposes. Such built spaces are used occasionally and hence not maintained regularly by owners. Eventually, such a situation leads to various types of conflicts that arise between urban residents, neighbours and the poor villagers. The development of a weekend home generates many disagreements and conflicts between developers and poor marginal farmers. The farms which are not part of the weekend home but located in the adjoining area receive rough treatment. The access to the farm is denied or blocked to non-participant individual farmland and therefore becomes a point of constant conflict. The new pattern of urban growth eventually pressurizes the marginal or poor farmers to egress the agriculture activities.

## *Urban Growth, Politics of Physical Planning and Shelter for the Poor*

The real estate developers from the emerging upper-middle class, through close connections with both the state government and development authority, can intervene and take advantage of vaguely defined borders (VUDA boundary and surrounding villages) in development planning and consequently shape the growing urban built environment in the favour of land nexus illustrated by the case study of weekend homes (Iscon Green weekend home at Bhaniyara, Vadodara), constructed at a strategic location before finalization of revised Development Plan (DP) at a much cheaper cost. A cricket stadium is proposed in front of the present weekend home location). Instant market alternatives are provided by outsourcing the facilities for a higher income group of stakeholders though; a low-income group of residents is not part of such an arrangement. It is the primary responsibility of the state to develop basic infrastructure; however, the basic infrastructure is developed for the powerful people, and poor people do not get the advantage of it though residing in the vicinity. The above conflicts negatively impact the basic livelihood sources (farming) of the poor.

The politics of drawing the boundary in a development plan (DP) changes the dynamics for the urban growth at two levels, one in silencing and making marginal farming community and low-income group of people invisible and misrecognized and second facilitating real estate developers in achieving lucrative land transaction deals (Vasudeva, 2018).

## *Urban Growth, Community and New Socio-spatial Structure for Residents*

While in Vadodara, the newly developed areas essentially belong to the Hindu community in which the role of state and [precondition of obtaining a no-objection certificate (NOC) from developers or prospective owners of the properties] developer [indirectly refuses home seekers from other than Hindu community is very obvious which led to the formation of territories based on religion particularly in post-liberalization period (Vasudeva, 2018)]. When the slum rehabilitation program was implemented, it was observed that VMC officials were pressurized against allotting newly built houses to a particular community people (Lobo, 2015). However, in Vadodara, the nearness and concentration of other built spaces where the majority of it was occupied by a similar religious group of people remains significantly important arrangement which is perceived to facilitate the human interaction possible within land nexus and forced social stratification.

### *Urban Growth, New Socio-spatial Structure and Economic Opportunities for the Poor*

The analysis results demonstrate that the new urban growth pattern shows a dualistic socio-spatial structure where a mix of lower-income (shifted to undeveloped peri-urban areas in the 1980s under resettlement housing schemes or slum rehabilitation scheme or homeless villagers) and upper-middle-income group (shifted in last one decade) of people live nearby. This new phenomenon, though not planned, is economically proving beneficial to the urban poor, as they get employed in proximity to their home. Their expenditure on transportation (daily commute to core city for work) also gets minimized. The family income rises as each family member gets an opportunity to work (Vasudeva, 2018).

### *Urban Growth, New Socio-spatial Structure and Shelter Insecurities for the Poor*

The study shows the improved condition of infrastructure facilities in one area created by new collaborations which originated with the vested interest of an influential developer and were furthered due to local authority's passive attitude. It can be stated that above processes and local politics manifested in terms of the highest rise in property prices in and developed uncertainty and insecurity among urban poor concerning their future period of stay as the area fetches high property prices as compared to any other infrastructure deficient developing area of the city. As a result, conflicts are also generated among small-time developers as they find it hard to sell their properties and are also not able to collaborate to develop infrastructure facilities. The whole process finally resulted in more infrastructure deficient, conflict-laden urban growth where state extract profit along with influential private sector and poor people remain in dilemma.

## Conclusion

This paper seeks to expand the understanding of urban growth dynamics and its impact on the shelter for the poor. During the last three decades, the rapid pace of urbanization and globalization has resulted in explosive urban growth throughout India. Rapid urbanization comes with the challenges of providing employment, housing and basic infrastructure to all stakeholders—a difficult task that is not being met as is evident in most of the cities including Vadodara. The effects of rapid urban growth are manifested in crowded slums in the centre of the city and mushrooming shanty habitation in the peri-urban region of a city. In the past, Government made various policies and programs such as JNNURM, Basic Service for the Urban Poor

(BSUP), Rajiv Awas Yojana (RAY) Pradhan Mantri Awas Yojana (PMAY) to make affordable housing available to poor and improve the condition of slum dwellers. Many EWS housing or slum rehabilitation sites were located in the peri-urban region far from the core of the city. In year 1990s, India moved towards economic liberalization which brought about changes in economic policies, political culture and consumption patterns of the people. By this time, the peri-urban regions of the cities have begun to be filled by the influx of upper-middle and middle-income groups of the population. The urban growth manifested in the form of new luxurious residential settlements, farmhouses and weekend homes in the peripheral region where EWS housing or slum rehabilitation sites were also located.

It can be concluded from the study that earlier the industrial and real estate development reciprocally strengthened each other; however, currently, the power of state and private sector is evident that influences the political rationality of speculative world of real estate by engendering new forms of informal planning and governance processes which intervene in poor people's shelter rights and livelihood practices in ways that add to their vulnerabilities, risks and conflicts. At the same time, the poor got better employment opportunities and access to basic infrastructure facilities. Close observation of research findings indicates the market forces, choice and constraints of stakeholders, and various informal processes play an equally important role in characterizing the shelter for the poor and pattern of urban growth. The research demonstrates that regardless of many challenges such as lack of understanding about the sociology of space and consideration in urban planning tools, the recent changes in the new economic order and vested interest of the powerful stakeholders, the current phenomenon of urban growth has brought a new position to the development of shelter for the poor. Such linkages and studies of other aspects have the potential to improvise the current system of urban development and providing shelter to the poor or homeless.

## References

Allen, A. (2003). Environmental planning and management of the peri-urban interface: Perspective of an emerging field. *Environment and Urbanization, 15*(April), 135–148.

Banerjee-Guha, S. (2002). Shifting cities: Urban restructuring in Mumbai. *Economic and Political Weekly, 37*(2), 121–128.

Banerjee-Guha, S. (2008). Space relations of capital and significance of new economic enclaves: SEZs in India. *Economic and Political Weekly, 22*(November), 51–59.

BCC. (n.d.). *Credit Connection-Baroda Citizens Council.* Retrieved dated April 4, 2020, from https://www.wsp.org/sites/wsp/files/publications/sa_creditbaroda.pdf

Behera, S. (2007). Role of urban poor in growth of cities and available support system for their sustenance in the context of state capital region, Orissa, Bhubneshwar. In *Conference Proceeding of 56th National Town and Country Planners Congress* (pp. 179–186). ITPI, India.

Castell, M. (1977). *The urban question—A Marxist approach.* Edward Arnold.

Census of India. (2011). Retrieved dated April 15, 2019, from https://www.censusindia.gov.in.

Cirvante, S.R. (1971). Prospects of private enterprise industries in Baroda. In *Profiles of a growing city socio-economic study of Baroda city*. Vadodara, India: Department of Economics, The M.S. University of Baroda.

Davis, K. (1965). Urbanization of the human population. *Scientific American, 213*(September), 40–55.

Development Plan 2031 (DP 2031). (n.d.). Retrieved on 12 July 2016 from http://www.vuda.co.in///organization.htm.

Dikshit, J. K. (Ed.). (2011). *The urban fringe of Indian cities*. Rawat Publications.

Gallion, E. (1986). *The urban pattern city planning and design*. CBS Publishers.

Harvey, D. (1978). The urban processes under capitalism: A framework for analysis. *International Journal of Urban and Regional Research, 2*(1–4), 101–131.

Harvey, D. (1985). *The urbanization of capital: Studies in the history and theory of capitalist urbanization*. John Hopkins University Press.

Jacob, J. (1961). *The death and life of great American cities*. Random House.

Jacobs, J. (1969). *The economy of cities*. Random House.

Jain, A. K. (2007). Integrating needs of the urban poor in the urban process. In *Conference Proceeding of 56th National Town and Country Planners Congress* (pp. 147–155). ITPI, India.

Lobo, L. (2015). What Vadodara's Slum Displacement Reveals-Case of Kalyannagar and Kamatipura. *Economic and Political Weekly, 50*(36), September 05, 2015, from https://www.epw.in/journal/2015/36/. Accessed on April 23, 2020.

Mohanty, L. N. P., & Mohanty, S. (2005). *Slums in India*. APH Publishing Corporation.

MOUD GOI (Ministry of Urban Development, Government of India GOI). (2016). *The Smart City Challenge Stage 2-Smart City Proposal*. Retrieved from https://smartnet.niua.org/sites/default/files/resources/Vadodara_SCP.pdf, dated December 12, 2019.

Rajyagor, B., & Tripathy, S. (1979). *Vadodara District Gazetteers*. The Director, Government Printing Stationery and Publications.

Ramchandran, R. (1989). *Urbanizations and urban systems in India*. Oxford University Press.

Rangaswami, V. (2011). Urban fringe of Vadodara: Development and transformation. In J. K. Dikshit (Ed.), *The urban fringe of Indian cities*. Rawat Publications.

Roy, A. (2005). Urban informality-toward an epistemology of planning. *Journal of the American Planning Association, 71*(2), 147–158.

Roy, S. (2007). Integrating needs of the urban poor in the development of process-key issues in Kolkata metropolitan area. In *Conference Proceeding of 56th National Town and Country Planners Congress* (pp. 172–178). ITPI, India

Roy, A. (2015). Governing the postcolonial suburbs. In Hamel, P., & Keil, R. (eds.). Suburban governance a global view, Toronto, Canada: University of Toronto Press.

Sassen, S. (2006). *Territory, authority, rights: From medieval to global assemblages*. Princeton University Press.

Shaw, A. (2005). Peri-urban interface of Indian cities growth, governance, and local initiatives. *Economic and Political Weekly*, January 8, 129–136.

Soja, E. W. (1989). *Postmodern geographies: The reassertion of space in critical social theory*. Verso.

Sridharan, N. (2011). Spatial inequality and politics of urban expansion. *Environment and Urbanization ASIA., 2*(2), 187–204.

UN-Habitat. (2015). *The State of Asian and Pacific Cities 2015: Urban transformations shifting from quantity to quality*. Retrieved from www.unescap.org, dated 10-1-2017.

UN-Habitat. (2007). *Enhancing urban safety and security: Global report on human settlement 2007*. Earth Scan.

Vasudeva, B. (2018). *Linkages between Built Environment Changes, Physical Planning, and Role of Stakeholders in the Peri-Urban Areas*. Unpublished Ph.D. Thesis. Ahmedabad, India: CEPT University.

Webster, D. (2002). *On the edge: Shaping the future of Peri-Urban East Asia*. Stanford University, Stanford, California.

Yadav, H. S. (2011). Expansion of urban capital and marginalization of fringe—An antipode of urban development. In J. K. Dikshit (Ed.), *The urban fringe of Indian cities*. Rawat Publications.

# Chapter 4
# Key Determinants of Urbanization in India: Evidence from State-Level Data

**Seema Joshi**

**Abstract** Urbanization and economic growth are closely related. Previous studies have revealed that there is a positive impact of economic growth on urbanization. As nations embark on the path of development, migration from rural habitats to urban centres in search of better livelihood opportunities and better incomes among other things is a key feature of structural changes. Workers' movement not only lifts their income and living standard but also thrusts them into the 'consuming class'. As higher purchasing power puts them in a better position to consume more, this huge consuming class adds fuel to the engine of industrialization and urbanization. Economic growth, of course, is very critical to urban growth, but studies focusing on certain other variables which on a priori can be expected to influence urbanization rate are very less in number. Our country-specific diverse data approach to understand and analyse the correspondence between urbanization rate and level of economic growth controlling for some other variables (as important determinants of urbanization rate) demarcates (constitutes) a clear-cut departure from the often done micro-level or country-level studies using panel data approach for identifying the determinants of agglomeration economies. The objective of the present paper is to ascertain the key determinants of urbanization in India. Taking cross-section data of 28 states and two Union Territories (UTs) of India for which data was available in case of all the variables and making use of multiple regression technique and ordinary least squares test procedure, we find evidence that economic growth, industrialization and infrastructure are the three key determinants of urbanization rate in case of India.

**Keywords** Urbanization · Economic growth · Industrialization · Infrastructure

## Introduction

Urban centres have always been engine of economic growth, industrialization and enhanced productivity. It is also equally true that economic growth stimulates urban concentration. Economic success and wealth creation in cities and sprawls contribute

S. Joshi (✉)
Department of Commerce, Kirori Mal College, University of Delhi, Delhi, India



A. Kateja and R. Jain (eds.), *Urban Growth and Environmental Issues in India*,
https://doi.org/10.1007/978-981-16-4273-9_4

to pulling out millions of people out of poverty. Cities attract skilled, semi-skilled and even unskilled labourers who contribute to productive business. 'The urban economic clout' is the most significant factor that attracts talented individuals to urban centres (Eggers & Skowron, 2018). As nations embark on the path of development, migration to urban areas causes rise in the income levels and living standards of the workers thrusting them into the 'consuming class'. As higher purchasing power puts them in a better position to consume more, this huge consuming class adds fuel to the engine of industrialization. The spiral process, realizing centrifugal force, makes these urban centres pockets of prosperity and growth.

There has been abundant literature (Abdel-Rahman et al., 2006; Bertinelli & Strobl, 2003; Henderson, 2003; Mooaw & Shatter, 1993) which postulates a strong positive correspondence between level of urbanization and economic growth. A World Bank Study (World Bank, 2015) of 750 cities across the world concluded that during 2005–2012, 72% cities experienced faster economic growth as compared to their respective national economies. Besides, the top 10% of cities registered 13.5% annual gross domestic product compared with 4.7% in an average city. According to Dobbs et al. (2012), nearly 65% of world economic growth in 2025 will be generated by the top 600 cities. In the year 2010, these 600 cities—a group called the City 600—home to just over 20% of the world population accounted for approximately $34 trillion, which was more than 50% of the global GDP. It is expected that the City 600s combined GDP would increase by 100% amounting $65 trillion during 2010–2025. Moreover, approximately 47%, i.e. half of expected global GDP growth will be accounted for by the emerging 440 cities in the City 600.

Expanding per capita income level in urban economies will drive rapid growth in annual household consumption. Mckinsey Global Institute report (Dobbs et al., 2012) projects the entry of one billion new buyers in the emerging market cities by 2025. The emerging 440 cities will account for 60% of the new urban buyers and $10 trillion rise in annual consumption that are projected to drive about $20 trillion a year in additional spending into the global economy. The increased demand is not limited to consumer goods and services, and cities attract heavy investment in infrastructure to keep the wheel rolling. Thus, it demands injection of massive investment further accelerating industrialization and also urban agglomeration. However, cities need to manage consumption and investment well to harvest the growth fruit, failing to do so can actually be big barrier to growth. If cities are planned to meet future needs, have well-functioning safe environment for consumers as well as producers, they may be called 'Smart Cities'. Sustaining environment and protecting ecosystems are two important issues associated with the rapid urbanization. Economic advances are of no/little value if accompanied with environmental degradation. In addition, ensuring the physical security and safety of residents from the increasing crime rates, unhealthy and unhygienic lives and civic unrest is a daunting task.

The present study has been undertaken to ascertain the key determinants of urbanization rate (URB) in India, which is so far a relatively less researched issue in economic literature. The degree of urbanization at 31% in India which used to be lowest in the world has got accelerated (Government of India, 2013). It is expected that 40% of India's population will live in cities and 70% of India's GDP by 2030 will be contributed by cities. The speed and scale of urban transformation that we will witness have been experienced by one country so far and that is China (McKinsey Global Institute-MGI, 2010). The urban population in India stood at 377.01 million out of the total population of 1210.02 million as on 1 March 2011. The decadal growth rate of people leaving in urban areas rose from 2.10% in the decade of 1990s (1991–2001) to 3.35% points in the first decade of the present century (2001–2011). The upsurge in the number of million plus cities/urban agglomeration (UA) from 35 in Census 2001 to 53 in Census 2011 (GOI, 2019) and a growing concentration of urban population in UAs have become quite noticeable over the decade of 2001–11). As per Census 2011, 53 million plus cities accounted for 43% of urban population. The growing concentration of population in urban centres has given rise to several questions. One vital issue pertaining to India (and to other developing economies) is whether rapid urbanization can help securing citizens of India improved standards of living and resolve related problems of unemployment and poverty without causing further destruction of the ecosystems. India aims at realizing its long-term vision of rapid and inclusive growth by managing the urbanization process properly. This phase of transition is likely to bring in opportunities and pose serious challenges.

There is enough empirical literature (Crozet & Koenig, 2007; Glaeser et al., 1995; Leitão, 2013; Vogel, 2012) emphasizing on the relationship between urban growth and economic growth. The cross-country analysis carried out by scholars like Henderson (2003) was first of its kind to estimate the impact of urbanization on economic growth (EG). By taking a bigger sample (of about 100 countries), Brülhart and Sbergami (2009) later revalidated and extended Henderson's analysis. In another study, Handerson (2010) observed high correlation between urbanization level and income per capita. Noteworthy fact is that the research hitherto is either country specific (Pandey, 1977 for India; Glaeser & Gottlieb, 2009 for United States; Chen et al., 2011 for China; Tripathi, 2013 for India) or specific to a country mix of developed and developing countries (e.g. Brülhart & Sbergami, 2009; Crozet & Koenig, 2007; Glaeser et al., 1995; Leitão, 2013; Vogel, 2012). There are certain other studies (Rani & Tripathi, 2016; Sridhar, 2010; Tripathi, 2013; Tripathi & Mahey, 2017) wherein attempt has been made to understand the factors which boost the process of urbanization in India. The present study stands apart because of its country-specific diverse data approach to examine the relationship between URB and EG controlling for some other variables (as important determinants of URB) unlike various other studies which are either micro-level in nature (Tripathi & Mahey, 2016) or make use of panel data approach for identifying the determinants of agglomeration economies (Hofmann & Wan, 2013) or rely on census data on all the cities or towns (Sridhar, 2010) or large agglomerations (Tripathi, 2013). In addition to EG, there are six control variables, viz. industrialization (INDUS), infrastructure (INFRA), literacy rate (LIT), environment (ENV), population growth ($POP_{GR}$) and population density ($POP_{DEN}$), which have been incorporated in this study after reviewing literature. The study has been carried out making use of cross-section data of 28 states and two UTs of India compiled from various sources. The study makes use of multiple regression technique and ordinary least squares (OLS) test procedure. The present study assumes importance as it can provide good understanding of the determinants of urbanization which can further enable urban planners to take informed policy decisions for future urban development and for building inclusive cities which are so essential for driving and sustaining India's EG. The paper has been organized as follows: Following the introduction, section on model specification specifies the model and also gives description of the explanatory variables/regressors that are anticipated to influence the URB in India. The data sources and methodology are discussed in the next section which is followed by a discussion on the empirical results in the light of the previous findings. The concluding remarks are given in Sect. 5.5, and the paper also draws some policy lessons in this section.

**Model Specification**

In the present study, we investigate the relationship between URB and the level of EG in case of India. For the same, dependent variable is URB. Besides EG, the present study has included six control variables. After careful review of literature, the choice of controls has been enumerated.

The following multiple regression model has been studied using OLS test procedure:

$$\begin{aligned} \text{URB} = {} & b_0 + b_1\text{EG} + b_2\text{INDUS} + b_3\text{INFRA} + b_4\text{LIT} \\ & + b_5\text{ENV} + b_6\text{POP}_{\text{GR}} + b_7\text{POP}_{\text{DEN}} \end{aligned}$$

where the intercept term is denoted by $b_0$ and the slope coefficients by 'b's'. The numerical values of these coefficients have been estimated using OLS technique. The partial effect of explanatory variables on URB while holding other variables constant gets reflected through each slope coefficient $b_1$. This also represents the rate of change in the dependent variable for a unit change in the value of the explanatory variables.

## Short Description of Explanatory Variables Used

**Level of economic growth**

State Net Domestic Product Per Capita (SNDPPC) at constant prices for the year 2011–12 has been used as a proxy for the variable EG. The data has been taken from various reports of Reserve Bank of India's (RBI) *Handbook of statistics on Indian economy*. The conventional wisdom (Chenery & Syrquin, 1977; Clark, 1940, 1960; Fisher, 1935, 1939; Kuznets, 1966, 1969) tells us that as per capita income of a country increases, structural shift from a predominantly agrarian economy to industry and service sector-based economy gets underway. Since industries and service activities are urban centric, therefore, expansion of secondary and tertiary sectors contributes to the process of urbanization (Joshi, 2004). As Hofmann and Wan (2013) point out that 'given that both localization economies and agglomeration economies are more likely to cause cost advantages in manufactured products than in agricultural goods, structural change may drive urbanization'. Or in other words, higher levels of EG will lead to urbanization. In later studies (Annez & Buckley, 2009; World Bank, 2009), urbanization has been regarded as a motor/or an engine of EG. We expect that the sign of the coefficient of the variable (EG) will be positive.

**Environmental effect**
The review of literature also reveals that there is an association between urbanization and environment (Turok & Mcgranahan, 2013; Wan, 2013). A study done by Population Reference Bureau (PRB) recognizes that urban populations interact with environment and cause change in it through their consumption of land, food, water and energy. Since urbanization can have negative implications on environment by changing the land-use pattern, therefore, to capture the environmental effect on urbanization, the percentage of geographical area of a state/UT under forest cover has been used as a proxy for factoring in the impact of environmental factors on urbanization. In respect of this variable, ENV, we take the hypothesis that urbanization, by changing the land-use pattern, can cause environmental problems and result in biodiversity loss, ecosystem collapse and climate change. It has been observed that the irrational and the inefficient use of urban land has brought about adverse consequences in the form of the loss of the forest cover. The forest cover declines as the demand for land increases for housing or commercial purposes in the urban centres. This notable change in land use can further damage the environment as the ability to filter air and remove carbon dioxide from the atmosphere gets stunted because of loss/decline of forest cover due to urbanization. The forest cover is expected to decline as the demand for land will increase for housing or commercial purposes in the urban centres with urbanization process gaining momentum. Therefore, the expected sign of the coefficient will be negative.

**Industrialization**
Industrialization has been measured as share of non-agricultural sectors in different states and UTs as percentage of State Net Domestic Product (SNDP) of the respective states/UTs for 2004–05. We have taken data for 2004–05 for this variable as our assumption is that industrialization will lead to urbanization, not immediately but after a time lag. The inclusion of this variable is in keeping with the contributions of structural economists like Fisher (1935, 1939), Clark (1940, 1960), Kuznets (1966, 1969) and Chenery and Syrquin (1977) who said that as an economy develops, there takes place a shift from an agrarian economy to an industry service-based economy (also see Henderson, 2005). Since industrial and services activities are generally urban activities, therefore, they are likely to add to the growth of urban centres. Moomaw and Shatter (1996) and Hofmann and Wan (2013) suggest that economic development impacts urbanization by agglomerating economic activities in urban areas by reducing transport costs and via industrialization. The World Bank (2009) very clearly says that industrialization and urbanization are a must for growth. According to World Bank (2009), *'No country has grown to middle income without*

*industrializing and urbanizing. None has grown to high income without vibrant cities. The rush to cities in developing countries seems chaotic, but it is necessary'.*

Further, Peng et al. (2000) considers it one of the biggest operators of the urbanization process as supported by the historical evidence. Accordingly, we expect the coefficient of industrialization to bear a positive sign.

**Road density**

Road density is defined as kilometre of roads per one square kilometre of land area. We have used it as a proxy for infrastructure. Some studies (Hofmann & Wan, 2013; Pradhan, 2007) clearly demonstrate that infrastructure availability facilitates rapid growth in urbanization. The availability of better infrastructure like road and transport facilities increases the productive capacity of households and can enable higher EG. In respect of this variable, we take the hypothesis that better infrastructure will attract more people into cities, and therefore, this variable is expected to bear a positive coefficient.

**Literacy**

There are studies (Black & Henderson, 1999; Henderson, 1988; Hofmann & Wan, 2013) which try to display the role of education in urban development. Therefore, LIT has been included as an explanatory variable. Higher LIT will encourage people to migrate to cities which are educational hubs. The coefficient of literacy is expected to bear positive sign.

**Population growth**

One of the hypotheses put forward for the rapid growth of cities and urbanization is that high $POP_{GR}$ rates can press on limited farm acreage and push landless labour into cities (Williamson, 1988). This hypothesis has been supported by labour surplus model of Lewis (1954) and probabilistic model of Todaro (1969). Hofmann and Wan (2013) point out that there can be direct effect of $POP_{GR}$ on urbanization via differential growth in urban vis-a-vis rural areas or indirect effect through the movement of population. The expected sign of the coefficient is positive.

**Population density**

A measure of $POP_{DEN}$, that is, population per square kilometre too has been included to account for states whose land areas are small relative to their population. Tisdale (1942) considered urbanization to be a process of population concentration. Wan (2013) postulated '*… for any given population, the high urban density is benign for the environment. The urban economics literature shows that compactness is a key determinant of energy use. High density can make public transport more viable and reduce the length of trips. Urban living encourages walking and cycling rather than driving'.*

Therefore, higher urban density can lead to greater urbanization. Accordingly, the sign of the coefficient of this variable will be positive.

**Table 4.1** Description of data sources

| Variables | Year | Data source |
|---|---|---|
| Urbanization | 2011 | Ministry of Statistics and Programme Implementation, GoI (2017) Selected Socio Economic Statistics India 2017. Central Statistics Office Social Statistics Division, GOI, New Delhi |
| EG | 2011–12 | RBI *Handbook of Statistics on Indian Economy*, various reports |
| INDUS | 2004–05 | RBI *Handbook of Statistics on Indian Economy*, various reports |
| INFRA | 2011 | Ministry Of Road Transport And Highways Transport Research Wing, GoI (August 2012) *Basic Road Statistics of India, 2012*. Ministry Of Road Transport And Highways Transport Research Wing, New Delhi |
| LIT | 2011 | Ministry of Statistics and Programme Implementation, GoI (2017) *Selected Socio Economic Statistics India 2017*. Central Statistics Office Social Statistics Division, Ministry of Statistics and Programme Implementation, New Delhi |
| ENV | 2011 | Ministry of Environment and Forests (2011) *State of Forests Report 2011*. Forest Survey of India, Dehradun |
| $POP_{GR}$ | 2001–11 | RBI *Handbook of Statistics on Indian Economy*, various reports, 2019 |
| $POP_{DEN}$ | 2011 | RBI *Handbook of Statistics on Indian Economy*, various reports |

*Source* Compiled by the author

## Methodology and Data Sources

The present study makes an attempt to empirically investigate a model in the Indian context by making use of cross-section data for 28 states and two UTs and applying multiple regression technique and using OLS test procedure. The following data sources have been utilized for the purpose (Table 4.1).

## Empirical Results and Discussion

We started with a limited number of explanatory variables, viz. EG, INDUS and INFRA, and gradually added more controls. It is reassuring to see that the coefficients of all these three variables remained highly significant in all the four equations (Table 4.2). They are bearing right expected signs. They are the only coefficients which are consistently significant at 1% level of significance, no matter which combination of explanatory variables worked out. However, the coefficient of the variable INFRA became insignificant in Eq. 4 with the introduction of the variable $POP_{DEN}$. But the coefficients of the other two variables (EG and INDUS) were found out to be highly significant even in Eq. 4.

In Eq. 1, three variables, namely EG, INDUS and INFRA, were regressed on the dependent variable URB. The value of $R^2$ is 80% implying that on the whole 80%

**Table 4.2** Multiple regression results for URB

| Equations | Intercept | EG | INDUS | INFRA | LIT | ENV | $POP_{GR}$ | $POP_{DEN}$ | $R^2$ | $R^2$Adj |
|---|---|---|---|---|---|---|---|---|---|---|
| Equation 1 | 3.17148 | 0.0003558 | 0.0182355 | 2.345712 | | | | | 0.8054 | 0.8 |
| | | (0.0000668)* | (0.0057523)* | (0.4732016)* | | | | | | |
| Equation 2 | 6.85412 | 0.0003623 | 0.020691 | 2.176123 | 0.0781478 | −0.08579 | | | 0.8203 | 0.8 |
| | 18.8315 | (0.0000871)* | (0.0066594)* | (0.4933256)* | (0.2920184) | (0.06295) | | | | |
| Equation 3 | -3.0516 | 0.00003659 | 0.0189778 | 2.090573 | 0.1755152 | −0.091 | 0.166412 | | 0.8226 | 0.8 |
| | 26.2924 | (0.00008886)* | (0.0074452)* | (0.5244199)* | (0.3454457) | (0.0645913) | (−0.30334) | | | |
| Equation 4 | −5.1956 | 0.0003516 | 0.0182211 | 1.39136 | 0.227429 | −0.08715 | 0.122033 | 0.00129 | 0.8233 | 0.8 |
| | 27.8455 | (0.0001033)* | (0.0080393)* | (2.485109) | (0.3959202) | (0.0672625) | (−0.34577) | (−0.0045) | | |

*Note* Figures in the parentheses refer to the standard errors of the estimates. N = 30
*Refers to 1% level of significance

of movements in the dependent variable (URB) were explained by the independent variables. All the variables are significant at 1% level of significance. The positive coefficients of these variables are as per our expectation and signify that there is a positive association between the urbanization and these three variables.

In Eq. 2, when an attempt was made to assess the impact of two more variables, namely LIT and ENV, on urbanization retaining all other variables of Eq. 1, the value of $R^2$ increased marginally to 82%. The coefficient of LIT was positive, whereas that of ENV turned out to be negative keeping in line with our hypotheses. The negative sign of the regression coefficient of ENV indicates the negative association of ENV with URB. This may be due to the reason that negative externalities may increase with expansion of cities (Turok & Mcgranahan, 2013: p. 468). This may result into the loss of forest cover/green areas to develop theme parks or recreation zones or to build low-cost houses in urban areas. This amounts to loss of biodiversity or city forests which help to maintain local ecology. Increased migration to urban areas and increased developmental activities affect the ecosystem, that is, soil ecology, water balance, etc. Human impact on the land increases through deforestation and increased use of pastures. Therefore, higher the URB, lower will be the forest cover. The sign of the coefficient on ENV is in keeping up with our expectations. But the coefficients of both the variables (LIT and ENV) turned out to be insignificant.

In Eq. 3, another variable $POP_{GR}$ has been added to Eq. 2 to observe how it impacts URB. The results very clearly show that impact of $POP_{GR}$ will be positive on URB. But the coefficient of the variable ($POP_{GR}$) is not significant.

In Eq. 4, another variable $POP_{DEN}$ has been included in Eq. 3. Though there is marginal improvement in the value of $R^2$, the coefficient on INFRA becomes insignificant which was otherwise highly significant in all other equations. The sign of the coefficient of $POP_{DEN}$ is as hypothesized. Higher $POP_{DEN}$ leads to higher URB.

To sum up from the regression exercise carried out, the three variables, viz. EG, INDUS and INFRA, emerge as the key determinants of URB.

The present study made an attempt to investigate the key determinants of URB by collecting cross-section data for different variables which are expected to influence the URB. The study makes use of the cross-section data collected from various sources in case of 28 states and two UTs. We used multiple regression technique and OLS procedure. We found evidence that EG, INDUS and INFRA are the three key determinants of URB in case of India.

The coefficients of these three variables not only yield the expected positive signs but also are highly significant at 1% level of significance in various equations. The coefficients of remaining all other variables, namely LIT, ENV, $POP_{GR}$, and $POP_{DEN}$, though were having expected signs but yet were not found to be significant.

The study very clearly highlights that growth is critical for India's urbanization process as the coefficient of EG bears correct positive sign as per our expectation. It is widely recognized now that for realizing the growth potential of India, urbanization is inevitable. Therefore, keeping in view the positive relationship between EG and urbanization exhibited in this study, steps should be initiated by the government to attain higher growth first, and then, attempts should be made to sustain that growth.

Higher growth can accelerate the pace of urbanization which in turn can result in significant sprawling of cities outwards or expansion of urban boundaries or growth of peri-urban areas. This will definitely make entire growth process more inclusive. But the development of peri-urban areas should become an integral part of urban planning process. Our study also demonstrates that there is considerable positive impact of INDUS and INFRA on URB. There is a need to turn around cities. There is an urgent need to address the issues of industrial and infrastructural development. Stepping up of the construction of infrastructure is required to erase infrastructure backlog. The 'Make in India' and 'Smart Cities Mission' need to be given a big boost in India. The challenge before the government is how to ramp up investment in industrial and infrastructure building in line with economic growth. Like China, India too needs an effective, systematic and mature urban planning regime. This highlights the needs for improving governance structure at the level of urban local bodies. The paper clearly reveals that ENV and URB are negatively associated. In other words, decline in forest cover is associated with urbanization. Therefore, there is an urgent need for effective and mature urban planning system in India, which is currently in poor state. A study done by MGI (2010) clearly brings out, 'India's planning is in very poor state. On paper, India does have urban plans, but they are esoteric rather than practical, rarely followed, and riddled with exemptions'. Therefore, there is a pressing obligation to revamp India's planning system. It has been pointed out in MGI report that India is having the potential to save more than six million hectares of potentially arable land over the next 20 years through revamping its planning systems at different levels (metropolitan, municipal levels, etc.).

The positive signs of coefficients of $POP_{GR}$ and $POP_{DEN}$ on URB are as per our expectations. But what is unique about the urbanization process of developing countries like India is the growth of slums which are the result and product of urbanization and changes in urban economy (Gavsker, 2011). With limited job opportunities and excess labour supply (a consequence of natural growth of urban population and migration to urban areas), the migrants get absorbed in urban informal sectors. The peculiarity of such jobs is low incomes, low quality of service, pathetic living conditions, restricted or limited legal ownership of land and no social security. So, this leads to the growth of slums and squatter settlements and urban poverty (Mitra, 1994). This points towards the need for raising rate of creation of jobs in the urban sector on the one hand and guiding the process of urbanization (by the government) and ensuring the basic services ranging from availability of sanitation, water supply services to housing facilities (GoI, 2013). The twelfth Five-Year Plan document warned about the consequences of improper handling of these challenges. It cautions that any mishandling of such issues may result in making India's cities chaotic and choking them too. Rural poverty can get converted into urban poverty, with no betterment in livelihoods for the rising population of India (GoI, 2013). Therefore, there is a pressing requirement for better and informed policies.

## Concluding Remarks and Policy Implications

For centuries, urbanization and economic growth have moved in tandem. Undeniably, the cities carry huge economic potential as they account for about 80% generation of the global GDP (United Nations, 2015). With enormous growth urban areas are likely to witness in the next few decades, there is an urgent need for more planned and well-managed cities and urban spaces.

In 2018, 4.2 billion people, 55% of the world's population lived in cities. By 2050, the urban population is expected to reach 6.5 billion. In coming decades, 90% of urban expansion will be in the developing world (United Nations, 2017). However, the fact that around 900 million people are estimated to live in slums today (Friesen et al., 2019) and the number is continuously rising calls for more planned and managed cities. Only 3% of the Earth's surface is occupied by cities, but these are responsible for 60–80% of power consumption and around 70% of greenhouse gas emissions (United Nations, 2015). The skewed consumption is heavily responsible for misutilization/overutilization of resources making development unsustainable. With two thirds of all humanity residing in pockets of cities, it can contribute to rising per capita GDP only if cities are made sustainable by creating greater opportunities for business and career, taking care of safe and affordable housing and building resilient societies and economies (MGI, 2010). It requires massive investment in public transport, green spaces, improved planning and management. For realization of a better quality of urban dwellers, reduced energy costs and emissions, safety and well-being of inhabitants must be given high priority. Increasing economic growth in urban centres demands spotlight to be shifted to better management of cities, environmentally conscious decisions, better utilization of resources and improved governance.

## References

Abdel-Rahman, A. N., Safarzadeh, M. R., & Bottomley, M. B. (2006). Economic growth and urbanization: A cross-section and time-series analysis of thirty-five developing countries. *International Review of Economics and Business, 53*(3), 334–348.

Annez, P. C., & Buckley, R. (2009). Urbanization and growth: Setting the context. In M. Spence, P. C. Annez, & R. M. Buckley (Eds.), *Urbanization and growth. Commission on growth and development* (pp. 1–45). The World Bank.

Bertinelli, L., & Strobl, E. (2003). Urbanization, urban concentration and eco growth in developing countries. Credit Research Paper, Centre for Research in Economic Development and International Trade, University of Nottingham, 3(14).

Black, D., & Henderson, V. (1999). A theory of urban growth. *Journal of Political Economy, 107*(2), 252–284.

Brülhart, M., & Sbergami, F. (2009). Agglomeration and growth: Cross-country evidence. *Journal of Urban Economics, 65*(1), 48–63.

Chen, Z., Lu, M., & Xu, Z. (2011). A core-periphery model of urban economic growth: Empirical evidence using Chinese city-level data, 1990–2006. Global COE hi-stat discussion paper seriesgd

11–206. Institute of Economic Research, Hitotsubashi University, Tokyo. Accessed on April 10, 2019, http://gcoe.ier.hit-u.ac.jp/english/research/discussion/2008/gde11-206.html.
Chenery, H. B., & Syrquin, M. (1977). *Patterns of development, 1950–1970.* Oxford University Press.
Clark, C. (1940). *The conditions of economic progress.* Macmillan.
Clark, C. (1960). *The conditions of economic progress* (3rd ed.). Macmillan.
Crozet, M., & Koenig, P. (2007). The cohesion vs growth trade: Evidence from EU regions. University of Paris, Mimeo. Accessed on April 10, 2019, http://www.parisschoolofeconomics.eu/docs/koenigpamina/paper_tradeoff.pdf.
Dobbs, R., Remes, J., Manyika, J. M., Roxburgh, C., Smit, S., & Schaer, F. (2012). *Urban world: Cities and the rise of the consuming class.* Mckinsey Global Institute.
Eggers, W. D., & Skowron, J. (2018). Forces of change: Smart cities, deloitte insights.
Fisher, A. G. B. (1935). Economic implications of material progress. *International Labour Review, 32*(July), 5–18.
Fisher, A. G. B. (1939). Primary, secondary and tertiary production. *Economic Record, 15*(June), 24–38.
Friesen, J., Hannes, T., Michael, W., & Peter, F. P. (2019). Size distributions of Slums across the globe using different data and classification methods. *European Journal of Remote Sensing, 52*(Sup 2).
Gavsker, K. K. (2011). Urban restructuring and change: Planning development practices and impacts on the urban poor. *Institute of Town Planners, India Journal, 8*(2), 24–44.
Glaeser, E., & Gottlieb, J. D. (2009). The wealth of cities: Agglomeration economies and spatial equilibrium in the United States. *Journal of Economic Literature, 47*(4), 983–1028.
Glaeser, E., Scheinkman, J., & Shleifer, A. (1995). Economic growth in a cross-section of cities. *Journal of Monetary Economics, 36*(1), 117–143.
Government of India. (2013). *Twelfth five year plan (2012–2017), economic sectors* (Vol. II). Sage Publications India Pvt Ltd.
Government of India. (2019). Ministry of Housing and Urban affairs. http://mohua.gov.in/cms/level-of-urbanization.php. Accessed on April 9, 2019.
Henderson, J. V. (1988). *Urban development: Theory, fact and illusion.* Oxford University Press.
Henderson, J. V. (2003). The urbanization process and economic growth: The so what question. *Journal of Economic Growth, 8*(1), 47–71.
Henderson, J. V. (2005). Urbanization and growth. In P. Aghion & S. Durlauf (Eds.), *Handbook of economic growth* (Vol. 1, pp. 1543–1591). Elsevier.
Handerson, J. V. (2010). Cities and development. *Journal Of Regional Science, 50*(1), 515–540.
Hofmann, A., & Wan, G. (2013). Determinants of urbanization. Working paper no. 355. Asian Development Bank, Manila.
Joshi, S. (2004). Tertiary sector-driven growth in India—Impact on employment and poverty. *Economic and Political Weekly, 39*(37), 4175-4178.
Kuznets, S. (1966). *Modern economic growth.* Oxford and IBH Publishing Co.
Kuznets, S. (1969). *Economic growth and structure: Selected essays.* Oxford and IBH Publishing Co.
Leitão, N. C. (2013). A panel data modelling of agglomeration and growth: Cross-country evidence. *Theoretical and Empirical Researches in Urban Management, 8*(1), 67–77.
Lewis, W. A. (1954). Economic development with unlimited supplies of labour. *The Manchester School of Economic and Social Studies, 22*(2).
Mahey, K., & Tripathi, S. (2016). Urbanization and economic growth in Punjab (India): An empirical analysis. MPRA_paper_71325.pdf. Available on Urbanization and Economic Growth in Punjab (India): An Empirical Analysis - Munich Personal RePEc Archive (unimuenchen. de).
McKinsey Global Institute. (2010). India's urban awakening: Building inclusive cities. https://www.mckinsey.com/~/media/McKinsey/Featuredpercent20Insights/Urbanization/Urbanpercent20awakeningpercent20inpercent20India/MGI_Indias_urban_awakening_full_report.ashx. Accessed on April 9, 2019.

Mitra, A. (1994). *Urbanization, slums, informal sector employment and poverty: An exploratory study*. BR Publishing Corporation.
Mooaw, R., & Shatter, A. M. (1993). Urbanization as a factor in economic growth. *The Journal of Economics, 19*(2), 1–6.
Moomaw, R., & Shatter, A. M. (1996). Urbanization and economic development: A bias towards large cities? *Journal of Urban Economics, 40*(1), 13–37.
Pandey, S. M. (1977). Nature and determinants of urbanization in a developing economy: The case of India. *Economic Development and Cultural Change, 25*(2), 265–278.
Peng, X., Chen, X., & Cheng, Y. (2000). Urbanization and its consequences. *Demography, II*. https://www.eolss.net/sample-chapters/C04/E6-147-18.pdf. Accessed on April 9, 2019.
Population Reference Bureau. Urbanization: An environmental force to be reckoned with. https://www.prb.org/urbanization-an-environmental-force-to-be-reckoned-with/. Accessed on April 10, 2019.
Pradhan, R. (2007). Does infrastructure play role in urbanization: Evidence from India. *Indian Journal of Economics and Business, 6*(1), 81–92.
Rani, C., & Tripathi, S. (2016). Determinants of urbanization in different size/class distribution of cities/towns in India. Munich personal RePEcarchive. Accessed at https://mpra.ub.uni-muenchen.de/74757/1/MPRA_paper74757.pdf, date of access April 9, 2019.
Sridhar, K. S. (2010). Determinants of city growth and output in India. *Review of Urban and Regional Development Studies, 22*(1), 22–38.
Tisdale, H. (1942). The process of urbanization. *Social Forces, 20*(3), 311.
Todaro, M. P. (1969). A model of labour migration and urban unemployment in less developed countries. *The American Economic Review, LIX*(1).
Tripathi, S. (2013). Do large agglomeration lead to economic growth? Evidence from urban India. *Review of Urban and Regional Development Studies, 25*(3), 176–200.
Tripathi, S., & Mahey, K. (2017). Urbanization and economic growth in Punjab (India): An empirical analysis. *Urban Research & Practice, 10*(4), 379–402.
Turok, I., & Mcgranahan, G. (2013). Urbanization and economic growth: The arguments and evidence for Africa and Asia. *Environment and Urbanization, International Institute for Environment and Development, 25*(2), 465–482.
United Nations. (2015). Sustainable Development Goals, https://www.undp.org/content/undp/en/home/sustainable-development-goals/goal-11-sustainable-cities-and-communities. Accessed on May 7, 2020.
United Nations. (2017). The World Urbanization Prospects 2018, Department of Economic and Social Affairs, United Nations, New York.
Vogel, J. (2012). Agglomeration and growth: Evidence from the regions of central and eastern Europe. ERSA Conference Papers ersa12p1089. European Regional Science Association. http://wwwsre.wu.ac.at/ersa/ersaconfs/ersa12/ersa12gmfinal01089.pdf. Accessed on April 10, 2019.
Wan, G. (2013). Why urbanization may benefit the environment? https://www.adb.org/news/op-ed/why-urbanization-may-benefit-environment-guanghua-wan. Accessed on April 9, 2019.
Williamson, J. G. (1988). Migration and urbanization. In H. Chenery & T. N. Srinivasan (Eds.), *Handbook of development economics* (Vol. I). Elsevier Science Publishers.
World Bank. (2009). *World development report 2009: Reshaping economic geography*. The World Bank.
World Bank. (2015). *Competitive cities for jobs and growth*. World Bank Document.

# Chapter 5
# Contextualizing Smart Cities and Their Impact on Social Relations

**Manish Yadav**

**Abstract** Improvement in the quality of life of its inhabitants is one of the major challenges being faced by the cities today. In order to cater to this challenge, cities are attempting to transform themselves into smart cities through the use of state-of-the-art technological tools and new urban designs. They are also taking into account the ecological standards in force while saving resources so as to achieve the desired output. However, it has been observed that man has always sought to control nature to ensure comfortable living conditions. In pursuit of this goal, techniques of means of production and mode of production have been developed giving rise to different and new social relations with new technology. Both, technology and social environment as resource, must be accorded equal consideration. However, in present context where technology becomes an inevitable part of our lives, human relations are becoming weaker. This article attempts to look at the ways in which smart cities have been evolved, defined and how they are impacting social relations in the Indian context.

**Keywords** Smart city · Social relationships · Technological development

## Introduction

According to MacIver and Page (1950) 'society is a web of relations, a complex system of usages and procedures, of authority and mutual aid of many groupings and divisions, of control of human behavior and of liberties'. Group living is essential for humans, for it not only ensures fulfilment of their needs but also provides mutual assistance and sentimental support. Sharing of thoughts, happiness and distress is instinctual to human nature. Human beings have developed various means ranging from gestures, language to modern means of communication to communicate with each other. These advancements and innovations have supported the human beings to expand their way of living and thus creating a life filled with more opportunities. Technological developments have led to many developments in society by altering the customs, institutions, laws and organizations in the society. As Ogburn (1927)

M. Yadav (✉)
Department of Sociology, University of Rajasthan, Jaipur, Rajasthan, India

 
A. Kateja and R. Jain (eds.), *Urban Growth and Environmental Issues in India*,
https://doi.org/10.1007/978-981-16-4273-9_5

stated, 'technology changes society by changing our environment to which we in turn adapt. This change is usually first in the material environment and the way in which society adjusts with these changes often modifies the non-material world consisting of customs and social institutions'. The significant areas where technological change takes place, among others, are production, communication and transportation making way for alteration of all types of social relations. In the primitive stage, the relations among the members of society were very intimate, simple and primary. These relations grew more complex as the society paced towards agrarian structure but still remained simple, personal and unmediated. The industrial revolution brought forth a new kind of society which for the first time had non-agricultural occupational base. This transformed the way society was structured. In the industrial age, the technological advancements happened at such great pace that they transformed the social relations profoundly making them complex, impersonal, professional and indirect. This age brought radical changes in the way production and transportation of goods happened. Simultaneously, there were changes in the information and communication system also, facilitating the capturing of time and space by former and the material capturing of the same due to changes in latter.

The advancements in communication technology and transportation have resulted in boosting international commerce and industry leading to interdependence of countries in the world. Today, the world is seen as a global village, one community or one world citizenship. These developments have brought all nations under one roof.

## Smart Cities: The Concept

The smart city concept came into existence with the upcoming need to handle the accelerated urbanization. This concept has matured from its initial phase termed as Smart City 1.0 to a one more developed, termed as Smart City 2.0. Smart City 1.0 was actually born out of a disaster. After 9/11 attacks, New York city and other cities installed 24 $\times$ 7 security cameras and AIs to secure public areas through Public–Private Partnerships (PPPs) funded by governments and businesses. Their primary purpose was providing an integrated network by inter-connecting the various points of public infrastructure and feeding the collected data in the integrated network. Smart City 2.0 brings the government, private enterprises and citizens together under one umbrella (Chakraborty, 2019). There is a continued growth in demand of smart cities driven by numerous socio-economic and technical innovations worldwide. It is because of the growing number of smart cities that electricity, infrastructure, houses and government providers are coming into the smart city market. At the same time, private players like startups are eyeing the emerging opportunities in the same sector. However, in India, this concept of was introduced by the government in 2014 for the comprehensive advancement of cities (MoUD, Government of India, 2016).

## What Is a Smart City?

A smart city is essentially centred on the idea of having urban planning incorporate information and communication technology (ICT) and Internet of Things (IoT) for better and safe management of urban infrastructure. The urban infrastructure includes communication systems, transportation, health delivery systems, power stations, water supply networks, public service delivery systems, waste management systems, educational institutions, governance systems and the like which all together work to enhance the quality of life of the citizens. The prime objective of developing smart cities is to enhance the quality of life of the citizens by making use of urban computing and technological base to enhance service efficiency and coping with the emerging needs of the population. ICT enables direct communication between city officials and community, provides services to track the activities in the area and engages directly with the way systems are evolving and how a better standard of life is feasible. Data from the people is obtained through software and stored. It is then evaluated by the use of sensors and combined with real-time monitoring systems. Thereafter, the collected information and knowledge becomes key to address the inefficiencies of the system and delivery mechanism (Fig. 5.1).

## Defining Smart Cities

The crucial question is how to define a 'smart city'? Scholars articulate the concept of smart cities in different ways. No globally accepted definition of a smart city has emerged so far; nonetheless, scholars agree on it being a city that is run through the interface of technology. The conception of smart city varies from place to place. It is contingent on the level of development of the country, its willingness to change and reform, the available resources and aspirations of the residents. Going by the tag of smart city, the technological aspect becomes overpowering; however, smart cities should not be limited to only technological aspects. They should be seen as responsive cities which provide its inhabitants the facility of good and smart governance. The use of technology in governance eases the delivery of public services.

Nam and Pardo (2011) review the meaning of the term 'smart' in the 'smart city context'. Smartness as a connotation is centred on a user perspective. In order to enhance the appeal for the citizens, smart has been preferred over the more elitist term 'intelligent'. According to Caragliu et al. (2011), 'a city is smart when investments in human and social capital and traditional (transport) and modern (ICT) communication infrastructure fuel sustainable economic growth and a high quality of life, with a wise management of natural resources, through participatory governance'. Giffinger et al. (2007) defines smart city as 'a city well performing in a forward-looking way in economy, people, governance, mobility, environment, and living, built on the smart combination of endowments and activities of self-decisive, independent and aware citizens'.

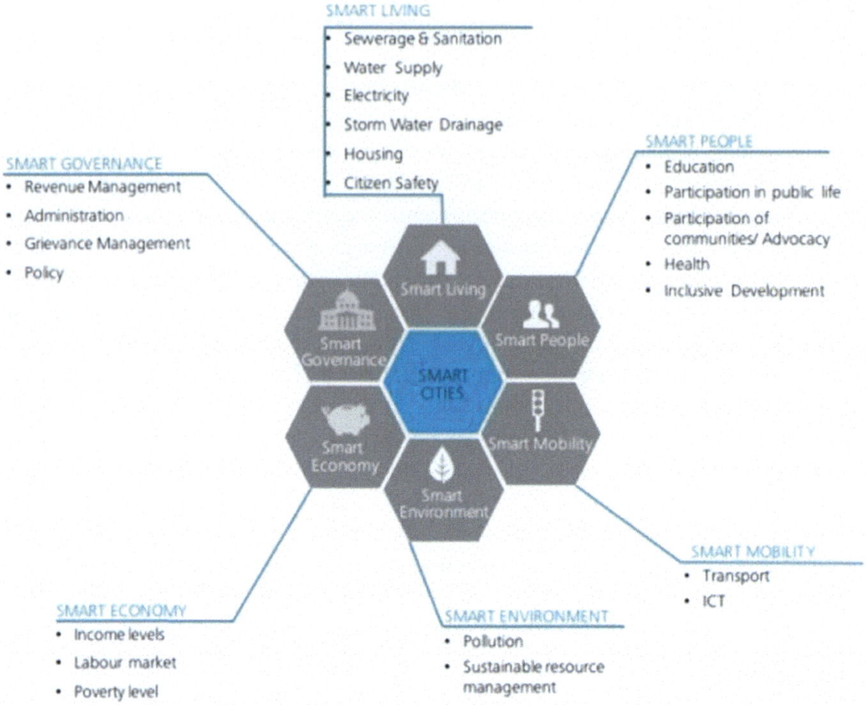


**Fig. 5.1** Integrated smart city framework comprising of various key enablers—smart living, smart governance, smart people, smart environment, smart mobility and smart economy. *Source* https://www2.deloitte.com/content/dam/Deloitte/in/Documents/IMO/in-imo-smart-cities-in-india-noexp.pdf

These definitions show that a smart city's aim is to create an ecosystem for the exchange of content, teamwork, interoperability and smooth interactions for all people anywhere in town. It is a sustainable city providing multiple possibilities where the human potential can be used to its maximum capacity and building a space for leading a creative life. However, as a result of significant increase in the number of urban residents and identified challenges, there is a growing need for new and innovative ways to manage the complexity of urban life. The smart city concept assumes that a city should be a sustainable, creative area which enables the improvement in the quality of life of its citizens, creates a friendlier environment and where the prospects of economic development are stronger (Lee et al., 2014). Thus, Smart city technologies not only provide substantial potential in the improvement in the quality of life by attempting to provide an environment with reduction in crime rate, improved public safety, optimized emergency responses, traffic safety through improved means of surveillance, better healthcare, clean and more sustainable environment (Mckinsey, 2018) but at the same time they also open the path for reducing the rampant socio-economic disparities and iniquitous hierarchies (Fig. 5.2).

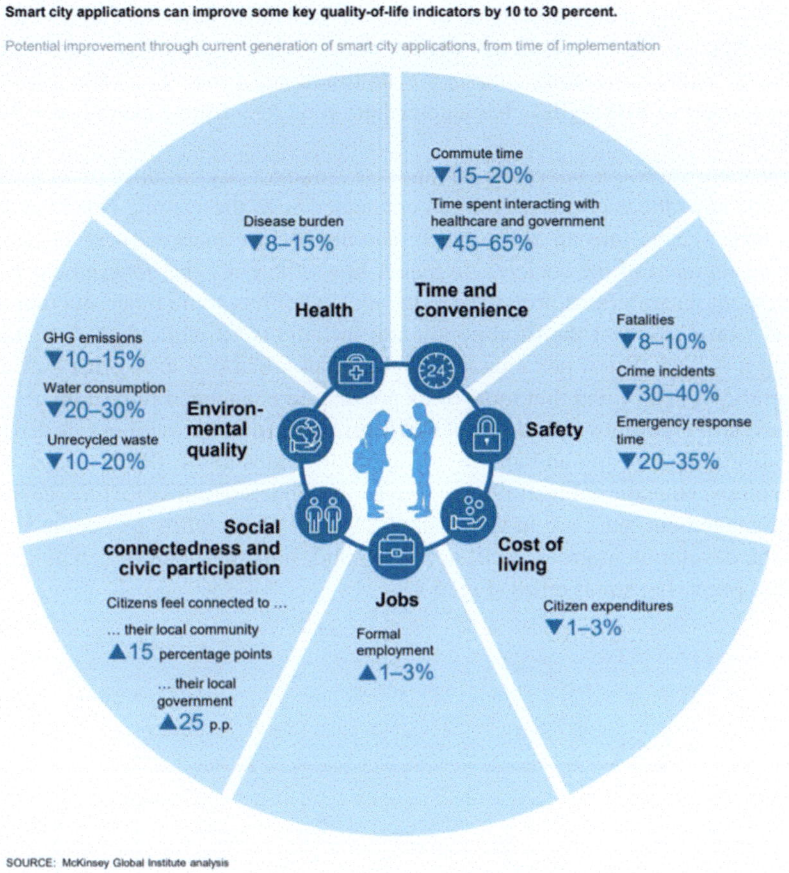


**Fig. 5.2** Potential of smart city applications. *Source* https://www.mckinsey.com/~/media/McKinsey/Business%20Functions/Operations/Our%20Insights/Smart%20cities%20Digital%20solutions%20for%20a%20more%20livable%20future/MGI-Smart-Cities-Executive-summary.pdf

## Smart Cities in India

India is building on the global growth of smart cities. The 'Digital India' Mission of Prime Minister Modi proposed to develop 100 smart cities throughout the country. Talking about smart cities, he said, 'Cities in the past were built on riverbanks. They are now built along highways. But in the future, they will be built based on availability of optical fibre networks and next-generation infrastructure' (Tolan, 2014). Digital India aims to make India a leader in the digital delivery of health, education, public services and financial services. Modi announced a $1.2-billion commitment in smart cities with further money coming from both the private and the foreign sectors. As reported by CNN, a number of new towns are now under development, particularly in the Delhi–Mumbai corridor. Some of the proposed cities have Special Investment Areas or Special Economic Zones (SEZ), which relax restrictions, reduce taxes, and generally make investing simpler for multinational firms. The Delhi–Mumbai corridor project of $100 billion has a contribution of 26% from Japan (Tolan, 2014) (Fig. 5.3).

While there are ample international funds for the smart city initiative of the government as of now, there is still a problem connected with the vision. This idea tends to turn urban scape into an elitist scape, leaving out the marginalized and underprivileged segment of the society. No matter how efficiently the government works to implement its ambitious programme, the idea itself has some inherent contradictions. The proportion of the rural population in India is 68 while that of the urban population is 32%. Rural percentage is higher than urban due to the fact that India is an agrarian country and that most of its population is engaged in farming.

India is still grappling with issues of basic infrastructure of providing safe drinking water, affordable housing and employment opportunities to all. Basic requirements like hospitals, educational institutions, provision of energy, infrastructure, adequate waste management, etc., are in short supply. Unless India fulfils its people's basic needs, the development ground will never be solid. India has a lot of catching up to do before preparing smart cities (Fig. 5.4).

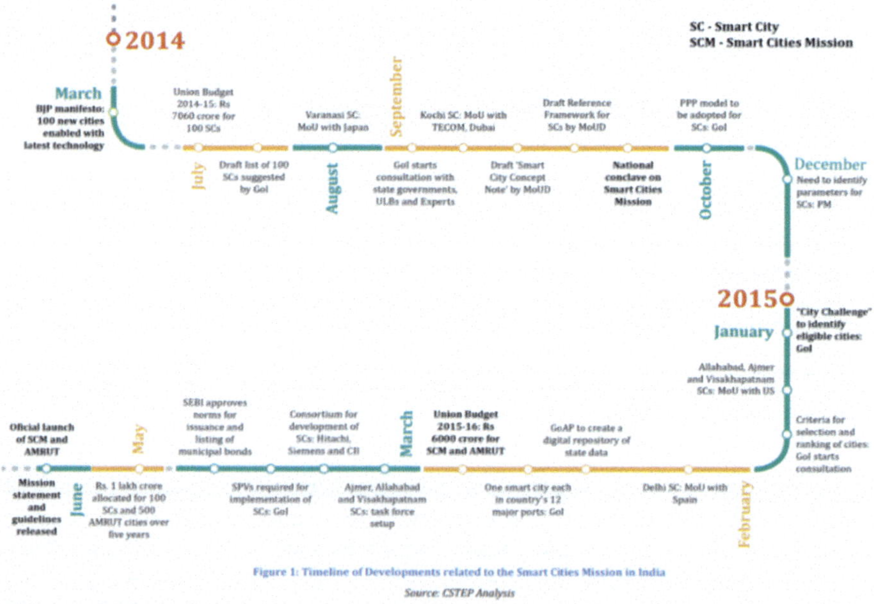


**Fig. 5.3** Timeline of developments and budget allocation related to Smart Cities Mission in India. *Source* https://niti.gov.in/writereaddata/files/document_publication/CSTEP%20Report%20Smart%20Cities%20Framework.pdf

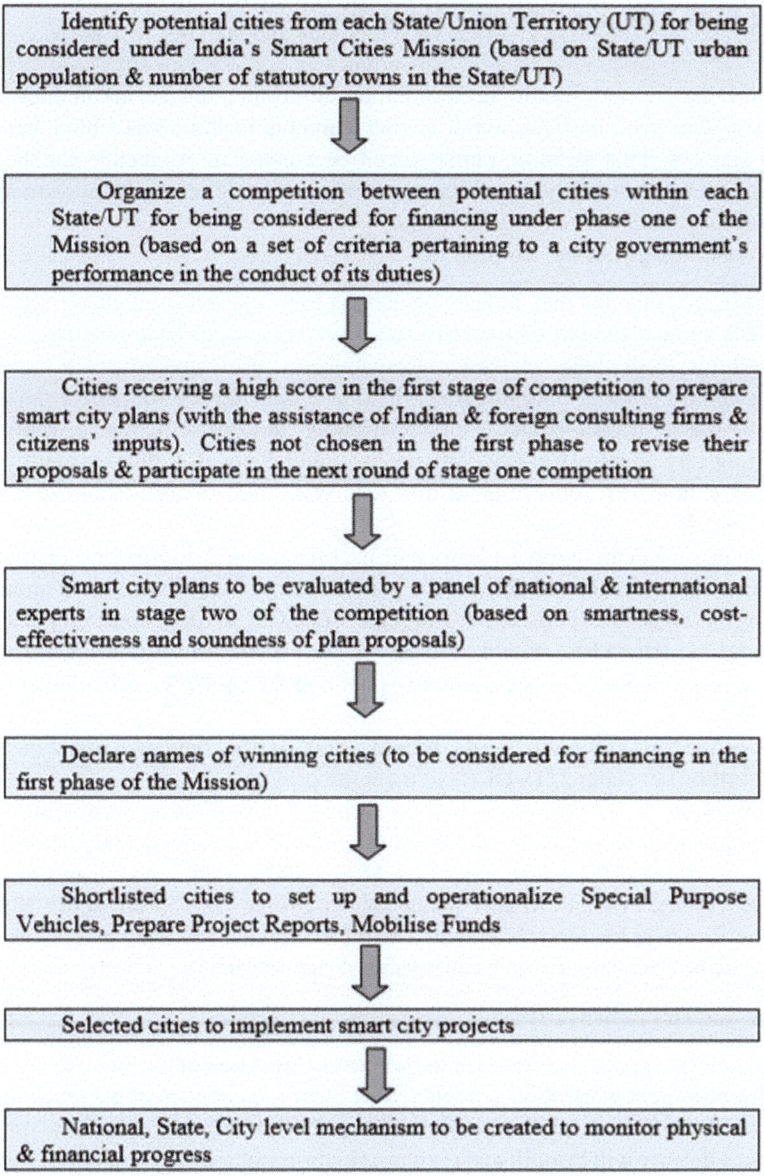


**Fig. 5.4** Smart city process. *Source* https://www.ifri.org/sites/default/files/atoms/files/av87_smart_cities_india_aijaz_0.pdf

## Smart City Mission (SCM): Its Relevance

Cities in India are in desperate need of enhanced urban planning and management. There is a dire need of fresh ideas. Earlier attempts in this regard have met with limited success. This mission, perhaps, can be viewed as an avenue for the local and state governments to plan and promote innovative solutions by addressing the obstacles they have encountered in the past.

The core strength of the Mission is:

- The Mission ensures that, at least one city is to be chosen from each Indian state and UT and a clear, straightforward selection process has been pursued.
- In addition to the basic characteristics suggested by Union Ministry that smart cities should possess (like integrated land use, housing for all, pedestrian zones, green spaces, transit choices, citizen-friendly and cost-effective governments, building city identity), it was left to the state/UT/local authorities and people to create their own conception of how they want their smart cities to evolve and function.
- The states/UTs can involve advisory firms, bilateral and multilateral institutions, foreign governments and domestic organizations in the preparation of smart city plans that possess experience in smart city renewal. In each smart city, a special purpose vehicle (SPV) would be used for the execution of the projects, against the conventional model of community planning driven by the urban local bodies (ULBs).
- The central government would provide half of the financial assistance (US $7.5 billion) to states/UTs/ULBs to meet the project expense.
- The mission, in its periphery, will compliment with other governmental urban development schemes such as Atal Mission for Rejuvenation and Urban Transformation (AMRUT), Clean India Task, Heritage City Development and Augmentation Yojana, Pradhan Mantri Awas Yojana-Urban, Jawaharlal Nehru National Urban Renewal Mission, National Urban Livelihoods Mission, programmes for capacity building and financial inclusion of the citizens.

Aijaz and Hoelscher (2015) list some areas of concern:

- Under this programme, instead of the whole city a portion of the city will be chosen for development work. Consequently, over the five-year span of the mission, only one portion of the city will experience renewal, while over the same time the other areas of the city will be built and regulated in the regular mode. This strategy could further broaden the gaps in growth.
- Improvement in sections of urban areas has to be done carefully. For instance, in order to provide drinking water or electricity $24 \times 7$, the facilities of other areas of the city should not be disrupted. Currently, there is evidence that, owing to the persistent demand of such facilities by commercial institutions (such as shopping malls, Cineplex's and other commercial centres); government departments frequently resort to the practice of load shedding. This disrupts supply in rural and suburban areas.

- Poverty-stricken and disadvantaged populations can be seen residing within the areas chosen from the region for development. However, the responsible agencies for smart city strategy do not provide adequate information about how to improve the housing and working environments of slum dwellers, beggars, disadvantaged people and persons with disabilities.
- Rapid irregular development in peri-urban areas is a negative consequence of urbanization. The peri-urban areas have a number of social, economic, development and environmental challenges. While these challenges should be addressed first, the Mission only provides for greenfield (new) development of vacant land around cities in order to meet the needs of the growing urban population. As time passes, conditions in peri-urban areas will deteriorate further, making it increasingly difficult to address the issue.
- Conventional progress and administrative instruments (i.e. parastatal agencies and municipalities) have been bypassed, and the proposed city-level SPVs will be used for all urban transformation work under the Mission. Although the SPV will be embodied by state and non-state players, they will have to demonstrate improved levels of efficiency in the receipt of project funds and in the implementation of the project and in the enforcement of the rules. These problems have severely hampered the advancement of sustainable planning in the past. Furthermore, efforts to improve the functioning of traditional institutions must continue because, on the one hand, SPVs will rely on them to fulfil their resource needs and, on the other hand, the areas of the city not protected by the Mission will need to be adequately controlled by traditional institutions.

## Smart Urbanization

Urbanization has been found to be a globally occurring phenomenon (Doytsher et al., 2010). In the last two decades, rapid urbanization has led to the proliferation of megacities worldwide (having population over ten million). It is favourable to be urbanized as it promotes well-being and augments economic development. However, the immense size and the scale of these cities put tremendous strain on the growth of infrastructure, the provision of public services and the protection of the environment. If we add economic, social and ethnic stratification as well as health, protection and security threats to the list of challenges, the problems faced by megacities appear daunting and are definitely those that technology alone cannot overcome. It is inappropriate to think that technology alone makes a city smart; it needs smart governance, smart enterprises and most importantly smart citizens. A smart city is one that can harness technology, infrastructure, public policies and community participation efficiently to create an urban ecosystem that fosters development, efficiency, creativity, civic changes, inclusiveness and sustainability. Thus, it is very important to embrace smart urbanization. 'Smart urbanization strategies shall look into smart growth, using the power of urbanized areas to increase GDP rather than getting affected by them' (Dobbs et al., 2012). Smart urbanization can be accomplished in several respects.

Smart neighbourhoods, sustainable and eco-sensitive growth and beginning with improvements to the environment through the green upgrading of current facilities are several moves towards this. Cities need green development in order to enhance the everyday lives of its inhabitants. 'Smart cities' can reconcile sustainability and growth (Hoornweg & Freire, 2013).

Urbanization refers to the process of change in the size of an urban area, density of population and subsequent changes in life style and the shift of majority from agriculture to non-agricultural occupations. It is predicted that around 500 million people will be urbanized by 2030 that is around 60% of the world's population will be living in cities (Unfpa.org; UN, 2014). One of the major pull factors of urban area is the availability of enormous economic opportunities. Economic drivers make cities the ideal places to work and live (Bouton et al., 2013). The journey to sustainable sustainability must be smart and adaptable to potential energy consumption demands. In this phase, cities play a key role. The cities are the driving force of world economy with around 70% of the global GDP coming from the urban areas. The economies of cities are improved, thus offering increased employment possibilities, better health and educational facilities and an enhanced quality of life. However, with increasing urbanization, many of these cities have been responsible for regional economic imbalances because of the high investment concentration, job opportunities, higher level of goods and services, as well as for urban areas with a population of over 10 million or more. Though megacities are economic hubs and have access to modern technologies, they also have better living standards, but at the same time, they also perpetuate wider economic disparities. The cities too face major challenges of urban poverty, poor health of the citizens, increasing crime rate, inefficient disposal of waste, other social issues and degradation of the environment. Therefore, the adverse impact of urbanization is seen as greater since the social fabric of urban areas is not close.

## Challenges of Urbanization

### *Increased Poverty and Decline in Quality of Life*

Grayson et al. (1994) remarked that 'Cities generally lack access to decent housing and sanitation, health care and education thus adding to urban poverty'. Massive increase in population has contributed to increase in the incidence of poverty. One major pull factor for people to migrate to urban areas has been the vast array of facilities and opportunities available in the cities. The anonymity provided by city life frees the people from the socio-cultural barriers experienced by them in rural areas. However, the huge amounts of uncollected waste, high levels of pollution, traffic congestion, lack of social networks, costly health services, out of reach educational facilities together lower the quality of life enjoyed by the citizens.

## *Climate Change and Associated Challenges—A Global Concern*

'Rapid urban growth is also contributing significantly to climate change as 20 largest cities consume 80% of the world's energy and urban areas generate 80% of greenhouse gas emissions worldwide' (Chauhan, 2008). Haphazard urban development can greatly increase the risk of environmental hazards such as famines, flash floods, melting of glaciers and greenhouse gas emissions. The reporting of melting of Atlantic glacier and its consequence for low coastal regions is self-explanatory in this regard.

## *Degradation of Ecosystem and Environmental Pollution*

Cities are major consumers of energy. The concentrated energy consumption contributes to higher air emissions with a major effect on public well-being. Automotive exhaust elevates the levels of lead in city air. Pollution significantly contributes in depleting urban green cover. Animal habitats are also threatened in cities by dumped toxic chemicals, waste products on road side, pollution on air, land and water. According to the air quality requirements of the World Health Organization, suspended particulate matter concentration (made up of airborne smoke, soot, pollen and liquid droplets from fuel combustion) should be less than 90 $\mu$g per $m^3$. Indian megacities like Delhi with 415, Calcutta with 375 and Mumbai with 240 $m^3$ of suspended particulates in the air are some of the highly polluted cities.

## *Housing Crisis*

Shelter is one of the primary needs of humans. Meeting the housing demands as a result of expanding population of the cities is a daunting proposition. The shortage of houses leads to overcrowding; insanitary conditions and it result in slums. As per report published in Business Standard (23 December 2020), the housing shortage in urban areas was around 29 million units in 2018 and most of the housing shortage lies in the economically weaker section (EWS) and lower income group (LIG). This shows how the urban poor are further marginalized in the cities.

## *Unmet Need of Infrastructure and Urban Services*

Surplus population in cities leads to problems such as water scarcity, electricity breakdowns, shortage of shelter, inadequate and inefficient transportation system and overcrowding. In many cities, water is available for short duration only, that too early morning or evening. The slum dwellers do not have access to piped water in their homes. Sometimes, due to leakage or mixing of sewer line, water gets contaminated. Keeping the city clean is also a thorny task. With a large number of footfall and plying of vehicles, it is difficult to keep the streets clean. According to a report, '2.5 billion people have unreliable or no access to electricity and 2.8 billion live in areas of high-water stress, by 2035 energy consumption will increase by 35% which will increase water consumption by 85%' (World Energy Council, 2010).

## *Challenges of Inequity*

Urban areas are seen as places where ideally there should be reduction in inequalities. But extreme differences are found between poverty and wealth. Besides, wealth differentials, inequity of social opportunities, gender equality is also found in urban areas. Cities witness higher rates of crime. Thus, unchecked urbanization in the towns causes many social and crime issues. Unemployment and long-term sickness issues are increasingly seen on rise.

'The challenge of rapid urbanization is to maximize social, economic and environment development through more effective and comprehensive land administration functions, supported by effective Infrastructures, resolving issues such as climate change, disaster management, insecurity, energy scarcity, environmental pollution, and extreme poverty' (Doytsher et al., 2010). The smart cities work towards providing technology driven solutions to these challenges. 'Smart urbanization' can make cities more effective and lower their total carbon footprint. The growth of cities offers an impetus for urban development to solve social and environmental challenges, including the elimination of greenhouse gas emissions. Combined with the reconstruction and upgrade of infrastructure and networks in existing urban centres, as well as effective planning and responsible governance, many cities could deliver education, health care and high-quality energy services more effectively and with less pollution than the sparsely populated areas, simply because of their advantages of size, proximity and lower geographic footprint (Falconer & Mitchell, 2012; Wagner.nyu.edu, 2014). Smart strategies like use of renewable energy, smart grids, smart planning and smart transport pave the path for development of inclusive and livable cities.

## Smart City and Social Relations

As already discussed earlier, the primary objective of smart city is improvement in quality of life of individuals through socio-economic development with the assistance of techno-managerial solutions. This ensures efficient and time-bound delivery of services at the door step which in turn will lead to better living conditions of the citizens. However, nothing much can be achieved by adopting technological innovation. The success of smart city plans will depend on active participation of the citizens in all its initiatives. The success of smart cities depends on smart citizens. The citizens are the driving force of smart cities. Unless the smart cities focus on engaging the citizens in its ambit, all exercises will be futile.

'Smart city project concentrates only on the economic aspect. The social aspect is altogether ignored or the extent to which it will help the community to improve human relations is debatable' (Jagannath, 2019). The government has enacted numerous social legislations to counteract social evils prevailing in the society. One such example is of Section 498A of Indian Penal Code which safeguards women from the abuse at the hands of in-laws and husband. Similarly, Senior Citizens Act ensures protection for the elderly and so on. These really make us ponder about the social relations which are arising from the advancement of technology.

## *Family Relations*

As a consequence of economic growth, the materialistic approach has deep ramifications for human relations beginning with family life. This has contributed to the breakdown of the traditional joint family (Kapadia, 1966; Mandelbaum, 1949; O'Malley, 1941; Ross, 1961), which not only served as the unit of production and consumption but also acted as a unit of social security. However, on the flip side, it did put a limit on the personal freedom. The demands of urban life gave rise to a new form of family, which was limited in size and included members consisting of parents and their children, termed as 'nuclear family'. This new modern family made it possible for not only men but also for women to assert their individuality. Young and Willmot (1975) opine that that nuclear family has proved to be functional. Due to strengthening of conjugal bonds, increasing level of education and participation of women in employment, the power equation within the family has become dynamic and egalitarian. Women now enjoy equal status as their husbands and equally participate in all decision-making processes. Modern means of communication and transport facilitated rural–urban migration and the change of family structure from joint to nuclear. This fact is observed by Morrison (1959) in his study of Badlapur that the acceptance of nuclear families among rural people was a response to the impact of modern education, urbanization, industrialization and modern values. In fact, the availability of sophisticated and advanced techno-solutions has made people believe that they can survive on their own. This is justified on the grounds of pragmatism. The hold of family's on regulating human actions has weakened. The traditional patriarchal family structure has turned more egalitarian under which each family member has the right to share his views; the opinion of women and children is often sought while making crucial decisions. Change can also be seen in terms of more importance being accorded to achieved status, particularly in non-traditional occupations (Roy, 2000). In the times of crisis, the families are forced to depend on their own power. Money is accorded prime value and has led to the development of the attitude that it can solve all problems (Diener & Biswas-Diener, 2002), as a result of which human ties are given less importance. Despite all these changes, the importance of being in deep relationships towards emotional illness cannot be undermined and to a certain extend joint family system fulfils the desire to love and be loved against the mental agony of complex social relationships of present times. Moreover, that joint household and kinship network helps in meeting the needs of economic development, and that education, modernization, economic development and other modern forces do not constitute a barrier to kin relationships and joint households. Instead, joint household and kinship network provide the needs of economic development (Conklin, 1976).

## *Marital Relations*

Marital relations have also undergone considerable shift in the society dominated by technology. The definition of love and marriage has changed (Uberoi, 2006); love is a private matter, and marriage is more of a legal and social arrangement. With the continued changes, privacy in relationships has taken on more significance at all levels of life. Overburdened with increased competitive nature of the society and pressure of professional commitments, the individuals are shying away from making family commitments and accepting responsibilities. Thus, newer forms of living arrangements like live-in relationships which ensures ample space, without compromising on the privacy are emerging. Live-in relationships also come without the added burden of family commitment (e.g. Brown, 2003, 2004; Hansen et al., 2007). Over commitment for professional excellence and dearth of emotional fulfilment have not only impacted social ties but also exacerbated mental deterioration and induced physical ill-health.

The smart city project is aimed to develop the infrastructure and provide a technological base that will benefit people of all levels of Indian society (Daily Hunt, 2018). However, it is also to be seen that technology at the same time should not restrict the potential of individuals and delimit community life. Mental tension, emotional distress and financial uncertainty have become part of everyday life. The state is responsible for its citizens; though, previously joint family shared most of the responsibilities of state. Bureaucracy, which is the by-product of the commercialized society, has greatly influenced people's working ties in public and private spheres. Bureaucratic structures have distinctive features of professional selection, obedience to rules, procedural objectivity and emotional neutrality. Such bureaucratic features contribute in building emotionless relationships between members of this setup.

## Conclusion

Smart city impacts human relations directly and society indirectly in both positive and negative manner. It is imperative that the smart city focuses on creating physical, economic and social infrastructure which in turn will benefit people in terms of making social and economic progress. However, caution must be exercised in ensuring that the desire to create a more convenient living should not be at the cost of compromising social and natural ecosystem of which we all are a part. It is therefore imperative to discuss the issues concerning smart cities and its implementation in creative ways which ensures all-inclusiveness. It is also important to study the changes that the modern advancements bring along with them and their probable consequences on social relationships, so that the established institutions which serve as a foundation to human relationships are not hampered and the harmony between the technology and social relationships is maintained. After all, it is important to assess the cost well in advance at which innovations are introduced in society.

## References

Aijaz, R., & Hoelscher, K. (2015). India's smart cities mission: An assessment. *ORF Issue Brief*, No. 124.

Bouton, S., Cis, D., Mendonca, L., Pohl, H., Remes, J., Ritchie, H., & Woetzel, J. (2013). *How to make a city great* (p. 35).

Brown, S. L. (2003). Relationship quality dynamics of cohabiting unions. *Journal of Family Issues, 24*, 583–601.

Brown, S. L. (2004). Moving from cohabitation to marriage: Effects on relationship quality. *Social Science Research, 33*, 1–19.

Caragliu, A., Del Bo, C., & Nijkamp, P. (2011). Smart cities in Europe. *Journal of Urban Technology, 18*(2), 65–82.

Chakraborty, M. (2019). https://medium.com/datadriveninvestor/the-smart-city-evolution-the-possibilities-in-the-future-c24cfe65baea. Accessed June 10, 2020.

Chauhan, S. (2008). *Climate change, disasters and security issues, concerns and implications for India.*

Conklin, G. (1976). The household in urban India. *Journal of Marriage and Family, 38*(4), 771–779. https://doi.org/10.2307/350696

Daily Hunt. (2018). The impact of smart cities on Indian infrastructure. https://m.dailyhunt.in/news/india/english/yourstoryepaperyourstory/the+impact+of+smart+cities+on+indian+infrastructure-newsid-87122667. Accessed June 16, 2020.

Diener, E., & Biswas-Diener, R. (2002). Will money increase subjective well-being? *Social Indicators Research, 57*(2), 119–169. https://doi.org/10.1023/a:1014411319119

Dobbs, R., Remes, J., Manyika, J., Roxburgh, C., Smit, S., & Schaer, F. (2012). *Urban world: Cities and the rise of the consuming class* (pp. 1–92). McKinsey & Company. Available at: http://www.mckinsey.com/insights/mgi/research/urbanization/urban_world_cities_and_the_rise_of_the_consuming_class\npapers2://publication/uuid/C1BBC14A-B51F-43FC-AE92-CAC2DB6E346E

Doytsher, Y., Kelly, P., & Khouri, R. (2010). Rapid urbanization and mega cities: The need for spatial information management. In C. Potsiou (Ed.), *Research study by FIG ..., (48)*. Available at: https://

www.fig.net/pub/fig2010/papers/ts01b\ts01b_potsiou_doytsher_et_al_4709.pdf. Accessed June 16, 2020.

Falconer, G., & Mitchell, S. (2012, September). Smart city framework: A systematic process for enabling smart connected communities. ... /web/about/ac79/docs/ps/motm/Smart-City-Framework. .... Available at: http://scholar.google.com/scholar?hl=en&btnG=Search&q=intitle:Smart+City+Framework+A+Systematic+Process+for+Enabling+Smart+++Connected+Communities#2

Giffinger, R., Fertner, C., Kramar, H., Kalasek, R., Pichler-Milanović, N., & Meijers, E. (2007). *Smart cities: Ranking of European medium-sized cities*. Centre of Regional Science—Vienna UT.

Grayson, L., Young, K., & London Research Centre. (1994). *Quality of life in cities: An overview and guide to the literature*. Keynote Series. British Library in association with London Research Centre. https://books.google.co.in/books?id=s19PAAAAMAAJ

Hansen, T., Moum, T., & Shapiro, A. (2007). Relational and individual well-being among cohabitors and married individuals in midlife: Recent trends from Norway. *Journal of Family Issues, 28*(7), 910–933. https://doi.org/10.1177/0192513X07299610

Hoornweg, D., & Freire, M. (2013). *Building sustainability in an urbanizing world* (p. 216). World Bank Publications. Available at: http://ideas.repec.org/b/wbk/wbpubs/15790.html

Jagannath, T. (2019). *Impact of smart cities on India*. THE Planning Tank. https://planningtank.com/city-insight/impact-smartcities-india. Accessed June 16, 2020.

Kapadia, K. M. (1966). *Marriage and family in India*. Oxford University Press.

Kline, G. H., Stanley, S. M., Markman, H. J., Olmos-Gallo, P. A., St. Peters, M., Whitton, S. W., & Prado, L. M. (2004). Timing is everything: Pre-engagement cohabitation and increased risk for poor marital outcomes. *Journal of Family Psychology, 18*(2), 311–318. https://doi.org/10.1037/0893-3200.18.2.311

Lee, J. H., Hancock, M. G., & Hu, M.-C. (2014). Towards an effective framework for building smart cities: Lessons from Seoul and San Francisco. *Technological Forecasting & Social Change, 89*, 80–99.

MacIver, R. M., & Page, C. H. (1950). *Society: An introductory analysis*. Macmillan.

Mandelbaum, D. G. (1949). The Hindu family. In R. N. Anshen (Ed.), *The family: Its function and density*. Harber and Brothers.

Mckinsey.com. (2018, June). *Smart cities: Digital solutions for a more livable future* (pp. 4–8).

Morrison, W. A. (1959). Family patterns in Badlapur: An analysis of changing institutions in a Maharashtrian village. *Sociological Bulletin, viii*, 45–67.

MoUD, Government of India. (2016). *The planning tank*. https://planningtank.com/urbanisation/smart-cityconcept.html. Accessed June 11, 2020.

Nam, T., & Pardo, T. A. (2011). Conceptualizing smart city with dimensions of technology, people, and institutions. In *Proceedings of 12th Annual International Conference on Digital Government Research*.

Ogburn, W. F. (1927). *Social change with respect to nature and original change*. Viking Press.

O'Malley, L. S. S. (1941). Family, anxiety and religion in a community of North India. In M. K. Opler (Ed.), *Culture and mental health* (pp. 273–289). Macmillan.

Ross, A. D. (1961). *The Hindu family in its urban setting*. Oxford University Press.

Roy, P. K. (2000). *The Indian family: Change and persistence*. Gyan Publishing House.

Tolan, C. (2014). *CNN*. http://edition.cnn.com/2014/07/18/world/asia/india-modi-smartcities/index.html#:~:text=%22Cities%20in%20the%20past%20were,and%20next%2Dgeneration%20infrastructure.%22. Accessed June 15, 2020.

Uberoi, P. (2006). *Freedom and destiny: Gender, family, and popular culture in India*. Oxford University Press.

Unfpa.org. (2014). *UNFPA—United Nations Population Fund. Urbanization*. [Online]. Available at: http://www.unfpa.org/pds/urbanization.htm. Accessed June 20, 2020.

Wagner.nyu.edu. (2014). *Intelligent cities: Technology, policy and planning*. NYU Wagner. Available at: http://wagner.nyu.edu/courses/courseDetail.php?nbr=URPL-GP.4611. Accessed June 6, 2020.

World Energy Council. (2010). *Survey of energy resources.*
Young, M., & Wilmott, P. (1975). *The symmetrical family*. Penguin.

# Chapter 6
# Management of *Rurban* Area Development: Selected Indian Experience and Challenges

**Yogesh C. Joshi**

**Abstract** Globally, *Rurban* has been understood as a region which is on the edge of urban areas or where urban as well as rural characteristics coexist, or residential areas where farming is being done. Different nations have unique conditions and priorities of development, and hence have adopted slightly varied approach to *Rurban* around the generally accepted concept of urban and rural areas. In India, Shyama Prasad Mukherji Rurban Mission (SPMRM) was launched in 2016 with an aim to developing such (Rurban) rural–urban areas by ensuring economic, social and physical infrastructure facilities. The mission aims at harnessing cluster approach to development to optimize benefits for the economy. This chapter attempts to study selected Indian Rurban experience and challenges involved in selected states of *Rajasthan and Gujarat*. This develops a case for critical gap funding for identified gram panchayats which were studied as part of a separate research study done for Ministry of Rural Development, Government of India. Author draws from experience of developed nations to compare and contrast Indian needs and experience till now. The policy implications are derived based on study of selected areas qualifying for *Rurban* area development in these two states of India. International evidence and experience indicate adoption of urbanization policy to encompass *Rurban* areas in India. There is need for appropriate adaptation to help achieve stated objectives of urbanization and *Rurban* area development.

**Keywords** Indian *Rurban* experience · Rural development · Urbanization · Critical gap funding · Policy implications · Cluster area development

## Introduction

Urbanization is a global phenomenon which is experienced along with economic development by all nations. It is attributed to many socio-economic and demographic factors which include improvement in education, increase in income levels, division of land among family members, advancement of technology and productivity, skill

Y. C. Joshi (✉)
Faculty of Management, Sardar Patel University, Vallabh Vidyanagar, Anand, Gujarat, India



A. Kateja and R. Jain (eds.), *Urban Growth and Environmental Issues in India*,
https://doi.org/10.1007/978-981-16-4273-9_6

development, attraction towards urban amenities, growth of secondary and tertiary sectors away from rural areas and many other such factors. The movement of people from rural to urban areas and expansion of urban areas invariably cause rural areas to shrink and get mingled with urban areas. Urbanization invariably leads to population congestion and excess pressure on urban amenities and public services in all nations. Managing continuous migration of population from rural to urban areas and expansion of urban areas is a challenge for economic planners and governments both in rural and urban areas.

It has been increasingly felt during last decades across nations that urbanization should take place without sacrificing many of the positive attributes of rural areas. Many efforts are being made to manage excessive pressure on urban amenities due to unplanned migration to urban areas and check rural environment degradation due to impact of urbanization. Every national government thus attempts to take a variety of measures to mitigate ill effects on human population as well as on ecosystem and environment due to urbanization process.

In an attempt to maximize the wellbeing of human beings during urbanization, one of the approach practised in developed countries of Europe, which is gaining focus in India and other countries globally, is *Rurban* area development. The definition and classification of *Rurban* area vary considerably from view point of regions and nations. Overall, the definition of *Rurban* revolves around the rural areas likely to be engulfed into the urban fold in near future. Such rural areas are either in the vicinity of existing urban areas or are those which have the potential to be classified as urban area in near future due to increased economic activity and development. The definition and interpretation may vary a bit from country to country, but objective is to provide urban amenities and opportunities in these areas while maintaining rural ecosystem and without spoiling environment. Globally, *Rurban* has been defined as regions on the edge of urban area where process of urbanization has just started or where urban as well as rural characteristics exist or residential areas where farming is being done. Different nations have unique conditions and priorities of development and hence have adopted slightly varied development approach for *Rurban* areas around the generally accepted concept of urban and rural areas development plans (Anon, 2020a).

In India also due to expanding population and increased economic development, urbanization is spreading across regions and states. Despite the fact that India is a fastest growing large economy, it is still a developing economy. There are large areas which need basic amenities and infrastructure, and there is lack of town planning and its effective implementation. We find that human settlements emerge before any planning for them is done, or even if plans exist, they are seldom implemented in reality. New human settlements emerge out of rural areas, agriculture activities give way to housing, trading, etc., without being preceded by appropriate planning even for basic amenities, like roads, education, drainage, health and others. During all these years, successive governments have taken initiatives, but they have not proved to be effective for appropriate development of such areas. The urban as well as rural in India are a picture of unplanned development and in some cases, a total chaos. In fact, people face a lot of challenges in their daily life like traffic congestion, potholed roads and clean drinking water to inadequate public services like drainage, sanitation, water logging, education and health. With the change in seasons, the population is forced to forget the daily tribulations as they prepare themselves to meet the challenges of changing seasons. Even the middle and rich classes face problems arising out of poor urban planning, transport, poor preventive health besides problems and challenges in rural areas. *Shyama Prasad Mukherji Rurban Mission (SPMRM)* aims to provide a solution to the burgeoning population increase in the urban areas by providing economic, social and physical infrastructure facilities in rural areas (Anon, 2020b).

The successive governments at the centre and state have been conscious regarding development of rural and urban areas in the country. However, more focus has been put on development of rural areas than urban areas, as is evident in the five year plans. Till the start of SPM Rurban mission, the focus had been on addressing the lack of basic facilities in the rural areas, and most of the funding was either being spent on poverty reduction or employment generation. The Rurban mission focuses on developing rural areas which have potential of development due to existence of economic activities, within an area of a few gram panchayats, identified based on certain population size in tribal and non-tribal regions. For such identified rural areas, cluster approach has been adopted in the form of SPM *Rurban* mission in 2016. The focus of this initiative is holistic development of identified clusters by giving a boost to the economic activities and providing the region with best of urban amenities for development of population leading to their improved wellbeing. The regions which are adjacent to urban areas, exhibiting significant characteristics of urban areas as well as rural areas, may be classified as *Rurban* according to internationally accepted definition.

In rural areas located on the fringes of urban areas and affected by urbanization, an array of socio-economic and demographic factors interacts, impacts and influences each other. Figure 6.1 provides a glimpse of interplay of various factors which act on one another and can be considered to be affecting nature of approximate '*Rurban*' development scenario as experienced in any country. These *Rurban* areas typically are rural areas which are influenced by urban development in their vicinity. The people here are still dependent on primary activities for their livelihood. Urban amenities like skill development, education, health and others are deficient but at the same time are also beginning to emerge and exist in a form which is a poor replication of urban services in terms of quality and sophistication. In underdeveloped countries like India, in such *Rurban* areas, people may have higher income but overall education, skills and fineness in the form of calibre to adopt modern technology and means of production is not their forte. These conditions indeed pose considerable challenge to administrative machinery and planners engaged in upliftment of population and development of rural areas.

During the course of urbanization, rural areas on urban fringes get engulfed by the urban amenities and this impacts the consumption and behaviour patterns of rural population. Improvement in transportation, information technology and others also contribute to this phenomenon which is further facilitated by the increase in automobile penetration, changes in financial sector and other social and economic imperatives. This phenomenon is more pronounced due to increase in income levels

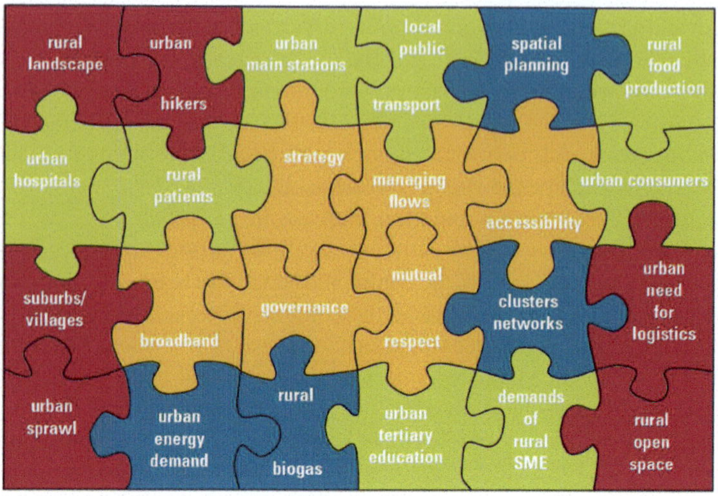


**Fig. 6.1** Interaction of rural and urban areas. Source Judith et al. (2012)

of population in rural areas due to escalation of land prices in India. All these factors have ensured that linkages of rural areas and their economic activities are much greater now than ever before with that of urban areas. The rural population aspires to have access to facilities and amenities at par with that of urban areas. At the same time, urban population aspires to spend time in rural ambience and areas during holidays and weekends in an attempt to be with nature and get away from hustle and bustle of urban areas. Thus, an increase in economic and social interaction among rural and urban areas and efforts for good governance by government is responsible for increasing focus of administration on efficient implementation efforts for rural area development. The experts in Europe, like in other developed nations have emphasized on linkages of urban and rural development and importance of regional planning (Anon 2010).

In India, the governments having realized the importance of planning in rural areas adopted *panchayati raj* and bottom-up approach for development. To give boost to rural area development near urban areas and in certain identified clusters of gram panchayats, the government launched '*Rurban* mission' in 100 selected clusters in the first phase under the name of *Shyama Prasad Mukherji Rurban Mission (SPMRM)* in 2016. The mission plans to identify 300 clusters during next few years for implementation SPM *Rurban* programme for implementation in the country (Anon, 2020b). Besides these identified clusters, there is need to implement SPMRM in areas most affected by urbanization for efficient planning and area development for benefitting population.

The paper attempts to study the provisions of SPMRM and Indian *Rurban* experience and challenges involved in its implementation of selected gram panchayats (GPs) in states of *Rajasthan and Gujarat*. These selected GPs are not among the Phase-I 100 clusters chosen under SPM *Rurban* mission, but are chosen based on definition of *Rurban* areas due to their proximity to urban areas. The paper develops a case for focusing efforts on development of selected GPs and for possible critical gap funding, which was studied as part of a separate research study commissioned by Ministry of Rural Development, Government of India. The paper draws from experience of developed nations to compare and contrast Indian needs and experience till now. The policy implications are derived based on study of selected areas qualifying for *Rurban* area development in the two states.

## Review of Literature

A brief review of literature is presented here to elicit experiences of a few leading countries in *Rurban* area development. It clearly reveals a variety of approaches adopted and followed in various countries for *Rurban* area development; however, the focus in each case remains on maintaining positive attributes of rural areas while amalgamating them in urban areas.

OECD (2013) study stressed that urban and rural areas enjoy different but often complimentary assets, and better integration between them is important for socio-economic performance. It provides a framework to understand the changing relationships between urban and rural areas. It is focused on an approach that can enhance and better manage rural–urban relationships—the use of rural–urban partnerships. It discusses the characteristics of these partnerships. It also examines the factors that can hinder as well as enable rural–urban cooperation. The study identifies and discusses different governance approaches to manage rural–urban relationships and gives recommendations to help national, regional and local policymakers to build effective and sustainable rural–urban partnerships for better economic development.

Busck et al. (2009) studied *Rurban* areas in Sweden, Denmark and the Netherlands. The authors opined that urban areas put pressure on rural areas leading to formation of '*Rurban* zones', where both *Rurban* and urban functions compete for multitude of land use. They studied one selected municipality of Sweden (Staffanstorp), Denmark (Roskilde) and the Netherlands (Werv-area) each in relation to European spatial development perspective (ESDP) focusing on the concept of compact city, rural urban linkages and conservation of nature, cultural heritage and proactive planning. The analysis provides insight on how *Rurban* areas are managed in order to avoid further urban encroachment and developing green landscape. The study found differences in approach in terms of public involvement in development, role of different administrative levels and use of zonation like prezonation based on local discretion (Sweden), zonation based on legislation and principles with little room for municipality (Denmark) and post-zonation with deliberate introduction of urban elements in rural areas and negotiations between stakeholders (Netherlands) (Busck et al., 2009).

The Policy Brief No. 1 of European Commission (2014) mentioned that the links between rural and urban people, activities and areas have great potential for socially equitable and environmentally sustainable economic development. For a very large proportion of urban dwellers, finding affordable shelter can be extremely difficult and many are forced to live in marginal locations that are often environmentally hazardous and do not have adequate basic infrastructure. The report shows that there are notable difference in the agricultural sectors of the selected countries due to their history, agro-ecological and socio-economic characteristics and governance systems.

Hans study focused primarily on the physical design aspects of *Rurban* design, excluding the social design aspects, which are no less important. The social design aspects must be realized in order to manifest the physical design aspects. The challenge of the new discipline is to understand the learning of the living system. The world's vast cornucopia of design dreams can be transmitted into every home on the planet.

Sudhira et al. (2008) stated that urban sprawls are a result of improper planning, inadequate policies and lack of good governance due to various reasons. They emphasized on the need for an integrated spatial planning support system. The administration is unable to capture and visualize the problems arising due to lack of governance and inappropriate policy decisions. The authors concluded that challenge for future research is to arrive at an integrated spatial planning support system to effectively plan, review and evaluate the different policy options while capturing the dynamics involved.

Kipnis (1989) discussed that rural renaissance and industrialization of the remote periphery of the metropolis in Israel evolved as a result of geopolitical needs to settle politically sensitive regions. The emerging *Rurban* communities, the industrial villages and the community settlements are attractive to middle class, educated, white-collar, middle-aged families who strive to secure their personal space and ex-urban lifestyle in a closed, intimate affinity community. Despite the high expense of such communities, which are financed almost entirely through public funds, the new forms of *Rurban* communities should continue to be considered as a viable vehicle for regional development under geopolitical constraints.

Barkley and Henry (1997) provide an overview of the advantages and disadvantages of promoting industry clusters as an industrial development alternative for rural areas. Advantages of successful cluster promotion are industrial reorganization, greater networking among firms and more efficient use of public resources. The industry cluster approach is hindered by the process of selecting industries to target, overcoming latecomer disadvantages and facilitating supportive institutions. It concludes that areas considering promotion of *Rurban* development should compare costs of initiating or expanding a cluster with the potential benefits of successful cluster development. Thus, industry cluster strategy is not appropriate for many rural communities.

Das and Das (2011) stated that in North East Indian states, there are many industries based on local resources and traditional skills which exist among local communities. The authors state that despite various attempts by central and state governments, these rural household industries are not performing up to the expectations. In order to give new life to these industries, government has adopted cluster approach of development. The authors concluded that this will provide competitive advantage to the industries by enhancing productivity, innovation and formation of new business firms and thereby boost rural economic and industrial development.

The study by Newman and Applebaum (1989) stated that the growth in the number of *Rurban* communities has been a key feature of the changing settlement space in Western societies. Due to the presence of mixed features, the classification of rurban areas becomes problematic. This problem is all the more compounded in Israel, where rigid institutional and organizational criteria provide an added dimension to the settlement definition. A study of 75 new *Rurban* communities in Israel shows that while the public agencies responsible for the development of the *Rurban* settlements have attempted to predefine idealized models according to traditional rural-based criteria, their models prove to be incompatible with the functional categories which best describe the operational norms of these communities.

Newman (1984) focused on the settlement activity in the West Bank and Galilee mountain regions of Israel. While both settlement location and settlement type are seen to have influenced Jewish colonization over the past 100 years, it is the former that provides the dominant framework for understanding decision-making in a situation of territorial conflict. The change in dominant ideology in Israel during the 1970s is evidenced by a process of modification in settlement policies, resulting in the development of new urban village types in a mountain environment.

Thus, the review of literature gives considerable insights and a fair idea about forms and variety of *Rurban* experiences and challenges experienced by various countries, in particular, the experience of selected developed and developing countries.

## The Indian Initiative: SPMRM

The development of rural areas has always been focus of public policy in India since a majority of population still lives there. Urbanization is increasing continuously during past decades. According to the last population census of India in 2011, a total of 833 million people constituting 68% of total population reside in rural areas. Further, during 2001–2011, the decadal growth of rural population has been a sizable 12% per annum. During this period a total of 2279 number of villages has increased in the country in absolute terms (Census, 2011). The successive governments have been making attempts from time to time to bring villages and rural areas in mainstream of development through various programmes and schemes. The state governments have adopted and implemented village-focused efforts through appropriate policy measures according to region specific requirements and geographical conditions. Some of the programmes launched include *Nirmal gram*, tribal area, *van Bandhu Kalyan* and many such other initiatives focusing on either, village, region, women, youth, skill development, agriculture activities, water resource provision and conservation, road and infrastructure, health, etc. in rural areas.

The policy initiatives are taken both by central and state governments for rural development. However, considering the overall magnitude of task across country the impact has been at best gradual and minimum in majority of the rural areas. Thus, even today it is imperative that focused and concerted efforts for rural development are made to keep pace with urban areas. The government has implemented SPMRM to focus specifically on developmental needs of such *Rurban* areas. The idea is to provide most modern amenities to these areas and simultaneously resolve urban problems also. It is argued that many parts of rural areas in India have the standalone capacity to develop due to their inherent characteristics, if they are provided with sufficient direction in terms of planning and resources. Such rural areas typically have potential for growth in terms of economic factors and location advantages for generating comparative advantages. Effective institutions of education and health are of considerable importance, obviously for rural development (Joshi & Dash, 2015; Joshi & Patel, 2006).

The programme aimed to identify such *Rurban* areas to develop them in centre of self-sustained clusters of economic activity with best of rural and urban conditions by provision of economic, social and physical infrastructure. Keeping in view such objective, NRuM has a vision of 'Development of a cluster of villages that preserve and nurture the essence of rural community life with focus on equity and inclusiveness without compromising with the facilities perceived to be essential urban in nature, thus creating a cluster of *Rurban* Villages'. The NRuM therefore aims at stimulating local economic development, enhances basic services and creates well planned Rurban clusters. The mission envisages various outcomes which include bridging the rural–urban divide in terms of economic, technological and availability of facilities and services. Secondly, it lays emphasis on stimulating local economic development with reduction of poverty and unemployment in rural areas. Thirdly, it ensures development in the region and lastly, attracting investment in rural areas. The SPMRM aims at developing 300 such clusters in the country in five years time and has identified first 100 *Rurban* areas to begin with.

The initiative under SPMRM is likely to focus on selected regions and therefore spread the spirit of development and sustainable economic activity in surrounding regions as well similar to one more novel initiative of *Sansad Adarsh Gram Yojana (SAGY)* which aims at developing model *gram panchayats* which can be emulated by other neighbouring gram panchayats as well. Therefore, such programmes and overall policy initiatives are likely to facilitate improving selected areas and give a push up to livelihood and wellbeing of population, incorporating good positive and sustainable development for people at their doorsteps itself. The SPMRM aims to draw funds from various existing schemes and programmes being implemented and tap Critical Gap Funds (CGF) if more funds are required for fulfilling the mission objectives in selected rural areas. These new initiatives will avoid sufferings for population as described in 'living in India's slums' by Schenk H. (2001) in the form of economic exploitation, social degradation, exposure to health hazards and insecurity of living illegally on the margin of and in Bangalore like in other urban centers.

A few selected regions of gram panchayats (GPs) have been identified purposefully for initiating SPMRM and likely challenges in fulfilling its objectives in selected GPs.

## Possibility of SPMRM Implementation in Selected GPs in Western India

The SPMRM refers to identification of clusters of economic activity, but it is equally imperative to focus policy on rural areas in immediate vicinity of urban areas and also qualify as *Rurban* areas for their planned development. There are many such *Rurban* areas existing in the country in all states near urban areas. The strength of rural area institutions becomes important in effective implementation (Joshi & Pathan, 2015). A few such cases are discussed for illustration here for consideration of stakeholders. The selected GPs are purposefully chosen, two each in states of *Gujarat and Rajasthan.* The GPs are namely *Vahelal, Ahmedabad* and *Panchot, Mehsana* in *Gujarat* while *Dhankya, Jaipur* and *Bara Padampura, Jaipur*. All four of these GPs are chosen due to their proximity to urban areas and convenience of accessing information. The VDPs of these GPs (Anon. 2015) prepared to initiate development in tune with impending urbanization by governments of Gujarat and Rajasthan is commendable and are likely to pave the way for planned development of these rural areas.

1. *Vahelal* GP is located in *Daskroi taluka* of Ahmedabad district, the largest district in terms of population, in Gujarat. It is a one village GP and has already been awarded as *Nirmal gram, Swarnim gram* and *Vrandavan gram* titles by state government. The GP is dominated by population of non-resident Indians. Besides it is a '*Satsangi*' Devotees village with a famous Swaminarayan temple. It has a population of 3074 (2011); the main crops grown are rice, wheat, castor and cotton. Almost 90% of population is also engaged in livestock rearing besides being engaged in service in nearby industries and business. The GP presents quite a promising status of development when compared to Indian rural areas in terms of infrastructure and other amenities.

The GP, however, needs better quality schools, which would provide quality education to the children and prepare them to excel in various walks of life at state and national level, both in private and government sectors. Improvement in amenities like sports ground and gardens, etc. which lead to better quality of life for people can further be improved in GP.

2. *Panchot* GP is located right on the edge of *Mehsana* city in *Mehsana* district, one of the largest agriculture crop producing and trading centres in the country. *Panchot* is one of most suitable case of *Rurban* area in India. Its population is 9214 (Census, 2011). A large majority of population is commuting to the city of *Mehsana* as they are engaged in trading of agricultural crops and edible oil. Its vicinity from city allows its population to make use of all amenities as enjoyed by people of *Mehsana.* Thus, equal per cent of population is engaged in agriculture, livestock rearing and trading and service. The main crops grown are cotton, castor, *bajra*, wheat, *guwar* (cluster beans) and millets. A very good sports complex is coming up in the GP itself.

The GP needs to focus on maintaining cleanliness, focusing on garden development and beautification of surroundings. Skill development activities among youth along with value addition to agriculture produce need to be focused by population in the GP for encouraging new economic activities. Education and health facilities improvement, in terms of engagement in prevention and development of positive aspects, rather than cure need to be focused, e.g. a large population is dependent on trading activities therefore focus should be on diversification of economic activities and training youth in skill development in other emerging areas of the economy.

3. *Bara Padampura* GP (11 villages) is located in *Chaksu tahsil* of Jaipur district. Its population of 4641 (Census, 2011) is engaged in agriculture and livestock mainly, and only 15% is dependent on service for livelihood. The main crops grown are wheat, maize, cluster bean—guwar, chicken pea and mustard. The GP also has a famous ancient *Digamber Jain* temple. The village needs to be developed under *Rurban* mission due to its location as well as need for involving all stakeholders in development process. A majority of population of GP is engaged in agriculture with Jain temple attracting devotees and tourists in large numbers. Due to the influx of religious tourists, it urgently requires attention of district and state administration for its planned development.

The *Bara Padampura* GP near Jaipur, which is on the verge of being engulfed and merged in urban area, presents an excellent case of a *Rurban* area. It presents an opportunity for planning and implementation for creating a more modern and vibrant urban suburb in the form of *Rurban* area. In the wake of an impending happening of being subsumed in Jaipur Municipal Corporation, there is an urgent need of programme like SPMRM being implemented in *Bara Padampura*. Not only the GP needs to be developed around famous Jain temple, there is scope of developing it by creating skill development and agribusiness opportunities in the region.

4. The GP (03 villages) of *Dhankya* is located in Jaipur *tehsil* of Jaipur district. It is very well connected with Jaipur and is about 27 km from Jaipur city. The total population of GP is 4626 (Census, 2011) with 90% dependent on agriculture and livestock. Main crops grown are wheat, millets, gram, maize, barely, groundnut and onion. The GP is renowned for its association with great thinker and philosopher Shri Pandit Deendayal Upadhyay. During the last few years, a grand memorial of Pandit Deendayal Upadhyay has been built. This is likely to promote tourism in the region further. However, the GP faces severe water crunch and is already categorized under dark zone for underground water deficiency.

The fact that GP is very near to a large urban area, it is imperative that appropriate planning needs to undertake in the form of developing quality education centres, providing for state of the art health facilities for population. A reasonably good transport is already provided for the focus should be on ensuring skill development among youth, job opportunities available near GP itself along with greater environment consciousness among population. Further efforts need to be made for improving and inculcating social values and citizenship among population to contribute to development of region and human beings together with development and modernization. Thus, an ideal *Rurban* area can hopefully emerge.

A further insight from study of these GPs reveals that efforts are needed on urgent basis if development in the mode of SPM *Rurban* mission is to have a meaningful contribution for development of selected regions and overall development of the country. The study of these four selected GPs also clearly shows the need for further attention towards modernization of basic education, health facilities. Further mechanization of agriculture and improving efficiency and productivity of agriculture practices is apparently required. There is need for promoting practices involving conservation of water and other natural resources and greening the environment for improving quality of life. The other area needing further improvement is use of renewable energy, waste disposal and treatment, which are so commonly used in modern developed *Rurban* areas across globe.

Quality of human life and wellbeing also depends on healthy lifestyle which is maintained by the people by adopting sporting activities and positive entertainment. Use of plastic needs to be judiciously regulated as it is already choking the cities and having harmful effects on public life. *Gutka, pan masala* chewing and liquor drinking and other social evils need to be addressed appropriately in the society. Other aspects need emphasis is steps to reduce pollution and carbon emission. These provide direct and indirect benefits in improving quality of human life and ultimately wellbeing of population. The garden and beautification of open space in *Rurban* area need to take up to help new urban marginalized sections, namely old and children in the society lead a decent and fulfilling life. This will contribute to sustainable development of rural areas (Joshi, 2017).

The population also needs to be trained in modern economic activities for transformation of rural population and to make them fit for taking advantage of *Rurban* area development initiatives. There is a need for skill development of people so that they can avail job opportunities and self-employment opportunities. Development of sustainable village-level micro and small firms will go a long way in productive engagement of population (Joshi, 2015). Sufficient training and support from government is need for this purpose. At the same time besides help and support in economic activity, improving environment (Joshi & Kurulkar, 2004), maintaining rural environment and traditional practices efforts for taking care of social evils is likely to contribute to developing the most modern and successful *Rurban* areas in the country through SPMRM.

The international experience is further useful in adopting successful and best practices for development of *Rurban* area development mission in the country. The international *Rurban* practices as presented briefly in review of literature provide ample of good examples for appropriate adaptation to suit individual *Rurban* areas while implementation of SPMRM in the country. It is therefore clear that proactive planning and its efficient implementation with well thought-out outcomes are the key to achieving appropriate *Rurban* development in the country. This is further likely to allow fulfilling objectives of a planned urbanization of new areas in the country providing excellent urban conditions for inhabitation of people in such areas.

## Conclusions

The above discussion reveals that SPMRM is a timely intervention and policy initiative adopted by the Government of India for focusing on area development issues emerging in the process of urbanization and economic development. It enables striking a balance between imperatives of sustainable environment. A concerted effort adapting to local realities and need of population in each of *Rurban* cluster allows efficient achievement of stated objectives of facilitating urban amenities in rural areas. International evidence and experience clearly favour adoption of urbanization policy to encompass *Rurban* areas in India as well. There is need for appropriate adaptation which will lead to further help in achieving stated objectives of urbanization and *Rurban* area development.

### *Suggestions and Policy Implications*

Urbanization is an inevitable process in all the countries without any exception. It is clear from experience of many developed countries that *Rurban* area development is an increasingly acceptable and practical means for effective planning and managing rural areas either on the edge of urban areas or having potential of development of diversified economic activity. In India too efforts in the form of appropriate policy initiatives are adopted from time to time for development of rural areas as their urbanization takes place. SPMRM is a recent initiative for development of selected rural areas in India.

The experience of four GPs indicates that the basic and primary needs for good *Rurban* development include education and health service available in GPs. Besides, a good and dependable transport system ensuring connectivity, drainage system, waste collection, electricity supply, garden and beautification efforts in GPs also need to be made available for the benefit of the people. There is a case for adopting preventive medicine and practices need to be aggressively adopted in GPs.

Education encompasses skill development aspect as well which need to be focused for sustained development. Activities of self-help groups (SHGs) need to be scaled up and meaningfully connected with local skills of population along with changing skill needs of the economy at present. Youth need to be empowered with adequate skills for their gainful employment. To strengthen harmony and greater bonding among population sports and other cultural activities along with festival celebrations needs to be encouraged and necessary facilities be made available in GPs. The role of village leadership and other opinion leaders like school teachers, hospital staff, etc. in conduct of various activities and sustainable best practices is of immense value in each GP.

The *Rurban* area development planning must incorporate modernization of agriculture as majority people are dependent on agriculture and allied activities. Adoption of horticulture, modern practices and scientific methods of cultivation in agriculture, agribusiness and value addition in GP, water conservation, use of solar and renewable energy need to be an essential part of the planning for development plan for these *Rurban* areas.

## References

Anon. (2010). https://ec.europa.eu/regional_policy/en/policy/what/territorial-cohesion/urban-rural-linkages/. Accessed May 20, 2020.

Anon. Census. (2011). *NSS*. Government of India.

Anon. (2015). Village development plan of *gram panchayats of Vahelal, Panchot, Bara Padampura and Dhankya*, in respective districts of Gujarat and Rajasthan.

Anon. (2020a). https://www.nationalgeographic.org/encyclopedia/urban-area/. Accessed May 20, 2020.

Anon. (2020b). http://rurban.gov.in/index.php/Public_home/about_us. Accessed January 15, 2020.

Barkley, D. L., & Henry, M. S. (1997). Rural industrial development: To cluster or not to cluster? *Applied Economic Perspectives and Policy, 199*(2), Fall–Winter. https://academic.oup.com/aepp/article-abstract/19/2/308/7228. Accessed January 15, 2019.

Busck, A. G., Hidding, M. C., Kristensen, S. B. P., Persson, C., & Søren, P. (2009). Planning approaches for *Rurban* areas: Case studies from Denmark, Sweden and the Netherlands. *Geografisk Tidsskrift-Danish Journal of Geography, 109*(1), 15–32. https://doi.org/10.1080/00167223.2009.10649593

Das, R., & Das, A. K. (2011). Industrial cluster: An approach for rural development of North East India. *International Journal of Trade, Economics and Finance, 2*(2), 161–165.

European Commission. (2014). *Rurban Africa, African rural-city connections* (Policy Brief No. 1). University of Copenhagen.

Joshi, Y. C. (2015). *Sustainable development of micro, small and medium enterprises (MSMEs) in Gujarat, India*. World Association of Sustainable Development (WASD). https://www.sudanknowledge.org/download/sustainable-development-of-micro-small-and-medium-enterprises-msmes-in-gujarat-india/. Accessed March 5, 2019.

Joshi, Y. C. (2017). Recent selected initiatives towards sustainable development in India. *Sanyojak, 8*(1), 1–4.

Joshi, Y. C., & Dash, S. (2015). Role of institutions in managing social sector: A perspective on education in Gujarat. *AMITY Journal of Management, 3*(2), 1–12. ISSN 23471832.

Joshi, Y. C., & Kurulkar, R. (2004). Greening the golden corridor: Exploring the possibilities in Ankleshwar and Vapi Industrial Estates of Gujarat, Western India. In *18th ECMSAS, EASAS, Conference Proceedings*, Lund, Sweden. Accessed from http://larseklund.in/Old.sasnet.lu.se-23.april/sasnet.lu.se/wwwroot/EASASpapers/12YogeshJoshi.pdf. http://citeseerx.ist.psu.edu/viewdoc/download?doi=10.1.1.579.9053&rep=rep1&type=pdf

Joshi, Y. C., & Patel, M. C. (2006). Emerging institutional responses in Gujarat for the fulfilment of the MDGs: Lessons for future. www.easas.org/19easas/panel46. Accessed April 2, 2019.

Joshi, Y. C., & Pathan, Y. (2015). Role of government in delivery of key services in rural sector. *Synergy—A Journal of Management, 17*(1&2), 01–11. ISSN: 0973922X.

Judith, A., Huttenloher, C., Kawka, R., & Scholze, J. (2012). *Partnerships for sustainable rural–urban development: Existing evidences*. Federal Institute for Research on Building, Urban Affairs and Spatial Development.

Kipnis, B. A. (1989). Untimely metropolitan field "Rurban" development—Rural renaissance as a geopolitical process in Israel. International aspects of rural and urban metropolitan systems. *Geography Research Forum, 9*, 45–66.

Newman, D. (1984). Ideological and political influences on Israeli Rurban colonization: The West Bank and Galilee mountains. *Canadian Geographer/le Géographe Canadien, 28*(2), 142–155.

Newman, D., & Applebaum, L. (1989). Defining the Rurban settlement: Planning models and functional realities in Israel. *Urban Geography, 10*(3), 281–295. https://doi.org/10.2747/0272-3638.10.3.281

OECD. (2013). *Rural-urban partnerships: An integrated approach to economic development* (OECD Rural Policy Reviews). OECD Publishing. https://doi.org/10.1787/9789264204812-en. Accessed March 21, 2017.

Schenk, H. (2001). *Living in India's slums: A case study of Bangalore*. Manohar Publishers and Distributors.

Sudhira, H. S., Ramachandra, T. V., & Bala Subrahmanya, M. H. (2008). *Urban sprawl management: Need for an integrated spatial planning support system* (pp. 1–30). http://wgbis.ces.iisc.ernet.in/biodiversity/pubs/ces_tr/TR119_HSS2007/HSS_Urbanpercent20Sprawlpercent20andpercent20SPSS,percent20Technicalpercent20Report_200707.pdf. Accessed May 20, 2020.

# Chapter 7
# Urban Health Infrastructure: Governance Issues

**Sandeep Narula and Narma Bhattacharya**

**Abstract** Employment generation initiatives of the Indian government have attracted rural populace to urban centres who face challenges relating to healthy living and access to affordable health care. The authors deep dive into the various facets of the disease burden being faced by this segment including the double whammy of non-communicable diseases (NCDs) and communicable diseases. The authors also find that given the very low public health allocation in GDP in India, the burden of unaffordable expenses falls squarely on the urban poor, often times pushing many families below the poverty line due to these health shocks. An attempt has been made to prescribe remediatory actions which include the need for an integrated health policy that orchestrates the participation of the government, private players and the citizenry to ensure easy availability of adequate and affordable health services. This will require increasing state expenditure on health, strengthening the urban infrastructure related to health services, greater involvement of medical colleges, proper triaging of diseases based on their prevalence among other initiatives described in this paper.

**Keywords** Governance · Urban development · Urban infrastructure · SDGs · Urban health

## Introduction

As rapid urbanization in India has set in, the need of infrastructure, renewed projects and sustainable living scopes are to be developed. Today when we are surrounded in the era of smart operational technology, it has also been observed that the expansion of urban infrastructure is not in sync with the exponentially growing urban population. Providing adequate facilities for the increased population influx in cities

S. Narula (✉)
IIHMR University, Jaipur, Rajasthan, India
e-mail: sandeep@iihmr.edu.in

N. Bhattacharya
Dr. B. L. Kapur Memorial Hospital, New Delhi, India



A. Kateja and R. Jain (eds.), *Urban Growth and Environmental Issues in India*,
https://doi.org/10.1007/978-981-16-4273-9_7

has become tougher, leading to inadequate infrastructure, overcrowding, pollution, lack of sanitation, lack of safe drinking water and several other issues which remain unaddressed. The overall wellbeing of the economical development demands for an inclusive growth to make all economic sectors of the society equally benefitting.

As per Census record (2011), 69% population lives in rural areas and only 31% in urban areas in India. It was for the first time a decline in rural population growth rate was noted. The difference between villages and cities is more than just beautiful buildings and infrastructure. The differences seep down to services, livelihood, income and consumption patterns too. These marked differences give rise to the term 'rural–urban divide'. It has been resulting in mass scale emptying of rural population and filing into urban cities seeking employment and better livelihood. In terms of health outcomes, there is a massive divide in urban and rural areas, leading many to believe that the urban health systems are far better.

The urban health in India has multiple stakeholders like Ministry of Health and Family Welfare, Urban Local Bodies and respective state governments. Health is a state subject. These overlapping jurisdictions and autonomy among the stakeholders have led to reduced accountability and ineffective planning and delivery of healthcare service to the urban poor. India is rapidly turning into a more urbanized nation; about 40% of the country's population will live in urban areas, i.e. around 600 million people will be living in towns and cities. The extent to which India's health system has the capacity to provide for this growing urban population will determine the country's success in achieving global health coverage and implementing national health indices.

In urban India, there is a lot of diversity in the healthcare system which ranges from private healthcare providers to public healthcare services to traditional healers (Butsch et al., 2012). The diverse healthcare systems in India make it very difficult to plan. The major challenges to healthcare system in the country are inadequate health infrastructure, lack of trained personnel, expensive treatment and inadequate health insurance. For the last 15 years, the public health spending has been about 1% of GDP for which has resulted into widespread inequities in distribution of health infrastructure and health personnel.

One out of four Indians is likely to die from non-communicable illnesses—cardiovascular infection, malignant growth, obesity and diabetes—before the age of 70. The World Bank report states that there are just 0.7 emergency clinic beds per 1000 individuals in India. This also varies across different states, and there are only 0.857 specialists per 1000 individuals (WHO). Further, limited accessibility and low quality services health facilities have prompted the individuals looking towards private health providers, which has a higher cost, for more care.

## Health Scenario in India

India has a well-established healthcare system that was evolved well before the beginning of modern science and British raj. Indian healthcare system is a mix of modern health practices and traditional health practices like Ayurveda, Yoga and Naturopathy, Unani, Siddha, and Homeopathy (AYUSH).

The Britishers laid the foundation of dispensaries at region level. Initially, they were meant for the treatment of soldiers. At the time of independence, there were around 7000 clinics in the country. In 75 years since India gained independence, a lot of progress has been made in terms of GDP, literacy rates and health indices and expenditures. Today right to health is a fundamental right. India has made remarkable progress in health sector in post-independence era. Health outcomes and health services utilization are found to be better in urban areas. The incidence of communicable diseases has reduced over time, but they continue to contribute to a significant proportion of the disease burden. Also, with urbanization there is a growing burden of non-communicable diseases along with communicable diseases that makes it more challenging.

India Council of Medical Research observes that the disease burden as measured using disability-adjusted life years (DALYs) dropped from 61 to 33% between 1990 and 2016 (ICMR, 2017). Similarly, a rise in disease burden from non-communicable diseases was observed from 30 to 55%. However, the epidemiological transition varied widely among Indian states, ranging between 48 and 75% for non-communicable diseases, 14–43% for infectious and associated diseases and 9–14% for injuries (ICMR, 2017).

The Indian healthcare delivery system is divided into two main domains—public and private. The public healthcare system comprises of primary, secondary and tertiary care institutions in key cities and focuses on providing basic healthcare facilities through Urban Primary health centres (U-PHCs) and community health centres (referral facility). There are 5190 functional U-PHCs in the country. There is a shortfall of about 44.4% of U-PHCs as per the urban population norms. Out of these 5190 U-PHCs, a total of 1734 PHCs have been upgraded as health wellness centres. The private sector provides majority of secondary tertiary and quaternary care institutions with major concentration in metros.

India has a 20% share in the global burden of diseases. In contrast to the share in the global diseases, India's contribution to the global healthcare infrastructure is highly inadequate. Globally, there are 9% global community health workers and 8% doctors (up from 50,000 in 1947 to 7.7 lakh in 2017). With 8% nurses, 6% beds for patient care (from 725 primary health centres (PHCs) in 1947 to 28,863 in 2017), and only 1% laboratory technicians in India, there is huge scarcity of human resources to tackle the mountain of diseases (Morgan & KPMG-FICCI, 2015). In 2015, our healthcare access and quality index at 44.8 were much lower than our neighbouring countries (Sri Lanka at 72.8, Bangladesh at 51.7, Bhutan at 52.7 and Nepal at 50.8, and it ranks above Pakistan at 43.1 and Afghanistan at 32.5) (Business Standard, 2017).

Table 7.1 depicts the different health missions started by government to tackle the health challenges.

**Table 7.1** National health missions in India

| Year | Name of mission |
|---|---|
| 1996 | Intellectual disability-related schemes (Vikaas, Samarth, Gharaunda, Niramaya, Sahyogi, GyanPrabha, Prerna, Sambhav, Badhte Kadam and Disha) |
| 2002 | Sarva Shiksha Abhiyan |
| 2005 | National Rural Health Mission (NRHM) |
| 2008 | National Mission on Medicinal Plants |
| 2012 | National AYUSH Mission |
| 2013 | National Urban Health Mission (NUHM) |
| 2014 | *Swachh Bharat Mission* (Clean India Mission) |
| 2015 | Affordable Medicines and Reliable Implants for Treatment (AMRIT) |
| 2018 | National Health Protection Mission (Ayushman Bharat Yojana/Pradhan Mantri Jan Arogya Yojana—PMJAY) |

*Source* Urban Health and Wellbeing, 2020, 251–256

## Constitution (74th) Amendment Act, 1992

The 74th Constitutional Amendment initiated the process of democratic decentralization and granted constitutional status to the municipal bodies and made the urban governance more responsive. The 74th amendment aimed at promoting people's participation in planning, provision and delivery of civic services. The municipal governance witnesses structural reforms with greater devolution of functions, more responsibility in planning and execution, and newer systems of fiscal transfers.

## Urban Lifestyle and Burden of Diseases

The modern lifestyle of an urban dweller is characterized by fast and sedentary living, stress and unhealthy eating habits. This leads to increased incidence of chronic diseases such as obesity, diabetes and high blood pressure. India is also known as the diabetic capital of the world, and at the same time seasonal outbreaks of diseases such as malaria, dengue, measles, diarrhoea and jaundice are also widespread here. Many of these diseases spread rapidly due to the large number of people in cities, especially due to overcrowding and slums.

Work stress, unhealthy eating and poor diet lead to a variety of uncontrollable illnesses including cancer. One million new cancer cases are reported every year. These are expected to rise rapidly and will increase five times by 2025. Due to a decrease in green cover and increased toxin levels in the environment, incidence of respiratory diseases has increased to 5 million since 2012 (European Respiratory Society, 2017). Seeking an effective and sustainable measure for permanent relief from air pollution-related health risks remains one of the most formidable challenges for city health planners, managers and public health workers today.

## Lack of Primary Healthcare Infrastructure

Since independence, a greater emphasis has been placed on public health in rural areas which can be garnered from Bhore report (1946). It has led to a disproportionate focus, lack of coordinated planning for health care infrastructure and the delivery of comprehensive health services to city dwellers. The rapid growth of urban population lands migrants and the urban poor in low cost and often dilapidated housing which puts them at a higher risk of catching diseases, increasing their need for affordable health care. Although a large number of secondary and tertiary care hospitals, public and private both, are located in cities, health infrastructure still remains inadequate in urban areas.

The health challenges in India require collective efforts and a multidisciplinary approach to address the situation. The growing proportion of urban poor and vulnerable population are found to be much worse than their rural counterparts on many health and development indicators. To address urban healthcare challenges and the unique needs of the urban population, there is a need to establish a primary healthcare system that connects the community with the UPHC, the CHC and the Higher Health Center. National Urban Health Mission (NUHM), launched in 2013, is a unique programme providing opportunities and mechanisms to improve urban health.

## Lack of Involvement of Medical Colleges in Urban Health Systems

Medical colleges play a major role to play in the development of evidence-based implementation of national health programmes (Kumar et al., 2018). However, their involvement in improving the health outcomes is less, ad hoc and unsystematic. Majority of medical colleges exist in urban areas. They are involved in the provision of education for future medical practitioners and the delivery of health care to the people. The medical colleges can act as link to primary and secondary institutions providing health care in urban areas. There is no developed mechanism to utilize the services of medical colleges in improving the service delivery and therapeutic compliance.

Most of the medical colleges are located in urban areas. They are valuable resources for providing educational, research and referral support to improve urban health delivery mechanisms and strengthen the various programmes initiated under NUHM. It will also be beneficial for the medical colleges as their engagement with national healthcare programmes will make their teaching and service programmes more relevant, effective, useful and sensitive to the changing health needs of society.

## National Urban Health Mission (NUHM)

NUHM was launched as a sub-division of National Health Mission (NHM) in 2013. It works towards reducing the cost of essential healthcare services and unaffordable expenditures for the treatment. This will be achieved by strengthening the existing healthcare service delivery systems, targeting the slum dwellers and synchronizing with various schemes of safe drinking water, sanitation, school education, etc.

The mission focuses on meeting the primary healthcare needs of the urban poor. The mission is being implemented in 779 cities and towns with more than 50,000 population and will cover about 7.75 crore people. The interventions will be targeted at reduction in IMR, MMR; providing universal access to reproductive health care and convergence of all health-related interventions. City-wise implementation plans based on baseline survey and felt need are prepared, and urban local bodies are to be fully involved in implementation of the mission.

The funding necessary for this scheme is split in 75:25 ratio between the central government and the state government in all states barring North-Eastern states including Sikkim and other special category states like Himachal Pradesh, Uttarakhand and J&K. For these states, the centre-state funding is to be split in 90:10 ratio. The Programme Implementation Plans (PIPs) sent by the states are appraised and approved by the Ministry.

### Inclusion of Community Participation Under NUHM

For greater accomplishment of objectives, NUHM aims to work with local communities of the area and local bodies. By devising a customized proactive approach for planning, implementation and monitoring, it partners with NGOs, for profit and non-profit health service providers and other stakeholders to provide quality health services. The NUHM proposes to leverage the existing non-government providers, who already are playing a significant role in community mobilization, to improve access to curative care. The NUHM proposes to forge partnership with this sector to promote active community participation and ownership (John et al., 2008).

## Urban Health Governance—And Related Issues

The urban health governance has been a contentious issue between centre and state governments. Being a state subject, the responsibility of providing adequate health care lies on the state governments, but the health sector is very tightly controlled by the central government. The COVID-19 pandemic uncovered various key issues plaguing the health services framework such as lack of hospital beds, lack of health equipment, unpreparedness of epidemic management systems, lack of trained human resource to name a few. The crisis exposed the mismanagement of health systems in the country. The public healthcare system was unable to provide for the rising cases and the health care provided by private hospitals was beyond the reach of the common man.

## *National Health Policy*

India rolled out its first National Health Policy in 1983 and thereafter in 2002 and again in 2017. India's National Health Policy, 2017 aims to provide quality healthcare and promote the wellbeing of all the citizens by making provision of preventive and promotive health care orientation in all developmental policies. It also envisions providing universal access to good quality healthcare services at affordable cost to the citizens. This involves strengthening of the healthcare institutions, reducing the unaffordable healthcare costs and raising the 'fiscal capacity' of the government to meet a widening healthcare financing deficit (Mohan, 2017).

Some of the specific goals and objectives as laid out by the policy are:

- Increase Life Expectancy at birth from 67.5 to 70 by 2025.
- Reduction in IMR.
- Increase utilization of public health facilities by 50% from current levels by 2025.
- Meet needs of family planning above 90% at national and sub-national level by 2025.
- Access to safe water and sanitation to all by 2020 (Swachh Bharat Mission).
- Increase health expenditure by government as a percentage of GDP from the existing 1.15–2.5% by 2025.
- Increase state sector health spending to >8% of their budget by 2020.
- Establish primary and secondary care facility as per norms in high priority districts (population as well as time to reach norms) by 2025.
- Ensure district-level electronic database of information on health system components by 2020.
- Strengthen the health surveillance system and establish registries for diseases of public health importance by 2020.
- Establish federated integrated health information architecture, Health Information.
- Exchanges and National Health Information Network by 2025.

The policy has been critiqued on the grounds that it does not provide any implementation plan. It elaborates on financial reforms in public healthcare systems but doesn't tell how it will be done. One of the major issues besides spending on public health care is poor management, inefficiency of administration and whole governance structure. It has been observed that some states with better management practices have been able to use the NHRM funds in a more prudent and effective manner than the others.

## Urban Health in Context with Sustainable Development Goals

The often used definition for sustainable development was given by Bruntland Commission, which says that the ultimate goal out of sustainable development is long-run stability of environment and economy. More importantly, it is only possible through decision-making process with social concerns and integrational equity. The major feature of developing sustainably is to make judicious use of resources, to conserve for the future generations to come.

Based on this approach, the framework of 17 SDGs and 169 targets was adapted by 193 United Nations Member States in 2015. Dedicated to create a synergistic effort to end poverty, hunger, ensure peace and aid global change of mindsets, SDG 3 aims towards good health and wellbeing for all ages, important for realizing full human potential. In these high population densities, several factors influence health as well as the economy. Addressing the aforesaid concerns SDG 11 aims at building safe, sustainable cities by virtue of improved air quality index, housing facilities.

The focus has been moderated towards low and middle income countries (LMICs) as United Nations have projected that in future more than 90% of world population will reside in these regions compared to 54% in 2015 (United Nations, 2014).

## Challenges in Healthcare

By 2030, 40% of Indian population will live in urban areas. It is crucial to understand the role of health care in ensuring a good standard of living for urban dwellers.

### *Barriers in Access to Healthcare for All*

Inaccessibility of healthcare service is a major concern among the urban poor population. Even though urban areas do not have geographical barrier as an issue, other factors such as inadequate health provider to patient proportion, overburdening of secondary and tertiary level services, and un-affordability of private health care are important. Other factors which lead to exclusion of population from seeking quality care are differently able or elderly, class or caste division (MoHFW, 2014) which culminates in seeking self-medication and care from informal health providers.

## *Disease Burden in Urban Areas*

In most parts of the world, the pattern of disease has shifted from communicable to non-communicable diseases (NCDs). In low- and middle-income countries, both co-exist and this phenomenon is known as 'double burden of diseases'. The shift of attention from child mortality, famine and infectious epidemics (e.g. tuberculosis, pneumonia) affecting large population to chronic, degenerative diseases causing diabetes, hypertension and obesity has exponentially increased.

WHO had released a list of 17 tropical diseases which are neglected and need immediate attention. The diseases that were considered to occur in developed areas have now been transferred to rural areas and vice versa. For example, lack of proper sanitation in urban areas has raised more than 893 million individuals in 49 nations viewed in danger of lymphatic filariasis, as now it has become a reality in urban areas of developing world (WHO, 2013).

The rapid urbanization has resulted in growth of slums and unplanned towns, where there is poor sanitation, overcrowding and poor housing (WHO, 2010b). Though migration and international travelling has squeezed the world small, it also acts as breeding ground for diseases like H1N1, Ebola, and more current COVID-19 to spread easily in high density populations.

A standard and more flexible method of measuring population health, other than Human Development Index (HDI), was published by WHO in 2014 and is called the Urban Health Index (UHI). It was used in a report analysis of 53 LMICs and 57 cities, including Delhi, Maharashtra, Kolkata and other cities from each continent. UHI has a range of 0 (lowest) to 1 (highest) and provided the following key analysis

i. Health conditions can greatly vary between same geographic region.
ii. Megacities in LMICs are worst affected.
iii. Health conditions at city level are not always determined by national level of wealth.

It highlights the challenges of health equity by adopting multi-sectoral responses (urban housing, transportation, power, etc.) which are both positively and negatively interlinked with each other. Thus, it also considers the challenges and introspection to possible solutions to resolve each of them.

## *Tailored Services for Urban Poor Population*

Due to huge migration in urban cities from different parts of the country, the population composition has a wide spectrum of health requirements. NUHM attempts to compose a uniform model for variedly urbanized areas, by supporting the designing of 'tailored' services for the population mix. This requires rigorous planning, technical and analytical system within health services which forms another set of challenges for ground implementation of the plans.

## *Reducing the Health Inequality*

Inequality in the socio-economic, political environmental health determinants is a common denominator across the global cities. Despite abundant resource availability, a large chunk of population is excluded from its consumption within same city due to scattering of population. It shows a very complex pattern and high inequalities exist within gender, wealth, caste and education levels. This is often defined as the inverse care law, which states that those who have greatest health need find it most difficult to access health services, often leaving their needs unmet. Within these massive populations, some urban poor households lag behind which increases the challenge of equitable delivery of health.

Equity in healthcare is attainable when factors affecting supply and demand of health services weighed on appropriate levels determine the accessibility. Examination of health equity in services through an integrated framework (Balarajan et al., 2017) of services has been used to describe demand and supply sides (Fig. 7.1).

In supply side, wide distribution of services between central and state governments, tiers of cities and geographical region determine the efficiency and physical availability of services to the population. In developing nations like India, inequity in access and poor service quality results as an outcome of inadequate supply among these areas. While health spending is roughly estimated at 1.28% of gross domestic product (GDP) in 2017–18, India has world's one of the highest out-of-pocket expenditure proportion at 61% (MoHFW, 2019).

In demand side, heterogeneous mix of large migrated population creates diversity in social, economic conditions, cultural and behavioural aspects of population thereby influencing their health behaviour. While this enriches the diversity, in LMICs, it also subsidizes the perception of availing better health services than rural residents. Age–sex composition mix is a big determinant in developed and developing nations. The

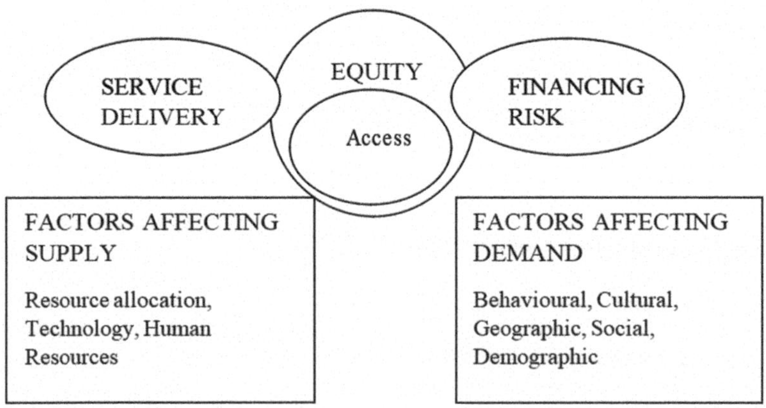


**Fig. 7.1** Model for conceptual understanding of healthcare-related challenges. *Source* Balarajan et al. (2017)

effect of migration in urbanizing cities paired with ongoing demographic transition impacts the life expectancy and health outcomes.

State and local governments play a crucial role in gaining information from their population including secluded communities or individuals. Formulating policies in collaboration with other sectors and organizations will help to reach out to wider cross section of their population. By empowering them with information and providing quality services, the burden of poor households will be lighter.

Improvement in knowledge of health system quality and researches on equity (Frenk & Horton, 2006) will help design interventions successfully. More independent organizations alongside NRHM (Bajpai & Dholakia, 2009) urgently need to evaluate specific programmes, with improved technologies to create multitudinal impact.

Urban health equity must be directed towards acknowledging social determinants of health. Use of equity metrics should be integrated in all allopathic and non-allopathic health systems (Chan et al., 2010). It can help to bring implementation of monitoring and evaluating health performances and its outcomes.

Public health expenditure plans coordinated by MoHFW, empowering of primary health service, technology systems, are the most significant way of ensuring health equity. The ultimate call of action required to translate this strength needs involvement of the affected communities and an intersectoral approach in both private and government infrastructure to build sustainable changes.

## *Universal Health Coverage (UHC)*

UHC is a step towards providing 'health-for-all' services under an umbrella for obtaining access to quality, promotive, preventive, curative, rehabilitative and palliative care. It is directed towards men, women and children of all communities with little or no access to these healthcare services. Other expected outcomes of UHC are increased job opportunities, improved health indicators, increased productivity and efficient health systems. With rapidly urbanizing and dynamic disease pattern changes, cities are crushing under the basic health needs of the population. UHC is also important part of United Nations' SDGs which is to promote wellbeing of all ages. Considering the condition of poverty already existing in LMICs, each year 100 million people in a population are pushed under the below poverty line (BPL) due to catastrophic unaffordable expenditures on health. Approximately only 17% of India's population is covered under some insurance (Reddy et al., 2011). This compounds the issue.

Health requirement in heterogeneity of rural and urban areas varies due to factors which affect the rates of mortality and morbidity. Thus, for achieving success utmost importance should be to cater priority health requirements of the masses. Aim of improving health services and protecting the disease and financial burden, there was increased push of involving UHC in SDGs.

The 3D approach towards Universal Health Coverage is an integration of population, services and financial factors of covering these service facilities. One of these dimensions is about covering population with needed health services through quality interventions in primary, secondary and tertiary levels to improve health status. It is to be based on criteria of prioritizing and monitoring of selected indicators like IMR, MMR, TB, malaria, etc. Second axis is about the equity in availing affordable services by people based on demographic, social, economic factors. Third axis is indicator of protection of financial risk projects the hefty out-of-pocket expenditures. These axes are quantifiable based on given indicators which are important part of the approach. Financing for UHC is facilitated through collecting insurance premiums from taxes, government budget allocations and insurance premiums from salaried employees to finance health coverage.

In 2018, publically financed Ayushman Bharat scheme was launched in two constituents, namely Pradhan Mantri Jan Arogya Yojana (PMJAY) and Health and Wellness Centres (HWCs). It accounts for insuring 40% of total population. The success of these schemes lies in strengthening of regulatory institution with increased budgets, modern technology and help of efficient systems of data management.

## *Managing Communicable Diseases*

Communicable diseases such as tuberculosis (TB), respiratory infections, chicken pox, etc. have direct modes of transmission through droplet infection or contaminated objects. The corona pandemic that the world is facing has also raised issues at the fragile healthcare systems, especially in India. In urban areas having high density of population, the population is more vulnerable of contracting and spreading these diseases rapidly. Transmission of diseases is affected by the inequity and diversity among population.

Communicable disease like TB is more likely to have high rates of incidence and prevalence in developing parts of the world. Similarly, respiratory disease outbreaks like Middle East Respiratory Syndrome (MERS), Severe Acute Respiratory Syndrome (SARS) propagate quickly through the crowded, malnourished population staying in poor housing conditions, resulting in severe outspread and high mortality.

Managing factors influencing the standard of lifestyle like vaccinations, hand hygiene, water hygiene will help to create individual stakeholders of creating better health status of population at large. It will help to regulate the pressure of double burden of diseases by decreasing the burden to a great extent. On national level, multidisciplinary approach should be given more focus in health sector. Nationwide implementation of these programmes can be done with involvement from local to central government plans.

Some of the following interventions can be adapted for better management of communicable diseases—prioritizing preparedness towards preventing specific infectious diseases, establishing epidemiological services in rurban areas, technical capacity building, information and risk communication awareness, awareness of hygiene and sanitation programmes. These measures will help increase preparedness and prompt response in situation of emergency.

### *Non-communicable Diseases (NCDs)*

The modern challenge of non-communicable disease epidemic is now in form of man-made diseases such as cancer, diabetes, mental health, coronary obstructive pulmonary disorders (COPD). They are creating double burden of disease along with diseases like influenza and flu-like diseases in rurban areas across the globe. Previously considered to be disease of the affluent, NCDs cause 63% of the global deaths which is alarmingly increasing with high premature mortality. In India, loss of 18 million lives is projected by 2030, due to premature deaths from NCDs (WHO, 2010a).

The NCDs are result of unhealthy diet consumption, high blood sugar levels, obesity, increased cholesterol, alcohol and tobacco consumption, etc. They can be controlled only by lifestyle and behavioural changes which is a big challenge in India with its diverse population matrix.

Apart from the loss of life, NCDs also increase the disease-adjusted life years (DALY) and economic burden on health systems of a nation. It represents increased expenditure on part of health systems, reduced productivity and fiscal risks of LMICs as well as high income countries. To quantify the representation in economic terms, on the basis of EPIC model, the economic burden of NCDs along with mental health conditions is estimated to be 4.58 trillion US dollars in between 2012 and 2030 in India (WHO, 2010a). Also in case of NCDs, the effect of catastrophic out-of-pocket expenditures keeps shooting up with time (Dwyer & Mitchell, 1999). Even within countries with UHC for its citizens, it creates an economic strain on health system and its resources. On individual level, it increases rate of absenteeism and reduced level of productivity.

Urban health systems can adopt integrated approach involving urban systems and environmental authorities, for better delivery of services. Utilization of current resources can help bring favourable modifications by adapting feasible, affordable care for all. Community participation should be included to work together for life-saving changes. Changing lifestyle and safe environment, warnings and providing information through dissemination of responsibility among communities, local, state government and, non-governmental institutions can provide safer urban living. In that manner, the common risk factors in population as well as individual can be mitigated to control and prevent diseases in sustainable manner.

## *Malnutrition*

Malnutrition is a huge cost absorber from the health system to tackle with reducing numbers. Intersectoral efforts to bring better urban services can help pay attention to cases of malnutrition globally. Inadequate supply of safe consumable water, unhygienic living conditions, improved food and nutrition can help reduce the chronic effects of deteriorated health on the health of an individual's life span. SDGs target towards achieving improved nutrition and ending hunger by promoting sustainable health nutrition for which role of primary health care is the propagating points on grass root levels.

The economic burden of all forms of malnutrition including anaemia, obesity costs 3.5 trillion dollar each year (UNICEF, 2018) Control over stunted growth among children and underweight women have been brought down to some extent in scattered manner over time. Phenomena of 'nutrition transition' including over and under nutrition is creating increasing problems within the same population. It is closely related to the trend of double burden of diseases due to more number of adult population who are overweight and younger children population being underdeveloped.

In LMICs, the picture has been very different, prevailing cases of under nutrition as well as over nutrition in its rurban areas is a major issue. Risk of increased rate of mortality, psychological and morbidity is associated with other costs like economic cost and loss of lives. Children and adult who are unable to utilize their full potential due to decreased quality of life and shortened life span is interlinked within the cities with poor health service infrastructure. Diarrhoea, worm infections are some diseases among the inhibitors of healthy physical cum mental development.

All communities should be through these steps—SMART objectives to attain equity in tackling population suffering with implications of poor urban food security and related outcomes. Smart prioritizing of information through integrated data processing is a promising road to attain the target. Several other roads to improving nutritional status through promoting healthy food habits by educating and food security can be taken to build sustainable improvements. Access to healthy and safe consumption of food, water and having clean living conditions have been attributed to healthy living.

## *Water and Sanitation for Safer Health*

In a WHO report (2008), there were estimates that globally one tenth of disease burden can be prevented by assuring safe water and proper sanitation. SDG 6 targets to attain good health through clean water and sanitation. The developing countries are making investments to build toilets, provide clean water supply, but the access is often limited to the privileged population. The less privileged in urban and rural cities do not have access to it. The use of toilets is inequitable due to either inaccessibility or as a result of behavioural patterns or poor maintenance. Community education programmes with proper action plans and initiatives between government companies, water body authorities and agricultural sectors should be implemented.

### *Developing Healthy and Sustainable Cities with Mobility*

Metropolitan cities, like Mumbai, Delhi have witnessed mass growth due to influx of people from all across the nation in search of better opportunities and standard of living. The cities were left with no other option but to grow horizontally. This resulted in narrow by-lanes, traffic filled streets, poor air quality all leading to severe health issues. Additionally, slums develop to provide housing solutions to the influx of migrants in the metros.

With the support of Government of India (GoI), housing boards in India are poised to manage the expansion of creating and establishing structured colonies. Several housing boards across India are accountable for optimal utilization of resources and ensure development in a sustainable fashion.

Healthy cities are described as cities with green friendly environment, dedicated passenger and bicycle pathways, public parks, clean and ecofriendly transportation facilities. Facilities like hospitals, schools, transportation, and housing facilities must be planned in the public infrastructure only then it would be advantageous in creating sustainability. Every city can take some steps to make it healthier where extensive makeovers are not feasible.

Climatic conditions have been changing causing environmental changes that is increasing further with fast urbanization. Climate change induced floods; heat waves affect the urban poor with lack of adequate housing facilities more. Such urban poor tend to stay in risk-prone zones.

## Conclusion

The dissolving barriers of development between rural and urban India is crucial to solving the issues of both worlds together. To reduce the burden of urbanization in the top tier cities, there's need of equity in developing policies and implementation for the rural and marginalized areas. Sufficient budget allocation, community partnerships, stronger infrastructures should be incorporated for a more sustainable growth. From this perspective, the solution to these aspects lies in the Pareto's 80/20 principle which would imply that there is need to address the 20% of aspects causing 80% of the issues. This will be necessary to ensure a bright future for the generations to come.

Remediatory actions which include the need for an integrated health policy that orchestrates the participation of the government, private players and the citizenry to ensure easy availability of adequate and affordable health services are suggested. This will require increasing state expenditure on health, strengthening the urban infrastructure related to health services, greater involvement of medical colleges, proper triaging of diseases based on their prevalence among other initiatives.

## References


Bajpai, N., & Dholakia, R. H. (2009). *Improving access, service delivery and efficiency of the public health system in rural India, mid-term evaluation of the national rural health mission.* The Earth Institute at Columbia University.

Balarajan, Y., Selvaraj, S., & Subramanian, S. V. (2017). *Health care and equity in India.* https://www.sciencedirect.com/science/journal/01406736/377/9764. https://doi.org/10.1016/S0140-6736(10)61894-6. Accessed January 22, 2019.

Bhore Committee. (1946). *National Health Portal of India.* https://www.nhp.gov.in/bhore-committee-1946_pg. Accessed November 15, 2018.

Butsch, C., Sakdapolrak, P., & Subramaniam, S. (2012). Urban health in India. *Internationales Asienforum, 43*(1–2), 13–32.

Chan, M., Kazatchkine, M., Lob-Levyt, J., Obaid, T., Schweizer, J., Sidibe, M., Veneman, A., & Yamada, T. (2010). Meeting the demand for results and accountability: A call for action on health data from eight global health agencies. *PLoS Medicine, 7*(1), e1000223. https://doi.org/10.1371/journal.pmed.1000223

Dwyer, D., & Mitchell, O. (1999). Health problems as determinants of retirement: Are self-rated measures endogenous? *Journal of Health Economics, 18*(2), 173–193.

European Respiratory Society. (2017). Forum of international respiratory societies. In *The global impact of respiratory disease*, 2nd edn.

Frenk, J., & Horton, R. (2006). Evidence for health-system reform: A call to action. *The Lancet, 368*(9529), 3–4.

Global burden of disease study: India at 154, lags behind Bangladesh. *Business Standard*. Available from: http://www.business-standard.com/article/current-affairs/global-burden-of-disease-study-india-at-154-lags-behind-bangladesh-117052000010_1.html. Accessed September 20, 2019.

ICMR. (2017). *India: Health of the nation's states: The India state-level disease burden initiative.* India Council of Medical Research, Public Health Foundation of India and Institute for Health Metrics and Evaluation.

John, D., Chander, S. J., & Devadasan, N. (2008). National urban health mission: An analysis of strategies and mechanisms for improving services for urban poor. https://doi.org/10.13140/2.1.2036.5443

Kumar, S., Kumar, S., & Gupta, B. (2018). Urban health: Needs urgent attention. *Indian Journal of Public Health, 62*, 214–217.

Mohan, D. (2017, March). New national health policy paves way for more hype and less action. *The Wire.*

MoHFW. (2014). *Making the urban health mission work for the urban poor*. Report of the Technical Resource Group, National Urban Health Mission. Ministry of Health and Family Welfare.

MoHFW. (2019). *Annual report 2019–20.* Government of India.

Morgan, J. P., & KPMG-FICCI. (2015, July). *Healthcare: The neglected GDP driver—Need for a paradigm shift*. Indian Healthcare Services, Asia Pacific Equity Research, via Thomson Research Database. Available from: https://www.gita.org.in/Attachments/Reports/Healthcare-the-neglected-GDP-driver-2015.pdf. Accessed September 20, 2019.

Reddy, K. S., Patel, V., Jha, P., Paul, V. K., Shiva Kumar, A. K., & Dandona, L. (2011) Towards achievement of universal health care in India by 2020: A call to action. *Lancet, 377*(9767), 760–768. https://doi.org/10.1016/S0140-6736(10)61960-5. Accessed February 15, 2019.

UNICEF. (2018). *Global nutrition report.* UNICEF. https://globalnutritionreport.org/reports/global-nutrition-report-2018/. Accessed February 17, 2019.

United Nations. (2014). *World urbanization prospects: The 2014 revision highlights*. United Nations Department of Economic and Social Affairs, Population Division. https://www.un.org/en/development/desa/publications/2014-revision-world-urbanization-prospects.html. Accessed March 12, 2020.

WHO. *Global health workforce statistics*. OECD. Available at: www.data.worldbank.org

WHO. (2010a). *Global status report on non-communicable diseases*. ISBN 978 92 4 156422 9. https://www.who.int/nmh/publications/ncd_report2010/en/. Accessed March 15, 2019.

WHO. (2010b). *Hidden cities: Unmasking and overcoming health inequities in urban settings*. World Health Organization.

WHO. (2013). *Sustaining the drive to overcome the global impact of neglected tropical diseases—Second WHO report on neglected diseases*. World Health Organization.

# Chapter 8
# Migration Pattern and the Emerging Trends of Migration in India

**B. K. Nagla**

**Abstract** People have been migrating all through history. They do it on account of various reasons. Migration at times is the result of the movement of people as a result of economic and political crises faced by them. They migrate from such places to other places where they get opportunities for their livelihood. Sometimes people migrate due to the natural calamities like drought or ensuing civil war or political repression, where exodus from their country is the only way out. The migration of displaced people is often the cause of political embarrassment for their home countries, the recipient countries and the countries they move through during transit. There is great pressure on developed countries like America, Canada and in Europe on account of immigrants from underdeveloped or developing countries. Brain drain is one of the major challenges faced by developing nations when these nations are not able to give jobs or incentives to good-qualified people and retain them in the country. Generally, nobody wants to leave from one place to another place if they get opportunities like developed countries in their own country. Migrants earn salaries and save money to send their home. The remittances constitute the second-largest flow of capital to developing countries, behind only governmental development aid. When the scenario of migration is viewed from domestic angle, it is observed that interstate and intra-state migration is a major issue within the countries. In a diverse country like India, the census data reports that only 1% of people migrate to other countries (2011). Rural-to-urban migration is an important contributing factor in rising levels of urbanization in developing countries. Migrants coming to urban areas compound the already existing challenges of urban areas like inadequate housing, employment, traffic, pollution, waste accumulation and disposal, squatters and slums, etc. In this context, the present paper attempts to examine human migration and migration patterns at the global and national front. It also discusses the process of adaptation and acculturation and how they aid in the process of urban cohesion.

**Keywords** Migration · Remittances · Interstate migration · Livelihood strategies

---

B. K. Nagla (✉)
Department of Sociology, M. D. University, Rohtak, Haryana, India

 
A. Kateja and R. Jain (eds.), *Urban Growth and Environmental Issues in India*,
https://doi.org/10.1007/978-981-16-4273-9_8

## Introduction

Migration (human) is the movement of people from their place of birth to another region, sometimes, across political boundaries, for taking up permanent or semi-permanent residence. When the movement is across national boundaries, it is termed as emigration and within national boundaries it remains migration. In India, we see farm labourers, who come to cities to earn extra income during non-farming time and go back to the villages during farming times. These are seasonal migrants who would have 'semi-permanent residence'. Generally, there are two types of migration, i.e. temporary and permanent. Temporary migration includes annual, seasonal or even daily movements of the population between two areas. It is also called 'commutation'. When people move from one place to another to reside permanently, it is known as permanent migration. People can either opt to move (voluntary migration) or be coerced to move (involuntary migration). Migration should not just be looked at as a simple movement of people across boundaries, but it should be looked as a much more complex issue that involves development, security and other social issues (Singh, 2016).

Migration has been found throughout human history, beginning with the movements of the first human groups from their origins in East Africa to their current location in the world. Migration takes place at a variety of scales: interregional, intra-continental and intercontinental. One of the most important migration patterns has been rural to urban migration—the movement of people from the village to towns and cities in search of better job opportunities. In some cases, it occurs as *stepwise migration* means the population moves from the countryside to small towns and then to a bigger city.

Migration occurs due to '*pull*' factors and '*push*' factors. Some '*push*' factors are unemployment, poverty, social insecurity, political instability and ethnic conflicts. On the other hand, movement of people owing to a number of attractions offered by a city is interpreted as migration due to '*pull*' factors. Consequently, people move out of villages due to '*pull*' factors—such as better opportunities for employment, education, recreation, healthcare facilities, business, etc. outside villages.

## Migration-Theoretical Perspectives

People migrate from their country to another country for a variety of reasons. The process of globalization has provided numerous opportunities to fulfil the demands of the workers to sustain national economies from other countries. Known as 'economic migrants', these migrants move to other countries to obtain sufficient income for survival (http/migration). They generally send savings from their income to family members in the form of remittances. This becomes an economic staple for developing and underdeveloped countries (Parle, 2007). Sometimes people also move or are forced to move in critical conditions of conflict, human rights violations, violence, or to escape persecution. According to World Migration Report (2020), the number of international migrants is estimated to be almost 272 million globally, with 164 million being migrant workers (ILO, 2018; UN DESA, 2019). Another reason for the migration is that people move in search of better opportunities within the country or to escape the vagaries of nature. This type of movement is usually from rural to urban areas. This is known as 'internal migration' (http/migration). Socio-cultural and geohistorical factors also play an important role in migration. Immigrating to developed nations like Europe, USA, Canada and Australia has been seen as a sign of social prestige. Often, the relatives, already living legally there, play the role of social network and help the new migrants to find job and accommodation.

## Migration and Remittances

According to the World Migration Report (WMR) 2020, 3.5% of the world's population is immigrant, up from 2.8% in 2000. 41.6% of these migrants are women who migrate alone or with their family members. India with a remittance receipt of USD 78.6 billion is on the top most recipient receiving country (WMR, 2020). Most of the emigrants from high-income OECD countries migrate to other high-income countries, and a substantial part (estimated at 43%) of emigrants from developing countries migrates to other developing countries. This shows that structurally, there is a substantial South–South and North–North migration, The United Nations Population Fund says that '[w]hile the North has experienced a higher absolute increase in the migrant stock since 2000 (32 million) compared to the South (25 million), the South recorded a higher growth rate' (Reips & Buffardi, 2012). In 2017, 68% of migrant workers were residing in high-income level countries—an estimated 111 million people. An additional 47 million migrant workers (29%) were living in middle-income countries, and 5.6 million (3.4%) were in low-income countries (WMR, 2020). The USA (51 m), Germany (16 m), Saudi Arabia (13 m), Russian Federation (12 m) and UK (9 m) were the top five destinations of international migrants, and the top five origin countries were India (18 m), Mexico (11 m), Russian Federation (11 m), China (10 m) and Syrian Arab Republic (8 m).

Migrant workers transferred their savings to their home country in the form of remittances from a substantial part of the economy in some countries. In 2018, India (78.61 USD), China (67.41 USD), Mexico (35.66 USD), the Philippines (33.83 USD) and Egypt (28.92 USD) were the top five remittance recipient countries.

Distinctions are found between voluntary and involuntary migration, or between refugees fleeing political conflict or natural disaster versus economic or labour migration. These distinctions are difficult to make as the motivation to migrate is quite often correlated. In the year 2018, 25.9 million migrants were refugees globally. In 2014, The United Nations reported that the highest level of forced migration on record, i.e. 59.5 million individuals, caused by 'persecution, conflict, generalized violence, or human rights violations', as compared with 51.2 million in 2013 (an increase of

8.3 million) and with 37.5 million a decade prior. In 2015, one of every 122 humans is a refugee, internally displaced or seeking asylum (Refugees, 2015).

Thousands of people are forced to seek refuge from war, poverty, political persecution and instability in their home countries. Europe has been flooded with a large number of migrants. Owing to this large influx, the European countries were under pressure to consider immigration reforms to accommodate the 'human tsunami' of people leaving Africa, the Middle East and Asia. In between January and July 2015, nearly 340,000 migrants entered Europe which is up from 123,500 in 2014. A large number of arrival of refugees ignited fear and xenophobic resentment in Europe. Different countries around the world have faced similar situations.

## Patterns in Global Migration

The number of international migrants globally was 272 million (3.5% of the world's population) in 2019, of which 48% were females [The World Migration Report 2020 (IOM)]. According to the IOM Report (2020), India had the highest largest number of migrants living abroad (17.5 million), followed by Mexico and China (11.8 million and 10.7 million, respectively). The top destination country (50.7 million international migrants) still is the USA. Between 2013 and 2017, there is a slight drop in migrant workers from 112.3 to 111.2 million in high-income countries. The highest increase from 17.5 to 30.5 million was experienced in upper-middle-income countries. Among the top 3 remittance recipients were India (USD 78.6 billion), China (USD 67.4 billion) and Mexico (USD 35.7 billion). Hence, international remittances increased to USD 689 billion. In 2018, the global refugee population was noticed as 25.9 million, and 52% of it was under 18 years of age.

The global migration pattern was examined by Mathias Czaika and Hein de Haas. They tried to study the volume and diversity of international migration during the second half of the twentieth century which was observed in their study published in International Migration Review in 2014. They used the World Bank's Global Bilateral Migration Database to identify the trends in the areas such as distance of migration and gaining and losing countries.

The important findings of the above study are as follows:

- Between 1960 and 2010, the countries that took more immigrants then what they lost through emigration decreased from 102 to 78.
- The number of countries that were net-exporters of people increased from 124 to 148.
- This finding exposes the common assumption that migration has become more diverse both in terms of the origin and destination of migrants.

Czaika and de Haas (2014) challenged the notion that improvements in communication and transportation have intensified migration. Technological progress has '*made the world more mobile, it has not necessarily made the world more migratory*'. They also analyse the migration patterns and found the importance of economic and demographic changes. International migration has not accelerated at a global level. The main shifts are linked to major geopolitical and economic situations which give rise to new migration hubs in Europe, Gulf and Asia. This boosts development-driven emigration in origin countries and the removing of emigration restrictions in former Communist and developing countries.

## Trends and Patterns of Migration in India

According to the Census of India (2011), migration arising out of various social, economic or political reasons was one of the important facets of the study. The movement of the population in different parts of the country helps in understanding the dynamics of the society better. During the present times, when the states are growing as a result of economic development, the study of migration data provides interesting insights.

A person is considered a migrant when he/she is enumerated in the census at a different place than his/her place of birth. This is because of migration due to marriage among females or for work as is the case generally among males, etc. This is also applicable when many return to their place of birth after living elsewhere. A study of the data collected by Census 2011 is important to understand the migration scenario in India. The following are the highlights of movement of people in India:

- About 453.6 million (38% of population) persons have been reported as migrants who settled in places of residence different from previous places of residence.
- Between 2001 and 2011, while population grew by 18% the number of migrants increased by 45%.
- In 2001, this number was 314.5 million (31% of population). Of these migrants by last residence, 268 million (85%) are intra-state migrants, those who migrated from one area of the state to another. In 2011, 99% of total migration was internal and immigrants (international migrants) comprised 1%.
- 41 million (13%) were interstate migrants.
- 5.1 million (1.6%) migrated from outside of the country.

The pull factors that attract migrants from rural to urban areas and from smaller towns and cities to larger urban areas are better opportunities in context to education, employment, etc. On the other hand, there is movement of people in the opposite direction from urban to rural areas due to various reasons. The Census of India 2011 reveals the following facts:

- Out of about 98 million, total intra-state and interstate migrants in the country during the last decade, 61 million have moved to rural areas and 36 million to urban areas.
- Migration stream out of rural areas (73 million) to another rural area was quite high (53 million) in comparison to from rural to urban areas (20 million).

- About 6 million migrants went to rural areas from urban areas. On the basis of net migrants by last residence during the past decade, i.e. the difference between in-migration and out-migration, in each state, Maharashtra stands at the top of the list with 2.3 million net migrants, followed by Delhi (1.7 million), Gujarat (0.68 million) and Haryana (0.67 million) as per 2011 census.
- Uttar Pradesh (−2.6 million) and Bihar (−1.7 million) were the two states with the largest number of net migrants migrating out of the state.
- Most of the female migrants have cited 'marriage' as the reason for migration, especially when the migration is within the state.
- For males, the major reasons for migration are 'work/employment' and 'education'.

The Indian National Sample Survey Office (NSSO) conducted the 64th round survey on 'Employment and Unemployment and Migration Particulars' from July 2007 to June 2008, covering 125,578 households and 572,254 persons. This revealed the following facts:

1. ~30% were found to be migrants, i.e. whose last usual place of residence (UPR) was different from the present place of enumeration.
2. In this survey, the UPR of a person was defined as a village/town where the person had stayed continuously for six months or more.
3. Among the migrants, the majority were found to be moving within the state (85%) as opposed to those moving across states (15%). Women formed a sizeable majority of this migrant population.

## Intra-state and Interstate Migration Patterns

The internal migration reflects the following pattern:

- There are two types of internal migration. One is the intra-state movement of people within the state itself. Other is interstate movement when the migrants cross the borders of a state and settle down in another state.
- The bulk of intra-state migrations are not caused by economic factors. Among them, about three-fourths of migrants were females for whom marriage is the prime reason for migration.
- The interstate migration pattern depicts that the states with the highest net outflow of migrants are Uttar Pradesh and Bihar. The top destinations for people leaving UP are Delhi, Maharashtra and Uttaranchal, respectively.
- The highest net inflows are Maharashtra and Delhi.
- A clear trend is the consistent preference of people across regions to migrate into states with geographical proximity.

To sum up, it may be viewed that the cause of migration in our country is the movement from underdeveloped areas to developed areas. Therefore, there is a need to develop the underdeveloped region to prevent migration.

## Migration Across Rural–Urban Areas

From the perspective of movement across rural–urban areas, the census data shows that more than half of migration in India happens from rural regions to urban areas. However, a contra-trend was found for the Union Territories and the north-eastern states as unlike the rest of India; over 70% of migration in these regions is towards the urban regions.

Recently, the World Bank Report (2020) stated that nearly 40 million internal migrants were affected by the nationwide lockdown in India as a result of COVID-19 pandemic. The volume of internal migration was about two and a half times that of international migration. The migrants faced crisis of food scarcity, shelter, loss of employment, increased risk of contagion, the uncertainty of the future and anxiety. Moreover, they also had to face the fear of loss of livelihood, and this became the main cause for the mass exodus of migrant labourers to their homes. This mass exodus became a point of the contest at the interstate borders as migrants walked barefoot to for their hometowns.

## Causes and Consequences of Migration

Some key trends observed in the analysis of reasons for migration in India are as follows:

- Majority of women migrate on account of 'marriage'. It is also found that a majority of the women who migrate for marriage end up doing domestic chores.
- The main reason for the migration of men is employment and education which happens in the age bracket of 18–40 years. It is seen that men who move for employment end up as wage employees/labourers.
- People also migrate due to forced reasons/disasters as in the case of Tripura.
- Housing-related reasons are cause of migration in Kerala and West Bengal.
- There are some other significant numbers for various reasons for migration after the 1990s. These are mainly: business, education, housing, post-retirement and healthcare.
- In India, globalization started with the new economic policy of liberalization more in line with the Indian economic development. Consequently, global linkages and global consciousness accelerated the movement for migration.
- Uneven development between regions is the major cause of migration. Currently, the major metropolitan cities of India like Mumbai, Delhi, Chandigarh, etc. witness heavy in-migration which increases population growth due to work opportunities available in these cities and also a movement towards developed regions like West Bengal, Maharashtra, Andaman and the Nicobar Islands.
- Out-migration witnesses generally from underdeveloped and backward states likes Uttar Pradesh, Bihar, Orissa, Madhya Pradesh and Rajasthan which are also known as the BIMARU states.

## Process of Adaptation and Acculturation: Problems and Prospects

Adaptation is a process of change and adjustment to new socio-cultural, environmental and political conditions (Berry et al., 2002). It is difficult to define and measure adaptation for migrants from one culture to another. Initially, migrants get 'culture shock' that affects their wellbeing (Ward et al., 2001). The 'culture shock' and ensuing adaptation may depend on motivation and expectations (Berry et al., 2002).

Migrants not only leave earlier well-established social networks behind but also experience at first a sense of loss, dislocation, alienation and isolation. This leads to the process of acculturation. Therefore, an adaptation of immigrants is the process of 'fitting in' to the society of settlement and adjusting successfully in a new environment functioning (Ward et al., 2001). There are two distinct aspects of intercultural adaptation (Ward et al., 2001):

1. Socio-cultural adaptation: This is based on the culture learning approach and ability to interact with a different culture. This also includes maintaining the culture of origin and preferred host culture which is to be adopted.
2. Psychological adjustment: This helps the individual's sense of general satisfaction with life.

On the basis of the interaction between two intercultural adaption, the migrants have to strategize four different acculturation processes: separation, assimilation, integration and marginalization as suggested by Berry et al. (2002).

1. Separation refers to keeping the culture of origin and not accepting the host culture.
2. In contrast, assimilation is the process of relinquishing the ethnic heritage and accepting new cultural identity.
3. Integration means maintaining the heritage together with the host culture.
4. Marginalization refers to a rejection of both the culture of the host and the culture of origin (Berry et al., 2002).

It may be viewed that the strategy of integration predicts more positive outcomes in psychological and socio-cultural adaptation (Berry, 1998). The three other strategies particularly marginalization are often associated with poor adaptation. This leads to serious psychological problems including depression, anxiety and other mental disorders (Berry, 1998).

## Migration and Acculturation

In the preceding section, we have discussed socio-psychological aspects of acculturation. It is essential to understand the process of acculturation for survival and the existential problems of individuals. Berry (2002) suggests the process of acculturation following psychological models of moving towards, moving against and moving away from a stimulation corresponding to adaptation or simulation, rejection and deculturation. The process of acculturation requires adjusting two cultures both the dominant host culture and the migrant's problem of identity. Gordon (1964) differentiates between structural, identification and behavioural assimilation. There are cognitive ways of individual behaviour in the process of acculturation in terms of language, religion and other habits of dress, food, etc. The concept of acculturation is closely intergrouped to the identity of the self as culture and personality.

## Migration and Cultural Identity

A particular group shares a common heritage that reflects the ethnicity of the group. This includes history, language, rituals and preference for a way of life. Cultural or physical criteria determine ethnic identity. It may overlap between race and ethnicity. Each has a different social meaning.

Racial identity means a sense of group or collective identity. This is based on the perception of the individual who shares a common racial heritage with a particular racial group in which the individual is seen as inside or outside of a particular biological group. In practice, the concepts of ethnic, racial and cultural identities are often not easily distinguishable. Berry (2002) recommends using the term ethnocultural identity.

## Conclusions

The above discussion for the analysis of reasons for migration in India and abroad reveals the reason for migration and the process of acculturation and adjustment. Although migration in general and rural-to-urban migration in particular is conducive for economic and urban transition, there are many challenges confronting migrants in urban areas. Though the Constitution of India guarantees the right to move as a fundamental principle under Article 19, there is often negative attitude and hostility towards migrants by the native population. The conflict between natives and migrants has been a core issue since the 1970s. The reasons for this hostility range from the success of migrant population in trade and commerce to the perceived loss of jobs for natives due to the influx of cheap labour from different parts of the country. The recent times has seen flaring up of the native card in Mumbai as a result of political backing.

Population explosion and the resultant unemployment, poverty, presence of social inequalities in society become the push factors in the process of migration. Secondly, uneven development across the states attracts the people of underdeveloped or developing regions like Bihar, U.P., M.P., Rajasthan and Orissa. This is the consequence of pull factors basically for better opportunities for education, employment, healthcare facilities, etc., and thirdly, lack of good governance is also a factor aiding migration. If the underdeveloped areas are also developed, the chances of people migrating from one region to another reduce or are more or less equal.

Although migration in general and rural-to-urban migration in particular is conducive for economic and urban transition, there are many challenges confronting migrants in urban areas. A negative attitude towards migration persists as well as a strong hostility towards migrants in spite of the fact that the Indian Constitution guarantees the right to move as a fundamental principle under Article 19. The conflict between natives and migrants has been a core issue since the 1970s.

Migration is more of a governance issue rather than one of development. Deshingkar and Sandi (2012) argue that there is untapped potential for human development if the positive impacts of migration can be harnessed properly, but this requires accompanying changes in attitude and institutional structures. While there is a lack of integration of migration into development planning, there are a host of labour laws that deal with the conditions of migrant workers. However, the Interstate Migrant Workmen Regulation Act, 1979 (ISMWRA), which deals with contractor-led movements of interstate migrant labour, is not enforced properly.

The World Migration Report 2015 states, 'Migration policy, however, should not be viewed merely as part of labour policy but needs to be embedded in urban development policy and planning as rural to urban migration is the predominant form of migration'.

## References

Berry, J. W. (1998). Acculturation and health: Theory and research. In S. S. Kazarian & D. R. Evans (Eds.), *Cultural clinical psychology: Theory, research, and practice* (pp. 39–57). Oxford University Press.

Berry, J. W. (2002). Conceptual approaches to acculturation. In K. M. Chun, P. B. Organista, & G. Marin (Eds.), *Acculturation: Advances in theory, measurement, and applied research* (pp. 17–37). American Psychological Association.

Berry, J. W., Poortinga, Y. H., Segall, M. H., & Dasen, P. R. (2002). *Cross-cultural psychology: Research and applications* (2nd ed.). Cambridge University Press.

Czaika, M., & de Haas, H. (2014). The globalization of migration: Has the world become more migratory? *International Migration Review*.

Deshingkar, P., & Sandi, M. (2012). Migration and Human Development in India: New Challenges and Opportunities in National Workshop on Internal Migration and Human Development in India, 6–7 December 2011, Workshop Compendium, Vol. 2, Workshop Papers, UNESCO/ UNICEF.2012.

Gordon, M. M. (1964). *Assimilation in American life: The role of race, religion, and national origin*. Oxford University Press.

Gürkan, Ç. In *Turkish review: Turkey pulls, The Netherlands pushes?*

http://esa.un.org/unmigration/wallchart2013htrn. Accessed December 25, 2019.

http://www.unfpa.org/migration. Accessed December 25, 2019.

International Labour Organization. (2018). *Global estimates on international migrant workers—Results and methodology* (2nd ed.). ILO. Available at www.ilo.org/wcmsp5/groups/public/---dgreports/---dcomm/---publ/documents/publication/wcms_652001.pdf
International Organization for Migration. (2019). *World migration report 2020*. Accessed November 11, 2020.
Parle, J. de. (2007, April 22). A good provider is one who leaves. *New York Times*.
Refugees. (2015). United Nations High Commissioner for "worldwide displacement hits all-time high as war and persecution increase". https://www.unhcr.org/news/latest/2015/6/558193896/worldwide-displacement-hits-all-time-high-war-persecution-increase.html. Accessed June 7, 2020.
Reips, U. D., & Buffardi, L. (2012). Studying migrants with the help of the internet methods from psychology. *Journal of Ethnic and Migration Studies, 38*(9), 1405–1424.
Singh, H. (2016). Increasing rural to urban migration in India: A challenge or an opportunity. *International Journal of Applied Research, 2*(4), 447–450.
UN DESA. (2019). *International migrant stock 2019*. UN DESA, Population Division. Available at www.un.org/en/development/desa/population/migration/data/estimates2/estimates19.asp. Accessed September 26, 2019.
Ward, C., Bochner, S., & Furnham, A. (2001). *The psychology of culture shock*. Routledge.
World Bank. (2020). *COVID-19 crisis through a migration lens* (Migration and Development Brief No. 32). World Bank. Available at: https://openknowledge.worldbank.org/handle/10986/33634. License: CC BY 3.0 IGO. Accessed March 5, 2021.
World Migration Report. (2015). Migrants and Cities, New Partnerships to Manage Mobility, International Organization for Migration.
World Migration Report. (2020). International Organization for Migration. Available at https://publications.iom.int/system/files/pdf/wmr_2020.pdf. Accessed March 15, 2021.
www.nationalgeographic.com/xpeditions

# Part II
# Environmental Challenges

# Chapter 9
# Environmental Kuznets Curve: A Revisit in the Case of SAARC

**Maniklal Adhikary and Chandrasekhar Hajra**

**Abstract** The Environmental Kuznets Curve is a hypothesized relationship between environmental degradation and economic growth. It states that in the early stages of economic growth environmental degradation and pollution increases, but beyond some level economic growth leads to environmental improvement. We empirically tested the hypothesis for six selected SAARC countries namely Bangladesh, Pakistan, India, Bhutan, Nepal and Sri Lanka using a panel data over a period of 1980–2014. The Composite Environmental Degradation Index (CEDI) has been calculated, and its relationship with per capita income has been tested to find out shape and to compute the turning point(s) of the Environmental Kuznets Curve in the SAARC region as well as in the selected countries of the region.

**Keywords** Environmental Kuznets Curve · Environmental degradation · Economic growth · Composite environmental degradation index · SAARC

## Introduction

In development economics, ideas of economic growth are gradually getting shifted into the ideas of sustainable economic growth. Environmental impact of the growth path is now being studied thoroughly to understand the relation between environmental damages and income growth. The Environmental Kuznets Curve (EKC) provides theoretical support and tools for such studies. Several indicators measuring environmental condition along with income are being studied in EKC literatures. The theory states that in the early phases of economic development, environmental degradation as well as pollution increases, but after certain level of per capita income, the trend starts reversing. Usually at a high per capita income level economic growth takes place with improvement in the environmental conditions. This implies that the income and economic growth relationship takes the shape of an inverted U-shaped

M. Adhikary
Department of Economics, Burdwan University, Burdwan, West Bengal, India

C. Hajra (✉)
Nistarini College, Purulia, West Bengal, India

 
A. Kateja and R. Jain (eds.), *Urban Growth and Environmental Issues in India*,
https://doi.org/10.1007/978-981-16-4273-9_9

function. The logical intuition of the Environmental Kuznets Curve is straightforward. During the early days of industrialization, people give more importance to growth of output, income and employment generations and tend to ignore air, water or soil quality (Dasgupta et al., 2002). To increase output in quick time people use cheap and easily available natural resources resulting in higher emission of pollutants, and putting more pressure on the environment. Vast majority of the people are very poor and disregard environmental consequences of growth. With time people become more conscious and capable and during the later stages of industrialization, environmental consciousness makes regulatory institutions more effective resulting into a declining pollution level.

The South Asian Association of Regional Cooperation (SAARC) was established in 1985. Member countries of SAARC are Afghanistan, India, Pakistan, Bangladesh, Sri Lanka, Nepal, Maldives and Bhutan. The region is housing more than 1.5 billion people and comprises some of the densely populated areas of the globe. This regional group was established to address some of the common goals of the member countries as improving the quality of life, accelerating economic growth, social and cultural development, etc. Economic dynamism of recent decades especially by India followed by Pakistan and Bangladesh is changing South Asia. But in the path of economic growth, forest covers have depleted, minerals were extracted at a brisk rate, water bodies were polluted, soils got contaminated by unscientific use of chemicals, rapid industrialization and urbanization led to polluted air around us. So it is very important to study how far this regional economic growth is affecting the environment around us, country-wise as well as region as a whole. As South Asia is housing nearly one fourth of total global population and nearly 40% of global poor, its economic performances significantly affects global issues.

We empirically tested the hypothesis for six selected SAARC countries, namely Bangladesh, Pakistan, India, Bhutan, Nepal and Sri Lanka using a panel data over a period of 1980–2014. The Composite Environmental Degradation Index (CEDI) has been calculated on the basis of the index of air pollution, water pollution, destruction of soil quality, extraction of ores and metals, deforestation, depletion of natural resources and urbanization. The relationship between CEDI and per capita income at constant price as of 2010 has been tested.

This paper has been organized as follows. The following section of this study states the objectives of the study. The next section reviews the literature for the theory behind the EKC and econometric methods used in EKC studies. Econometric modelling, methodology and data sources of the study have been discussed in the section after review of literature. The next section describes the empirical estimates of our study, followed by the summary of our findings with conclusions in the last section.

## Objectives of the Study

South Asia is a vast region with a huge population. It is very rich in natural resources and is considered as one of the few important parts of the globe where human civilization started its journey. Huge cultural as well as physical differences among different parts of this area exist. For many centuries, fertile plain lands, abundant water resources, forest resources, and huge mineral resources were used by the people in a need-based approach. But industrialization through the path of privatization and liberalization has made most of these resources as commercially accessible to investment capital. To address hunger and poverty, agriculture has become modern through the use of modern inputs. Electrification, use of machine tools, chemical fertilizers, pesticides, improvement in storage facilities and marketing has made agriculture more productive. Mineral resources were used by industries. New urban settlements had been developed across nations indicating new levels of development. Poverty and hunger have reduced in general. But economic prosperity has invited new challenges in the name of environmental pollution.

Increased pollution level has increased morbidity of new kind. Path of economic development itself is now under question. Global policymakers are advocating sustainable growth strategies rather than simple economic growth strategies. Testing of EKC hypothesis is particularly relevant in this perspective. This will give us an idea about the current situation of environmental pollution in the area as well as in specific countries in this region along with the course of pollution with country-specific results. It is advocated by the EKC theory that an increase in per capita income is accompanied by an increase in pollution first. The pollution will start declining. Our objectives in this study are as follows.

1. To test the shape of the EKC in the SAARC region as well as in the selected countries of the region.
2. To compute the turning point(s) in the EKC for SAARC as well as selected countries of the region.

## Literature Survey

The EKC theme was popularized by the World Bank's *World Development Report 1992* (IBRD, 1992), which argued that: 'The view that greater economic activity inevitably hurts the environment is based on static assumptions about technology, tastes and environmental investments' and that as incomes rise, the demand for improvements in environmental quality will increase, as will the resources available for investment. Beckerman (1992) claimed that 'there is clear evidence that, although economic growth usually leads to environmental degradation in the early stages of the process, in the end the best- and probably the only- way to attain a decent environment in most countries is to become rich'. According to Panayotou (1993), the traditional view that economic development and environmental quality are conflicting goals reflects the scale effect alone. It simply states that if economic growth is not accompanied by corresponding structural and technological changes then it will generate proportional increase in pollution with many other environmental issues. As per the supporters of EKC hypothesis, it is argued that with increase in income countries gradually shift towards better technology and service-based industries with better environmental regulatory measures and awareness which ultimately helps in gradual declining of environmental degradation.

The standard EKC regression model is as

$$\ln\,(E/P)_{\text{it}} = \alpha_i + \gamma t + \beta_1\,\ln(\text{GDP}/P)_{\text{it}} + \beta_2\,(\ln(\text{GDP}/P))^2_{\text{it}} + \varepsilon_{\text{it}},$$

where $E$ is emissions, $P$ is population and ln indicates natural logarithms. The first two terms on the RHS are intercept parameters which vary across countries or regions $i$ and $t$ years. Time-varying omitted variables and stochastic shocks are common to all countries and to account that time-specific intercepts have been used in the model. Generally, panel data is used to estimate the model. Fixed effects and random effects models both are attempted by most of the researchers to estimate the parameters. It is found that $\alpha_i$ and $\gamma t$ are treated as regression parameters in the fixed effects model, and in the random effects model, they have been treated as components of the random disturbance. The random effect model cannot be estimated consistently when the effects of $\alpha_i$ and $\gamma t$ are found correlated with the explanatory variables (Hsiao, 1986; Mundlak, 1978). If no other statistical problems are assumed then the parameters can be estimated consistently applying fixed effect model though the estimated parameters are conditional with the country or region and time in the selected sample of data (Hsiao, 1986).

Using a panel dataset from a number of locations around the globe Grossman and Krueger (1991) estimated EKCs for $SO_2$, dark matter (fine smoke) and suspended particles (SPM). Their regression model involved a cubic function in levels of per capita GDP adjusted Purchasing Power Parity, a time trend and a trade intensity variable. They found different income levels for the turning points of the EKC for different pollutants. The turning points for $SO_2$ and dark matter are at around $4000–5000, while the concentration of SPM appeared to decline even at low income levels.

Three different functional forms and ten different indicators were used by Shafik and Bandyopadhyay's (1992) study to estimate the EKC relationship. They found with increase in income and time shortage of clean water and that of urban sanitation decline uniformly. They found no relationship between increase in income and deforestation but a positive relationship between worsening of river water quality and increase in income. As per this study, local pollutant concentration and income support the EKC hypothesis with incomes $3000 and $4000 as the turning point. Per capita carbon emissions and municipal waste both clearly increased with increase in income.

One study on EKC using longitudinal data was done by Selden and Song (1994). They estimated EKCs for four emissions series: $SO_2$, $NO_X$, SPM and carbon monoxide (CO) (WRI, 1991). Their data was primarily taken from the developed countries, and the corresponding turning points are higher than earlier studies.

Additional explanatory variables such as political freedom (Torras & Boyce, 1998), output structure (Panayotou, 1997) or trade (Suri & Chapman, 1998) were introduced by researchers and testing the modes found them significant at traditional level of significance as well. Testing individual variable is subject to the problem of potential omitted variable bias and along with this it is also not clear whether this was going to be a problem or not.

Impact of foreign trade on EKC relationship is highlighted by some studies like Arrow et al. (1995) and Stern et al. (1996). They argued that trade can have an effect on the distribution of pollution generating industries, and this can be one of the causes of the EKC-type relationship. As per the famous H–O trade theory, countries need to specialize in the production of goods which use intensively their abundant factors and accordingly will export those commodities. The developed countries would specialize in human capital and manufactured capital-intensive activities and the developing countries will specialize in primary and labour intensive activities. The specialization pattern in the developed and developing countries partly explains the gradual fall in environmental degradation in economically advanced countries and the corresponding rise in the developing ones (Lucas et al., 1992; Suri & Chapman, 1998). Enactment and enforcement of strict environmental regulation in developed countries make some polluting industries to relocate their production activities towards the developing countries (Lucas et al., 1992).

In the estimated EKC models are mostly criticized from four major econometric issues as heteroskedasticity, simultaneity, omitted variables bias and counteraction issues. The issue of heteroskedasticity was raised by Stern et al. (1996). This problem may be important one for analysis which uses grouped data for cross-sectional regressions. In some studies, regression residuals of OLS models were found heteroskedastic (Schmalensee et al., 1998).

The simultaneity issue is addressed by Holtz-Eakin and Selden (1995), and they carried out Hausman tests for regressor exogeneity. But in their study, they found no evidence of simultaneity bias. According to Perman and Stern (2003) traditional stationary econometric model suffers more from simultaneity bias than models with integrated variables. Relation between $CO_2$ emission and income was tested across different countries by Coondoo and Dinda (2002). They carried out a Granger Causality test, found no significant relationship in developing countries but for the developed countries the causality runs from emissions to income. They found a positive relationship without any EKC-type effect but a very high level of serial correlation probably indicated towards misspecification.

Few studies showed that environmental degradation will be lower in countries that developed earlier (Dasgupta et al., 2002). They also support the view that improvement in environment is also possible in developing countries. They show that though regulation of pollution increases with income, the greatest increases take place from low to middle-income countries. Further there would be expected diminishing returns to increased regulations, although better enforcement takes place at higher income levels.

According to Smulder and Bretschger (2000), cyclical patterns arise in technologies, which first diffuse, then become regulated, and are finally phased out by the next generation of technologies. This suggests existence of an inverted U shape with reference to each technology. With the change of innovation, income growth and pollution over cycles, a sequence of EKCs emerge related to each technology which may produce an envelope of EKCs, with inverted U- or N-shaped or an inverted L curve Dinda (2003). In a period with increasing income, one pollutant may decline but the others may increase.

Instead of carbon emissions, the ecological footprint is used as a proxy for environmental degradation by many researchers as Destek et al. (2018), Ozcan et al. (2018) to investigate the presence of Environmental Kuznets Curve hypothesis. The results show that there is U-shaped relationship between the real income and ecological footprint. Usman et al. (2019) studied the EKC relationship on energy consumption and democracy in India. They found the effect of democracy in reducing environmental degradation is weak (statistically insignificant) in the long run but strong (statistically significant) in the short run.

## *Econometric Model, Methodology and Data*

In order to examine the relationship between environmental pollution and per capita income in selected SAARC countries, we have computed a Composite Environmental Degradation Index (CEDI). CEDI has been calculated on the basis of the index of air pollution, water pollution, destruction of soil quality, extraction of ores and metals, deforestation, depletion of natural resources and urbanization. The relationship between CEDI and per capita income at constant price (of 2010 at US$) has been tested. Different constituent pollution index of CEDI has been defined using the following max–min formula:

$$\mathrm{IP}_X = 1 - \frac{\text{Actual measure of } X - \text{Minimum measure of } X}{\text{Maximum Measure of } X - \text{Minimum Measure of } X}$$

where $\mathrm{IP}_X$ stands for index of pollution of $X$ and $X$ = air pollution, water pollution, soil quality, extraction of ores and metals, deforestation, depletion of natural resources and urbanization. Thus, we have seven constituents of environmental degradation. CEDI has been defined as an average of all these seven indicators. Value of $\mathrm{IP}_X$ lies between 0 and 1. The value of $\mathrm{IP}_X = 0$ (or $= 1$) indicates maximum (or minimum) level of pollution. This implication is therefore true for CEDI. The following model is used to study various possible relationships between pollution level and income.

$$\begin{aligned} \mathrm{CEDI}_{\mathrm{it}} &= \alpha_0 + \gamma t + \delta \mathrm{POPUG}_{\mathrm{it}} + \sum_{i=1}^{5} a_i D_{\mathrm{it}} + \beta_1 \mathrm{GDPC}_{\mathrm{it}} \\ &\quad + \beta_2 \mathrm{GDPC}_{\mathrm{it}}^2 + \beta_3 \mathrm{GDPC}_{\mathrm{it}}^3 + u_{\mathrm{it}} \\ t &= 1, 2, \ldots, 35;\; i = 1, 2, 3, 4, 5 \end{aligned} \quad (9.1)$$

where $CEDI_{it}$ is the CEDI for *i*th country at *t*th period is the per capita gross domestic product,

POPUG is population growth rate, and

$u_{it}$ relates to random variables of influence on environmental degradation.

$D_{it}$ refers to the dummy variable for the country *i* at *t* point of time, where

$D_{it} = 1$, if we consider the country *i* and zero otherwise.

The aforementioned model provides us to test several forms of environment–economic development, growth relationship or what is known as EKC. In contrast to the traditional inverted U-shaped EKC, our EKC which reflects the relationship between CEDI and per capita GDP will look like normal U with several other possibilities as discussed below.

1. If $\beta_1 < 0; \beta_2 = \beta_3 = 0$, the equation of CEDI will be linear, and it shows a monotonically decreasing relationship between environmental degradation and per capita GDP.
2. If $\beta_1 > 0; \beta_2 = \beta_3 = 0$, the equation of CEDI will be linear, and it shows a monotonically increasing relationship between environmental degradation and per capita GDP.
3. If $\beta_1 < 0; \beta_2 = \beta_3 = 0$, the equation of CEDI will be quadratic. The traditional inverted 'U'-shaped EKC is replaced in the present situation by normal U-shaped EKC. At the turning point on such EKC, the level of per capita GDP is found at $\frac{dCEDI_{it}}{dGDPC_{it}} = 0$. Thus, $\beta_1 + 2\beta_2 GDPC_{it} = 0$ and under the condition that$\beta_3 = 0$, we get $GDPC_{it} = \frac{\beta_1}{2\beta_2}$ after which the environmental degradation starts declining with rising per capita GDP.
4. If $\beta_1 > 0, \beta_2 < 0, \beta_3 = 0$, the equation of CEDI will be quadratic. The traditional inverted 'U'-shaped EKC will exist in the present situation with a reverse interpretation. At the turning point on such EKC, the level of per capita GDP is found at $\frac{dCEDI_{it}}{dGDPC_{it}} = 0$. Thus, $\beta_1 + 2\beta_2 GDPC_{it} = 0$ and under the condition that$\beta_3 = 0$, we get $GDPC_{it} = \frac{\beta_1}{2\beta_2}$ after which the environmental degradation starts declining with rising per capita GDP.
5. If $\beta_1 > 0, \beta_2 < 0, \beta_3 = 0$, the equation of CEDI will be cubic. We note that with the increasing per capita GDP, the environmental degradation after its increase (first stage) takes an upward turn (second stage) followed by another increase (third stage). This implies that the EKC will be reverted '*N*' shaped. The level of per capita GDP at which the second turning point is found at $\frac{d(dCEDI_{it}/dGDPC_{it})}{dGDPC_{it}} = 0$. Thus, $2\beta_2 + 2\beta_3\ GDPC_{it} = 0$ and we get $GDPC_{it} = -(2\beta_2/3\beta_3)$ after which the environmental degradation starts increasing again with rising per capita GDP.
6. If $\beta_1 > 0,\ \beta_2 < 0,\ \beta_3 = 0$, the equation of CEDI will be cubic. We note that with the increasing per capita GDP, the environmental degradation after its decrease (first stage) takes a downward turn (second stage) followed by another decrease (third stage). This implies that the EKC will be a regular '*N*'-shaped curve. The level of per capita GDP at which the second turning

point is found at $\frac{d(d\text{CEDI}_{\text{it}}/d\text{GDPC}_{\text{it}})}{d\text{GDPC}_{\text{it}}} = 0$ Thus, $2\beta_2 + 3\beta_3\text{GDPC}_{\text{it}} = 0$ and we get $\text{GDPC}_{\text{it}} = -(2\beta_2/3\beta_3)$ after which the environmental degradation starts increasing again with rising per capita GDP.

Let us now proceed to calculate different indexes of pollution. Variables as follows have been used in this study.

1. *Index of Air Pollution (IAP)*: Total greenhouse gas emissions (kilo ton of $CO_2$ equivalent) have been used to calculate it. It captures all major types of gaseous air pollutants as $CO_2$, $SO_2$, $CH_4$, other greenhouse gases and nitrogen oxides.
2. *Index of Water Pollution (IWP)*: Total aquaculture pollution in metric tons has been used for calculation this index of water pollution.
3. *Index of Soil Quality (ISQ)*: Soil quality through soil contamination has been measured in hectares of arable land.
4. *Index of Deforestation (IDF)*: Deforestation has been measured by the adjusted savings on net forest depletion measured on current US dollars.
5. *Index of Depletion of Natural Resources (IDNR)*: Depletion of natural resources has been measured in current US dollars.
6. *Index of Ores and Metals Extraction (IOME)*: Ores and metals' extraction has been captured by the adjusted savings on energy depletion measured on current US dollars.
7. *Index of Rate of Urbanization (IRU)*: Rate of urbanization has been calculated by the ratio of urban population to total population of a country in a given year.

Panel regression has been done to carry out this study. We have empirically tested the hypothesis on EKC (CEDI–GDPC relation) for six selected SAARC countries namely Bangladesh, Pakistan, India, Bhutan, Nepal and Sri Lanka using a panel data over a period of 1980–2014. This study has used water and air pollution data from United Nation and World Bank's database, UN Statistical Yearbook, compendium of the OECD, FAO Production Yearbook, WHO's health database, etc. Economic data (GDP per capita, trade, urbanization, etc.) are taken mostly from the World Tables or the World Bank.

## Empirical Estimates

Based on our data and the econometric models as aforementioned, we now proceed to estimate the CEDI–GDPC relationship.

## Descriptive Statistics

Comparative position of different countries with respect to different pollutions is shown in different radar diagrams. It has been seen that among the selected countries, India has the highest value for most of the pollutions followed by Pakistan,

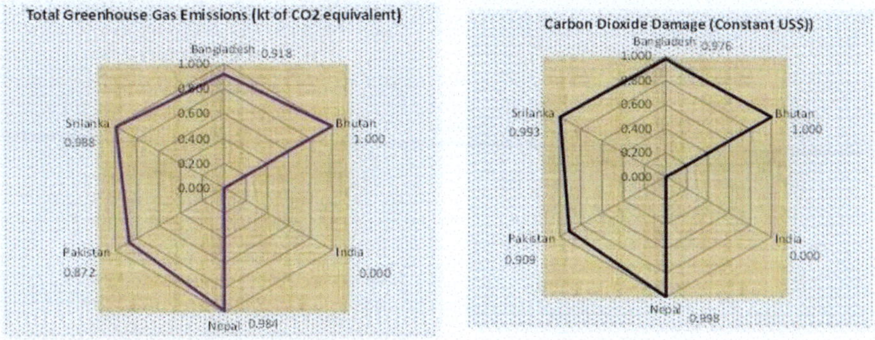


**Fig. 9.1** Greenhouse gas emissions ($CO_2$ equivalent kt.) in SAARC during 1980–2014

Bangladesh, Nepal, Sri Lanka and Bhutan. Different pollution values have been plotted in separate radars for a comparative view.

As per construction, values closer to the origin indicate higher levels of pollution and with distance from the origin pollution level decreases gradually. In our study period, the index of air pollution is highest in India followed by Pakistan and Bangladesh as is shown in Fig. 9.1. For India, the value of the index is very close to zero, signifying actual pollution was very close to the maximum pollution level. The second highest air pollution level was found in Pakistan with an average value of air pollution of 0.872. Pollution level in Bangladesh during our study period was found to be 0.918. Chronologically, Nepal, Sri Lanka and Bhutan positioned in the last three slots with respect to respective air pollution levels.

$CO_2$ is considered as the major pollutant in air pollution. If only $CO_2$ emissions are considered to calculate air pollution, then values of the air pollution level change but comparative positions of different selected countries remain the same, which are shown in the radar beside the first one.

Energy requirement as well as sources of energy used changes with the change in levels of development. At an initial stage, per capita use of energy remains low, and with development it gradually increases. Initially, primary sources of energy were used such as timber, coal, etc. which are more environmentally polluting, and with development people shift to other sophisticated sources of energy like petroleum or nuclear energy.

Comparative positions of different countries are shown in Fig. 9.2. India has occupied the highest polluter's place followed by Pakistan and Bangladesh. In other three countries, this type of pollution remained insignificant. Deforestation is another important cause of pollution. More and more deforestation indicates higher levels of environmental pollution. Second half of Fig. 9.2 plots the same. India showed the highest level of deforestation followed by Bangladesh, Pakistan, Nepal, Sri Lanka and Bhutan. Forest depletion in other countries is far lower than in India. Second highest level of forest depletion is seen in Bangladesh which is more than ten times lower than that of India.

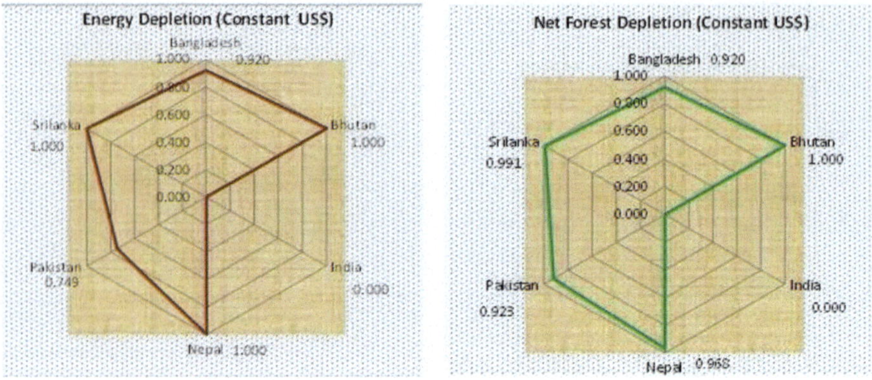


**Fig. 9.2** Energy and net forest depletion in SAARC during 1980–2014

Land/soil is the primary input of production. With more and more pressure on land, soil gets contaminated for different reasons. Soil pollution across selected SAARC countries is shown in Fig. 9.3. India again is the most polluted country followed by Pakistan, Bangladesh, Nepal, Sri Lanka and Bhutan. Water pollution level is highest in India followed by Bangladesh, Pakistan, Nepal, Sri Lanka and Bhutan.

This is quite consistent with the water body's distribution in the SAARC area also. After India, Bangladesh uses its water bodies more commercially than Pakistan. Nepal and Bhutan are landlocked countries and water resources are used less commercially except for some hydroelectricity projects which explain low levels of water pollution in those countries. Comparatively, more industrialization as well as commercialization of agriculture has taken place in Nepal creating greater water pollution in the country than Bhutan. Sri Lanka uses its coastal areas more commercially than its inland water resources. Low industrial and commercial use of inland water bodies has kept the water pollution at a level even lower than Nepal. Second part of Fig. 9.3 plots water pollution level at different selected SAARC countries

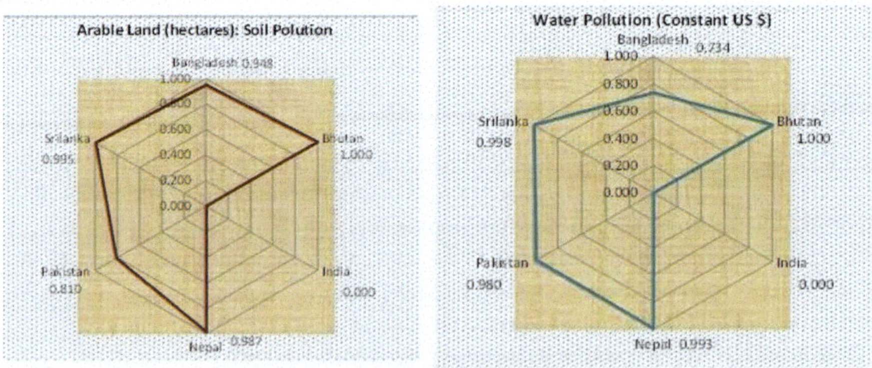


**Fig. 9.3** Soil quality and water pollution in SAARC during 1980–2014

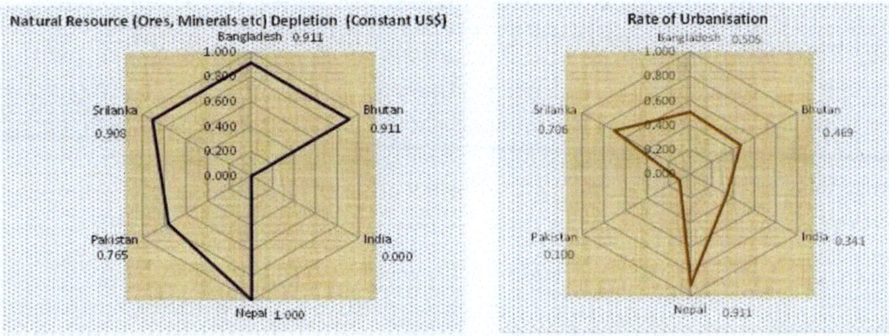


**Fig. 9.4** Natural resource depletion and rate of urbanization in SAARC during 1980–2014

during 1980–2014. Water pollution in India is much higher than all other SAARC countries, and the immediate second highest water pollution level is in Bangladesh which is four times lesser than that of India.

All the SAARC countries are endowed with different types of natural resources. These natural resources are used for various economic activities in this region. Most of the export baskets from this region are based on the natural resources or crude intermediate products used by sophisticated industries in developed countries. With respect to natural resources depletion, India takes the first position again followed by Pakistan, Sri Lanka, Bangladesh, Bhutan and Nepal (Fig. 9.4). Level of urbanization is regarded as an index of economic development. At the same time, urbanization puts huge pressure on the environment when it takes place in an unplanned way.

This is to be noted that in our area of study, most of the urbanization is unplanned. We have taken the rate of urbanization as one of the determinants of CEDI in our study area. In the second half of Fig. 9.4, rates of urbanization in different SAARC countries are shown. Pakistan showed the highest level of urbanization followed by India, Bhutan, Bangladesh, Sri Lanka and Nepal.

## Regression Results

Following the basic regression model in Eq. 1, we have estimated three separate regression models as follows.

Model A:

$$\mathrm{CEDI}_{\mathrm{it}} = \alpha_0 + \gamma t + \delta \mathrm{POPUG}_{\mathrm{it}} + \sum_{i=1}^{5} a_i D_{it} + \beta_1 \mathrm{GDPC}_{\mathrm{it}} + u_{\mathrm{it}} \tag{9.2}$$

Model B:

$$\mathrm{CEDI_{it}} = \alpha_0 + \gamma t + \delta \mathrm{POPUG_{it}} + \sum_{i=1}^{5} a_i D_{it} + \beta_1 \mathrm{GDPC_{it}} + \beta_2 \mathrm{GDPC_{it}^2} + u_{\mathrm{it}} \tag{9.3}$$

Model C:

$$\mathrm{CEDI_{it}} = \alpha_0 + \gamma t + \delta \mathrm{POPUG_{it}} + \sum_{i=1}^{5} a_i D_{\mathrm{it}} + \beta_1 \mathrm{GDPC_{it}} + \beta_2 \mathrm{GDPC_{it}^2} + \beta_3 \mathrm{GDPC_{it}^3} + u_{\mathrm{it}} \tag{9.4}$$

In Model A, estimated coefficients (Refer to Table 9.1) show that environmental degradation is increasing with increase in per capita income, and the coefficient is found to be significant at 7.5%, implying the rejection of null hypothesis. Thus, EKC shows a linear and monotonically decreasing relationship between CEDI and per capita GDP for SAARC. Population growth has been found to be significant in affecting environmental degradation. With time which is positively associated with CEDI and significant at 8.5% level, environmental degradation is decreasing. This is most natural because of the fact that people conscious of adverse effects of pollution take appropriate measures.

All the country dummies are significant at 1% level and negatively associated with CEDI. This implies that at given income, population growth, and time, we

**Table 9.1** Determinants of composite environmental degradation in the SAARC 1980–2014

*Model A*: Linear model: Linear in GDPC
Dependent variable: CEDI; Method: Panel least squares; Sample: 1980–2014
Periods included: 35; Cross sections included: 6; Total panel (balanced) observations: 210

| Variable | Coefficient | Std. error | *t*-Statistic | Probability |
|---|---|---|---|---|
| GDPC | −4.87E−06 | 2.72E−06 | −1.792 | 0.075 |
| POPUG | 2.94E−03 | 7.11E−03 | 0.413 | 0.680 |
| TIME | 9.00E−04 | 5.52E−04 | 1.695 | 0.085 |
| DBANG | −0.792 | 0.009 | −88.909 | 0.000 |
| DBHU | −0.859 | 0.009 | −95.372 | 0.000 |
| DNEP | −0.934 | 0.009 | −104.007 | 0.000 |
| DPAK | −0.693 | 0.010 | −70.505 | 0.000 |
| DSRI | −0.868 | 0.013 | −66.520 | 0.000 |
| C | 0.941 | 0.019 | 48.350 | 0.000 |

Diagnostic statistics

| | |
|---|---|
| *R*-squared = 0.99 | *F*-statistic = 2114.211 |
| Adjusted *R*-squared = 0.99 | Prob (*F*-stat) = 0.000 |

*Source* Authors calculations

**Table 9.2** Determinants of composite environmental degradation in SAARC, 1980–2014

| *Model B:* Nonlinear model: Quadratic in GDPC<br>Dependent variable: CEDI; Method: Panel least squares; Sample: 1980–2014<br>Periods included: 35; Cross sections included: 6; Total panel (balanced) observations: 210 | | | | |
|---|---|---|---|---|
| Variable | Coefficient | Std. error | *t*-Statistic | Probability |
| GDPC | −4.63E−06 | 9.09E−06 | −0.509 | 0.611 |
| GDPC^2 | −1.83E−11 | 6.63E−10 | −0.028 | 0.978 |
| POPUG | 2.92E−03 | 7.15E−03 | 0.408 | 0.684 |
| TIME | 8.86E−04 | 7.33E−04 | 1.209 | 0.228 |
| DBANG | −0.792 | 0.010 | −79.506 | 0.000 |
| DBHU | −0.859 | 0.009 | −90.768 | 0.000 |
| DNEP | −0.934 | 0.011 | −88.726 | 0.000 |
| DPAK | −0.693 | 0.010 | −66.459 | 0.000 |
| DSRI | −0.868 | 0.015 | −59.103 | 0.000 |
| C | 0.940495 | 0.020165 | 46.64038 | 0 |
| Diagnostic statistics | | | | |
| *R*-squared = 0.99 | | | *F*-statistic = 1869.956 | |
| Adjusted *R*-squared = 0.99 | | | Prob (*F*-statistic) = 0.000 | |

*Source* Authors calculations

can calculate expected values of CEDI for all six countries. Pakistan tops in its environmental degradation in the SAARC and India lies at the bottom. The regression result shows that in the order of composite environmental degradation, Pakistan is followed by Bangladesh, Bhutan, Sri Lanka and Nepal.

Estimated results of Model B are presented in Table 9.2. We introduced square of income to assess the marginal impact of income per capita on environmental degradation. This time none of the previous variables as income, square of income, population growth and time are found to be significant accepting the only country dummies. Therefore, the EKC in this situation cannot be confirmed. So the EKC hypothesis is not explained for the SAARC in the given period with quadratic specification of the model.

Estimated results of model C are presented in Table 9.3. We have used a cubic polynomial in income to assess the marginal impact of income per capita as well as change of marginal impacts on the environmental degradation. All the explanatory variables as income, square of income, cube of income and time along with all country dummies are found to be significant at 1% level.

As in the linear model, we find time as significant at 1% level implying the reduction in the environmental pollution for SAARC. Pakistan tops in the list of environmentally degraded countries with India in the bottom-most position. All other countries, namely Bangladesh, Bhutan, Nepal and Sri Lanka, come in order in between.

**Table 9.3** Determinants of composite environmental degradation in the SAARC 1980–2014

*Model C:* Nonlinear model:Cubic in GDPC
Dependent variable: CEDI; Method: Panel least squares; Sample: 1980–2014
Periods included: 35; Cross sections included: 6; Total panel (balanced) observations: 210

| Variable | Coefficient | Std. error | *t*-Statistic | Probability |
|---|---|---|---|---|
| GDPC | 1.08E−04 | 2.01E−05 | 5.363 | 0.000 |
| GDPC^2 | −2.01E−08 | 3.33E−09 | −6.054 | 0.000 |
| GDPC^3 | 1.12E−12 | 1.82E−13 | 6.152 | 0.000 |
| POPUG | 7.85E−04 | 6.58E−03 | 0.119 | 0.905 |
| TIME | −2.16E−03 | 8.36E−04 | −2.586 | 0.010 |
| DBANG | −0.763 | 0.010 | −74.082 | 0.000 |
| DBHU | −0.874 | 0.009 | −96.997 | 0.000 |
| DNEP | −0.896 | 0.011 | −78.541 | 0.000 |
| DPAK | −0.723 | 0.011 | −67.520 | 0.000 |
| DSRI | −0.917 | 0.016 | −58.609 | 0.000 |
| C | 0.856 | 0.023 | 37.207 | 0.000 |
| Diagnostic statistics | | | | |
| *R*-squared = 0.99 | | | *F*-statistic = 1996.839 | |
| Adjusted *R*-squared = 0.99 | | | Prob (*F*-statistic) = 0.000 | |

*Source* Authors calculations

The CEDI first increases (fall in environmental degradation), then falls (increase in environmental degradation), and after that it takes an upward turn (fall in environmental degradation) with the increasing per capita GDP. This implies that the EKC will be '*N*' shaped for SAARC. Exactly this is found in model C.

This should be noted that in the context of estimating the EKC using the time series and panel data, the problem of non-stationarity or the existence of unit root is very much common. We have taken care of this issue by testing the co-integration of CEDI and GDPC. Our analysis shows that there is no such problem of non-stationarity in the data used for the estimation of EKC.

## Environmental Kuznets Curve for SAARC and Its Members

Let us examine the relationship between the CEDI and the GDPC for SAARC as well as its member countries on the basis of the observations available in the World Tables. Table 9.4 shows the estimated coefficients with respect to per capita income, square of per capita income and cube of per capita income on CEDI.

All the coefficients are found to be significant and alternate in sign. Graphically, we have plotted the EKC for the SAARC in Fig. 9.5. It clearly makes an N-shaped curve with two turning points. The first turning point takes place at an income level of

**Table 9.4** Environmental Kuznets Curve in SAARC during 1980–2014

| Dependent variable: CEDI<br>Method: Panel least squares; Sample: 1980–2014; Periods included: 35<br>Cross sections included: 6; Total panel (balanced) observations: 210 | | | | |
|---|---|---|---|---|
| Variable | Coefficient | Std. error | $t$-Statistic | Probability |
| GDPC_SAARC | 0.009257 | 0.002419 | 3.827 | 0.062 |
| GDPC_SAARC^2 | −3.42E−06 | 9.20E−07 | −3.715 | 0.065 |
| GDPC_SAARC^3 | 3.77E−10 | 1.05E−10 | 3.593 | 0.070 |
| C | −6.99014 | 1.865704 | −3.747 | 0.064 |
| Diagnostic statistics | | | | |
| $R$-squared: 0.90<br>Adjusted $R$-squared: 0.75 | | $F$-statistic: 5.920<br>Prob ($F$-statistic): 0.015 | | |
| | | Durbin-Watson stat: 2.091 | | |

*Source* Authors calculations

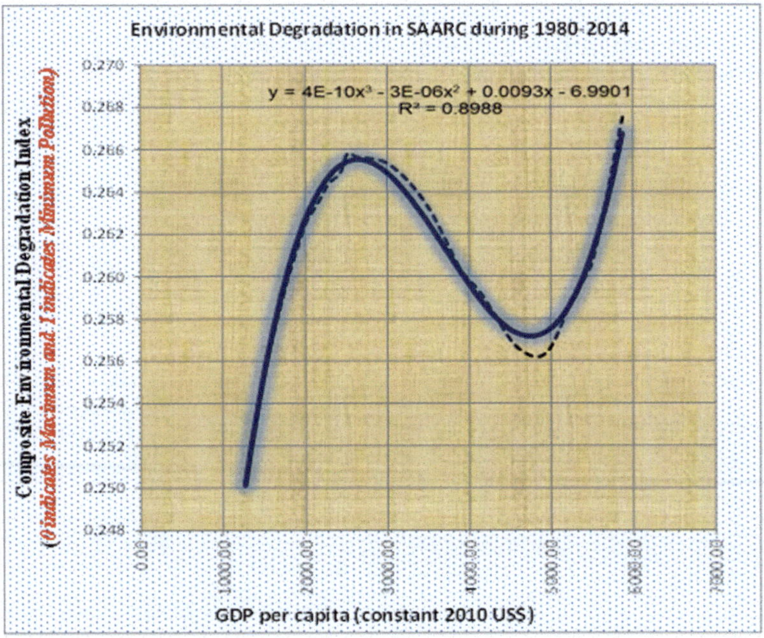


**Fig. 9.5** CEDI in SAARC during 1980–2014

US$ 2500 per annum. After this per capita income level, CEDI decreases (increase in environmental pollution) with rise in income per capita and then it starts increasing (decrease in environmental pollution). At around an income level of US$ 4000 per annum, environmental degradation further starts falling.

Individually country-wise EKCs were studied and the estimation results along with shapes of country-specific EKCs are discussed further. Individual country-wise structure of the EKCs is quite different for different countries indicating different types of EKC relationships. We found U-shaped, inverted U-shaped, N-shaped and L-shaped EKCs among our chosen countries in the SAARC.

## Environmental Kuznets Curve for Nepal

Figure 9.6 plots the CEDI–GDP relationship for Nepal. It looks like a reverted L-shaped curve. Environmental degradation remains high during early increase in per capita income ($700–$1500).

Turning point took place around per capita income of $1500. Environmental degradation of the country has shown a steady and sharp falling trend after that. Value of the CEDI was close to 0.01 at the turning point, and then it has increased to 0.17. It is important to note that Nepal has decreased environmental degradation almost 35 times, during its increase in per capita income from $1500–$2300. Among selected SAARC countries, environmental degradation in Nepal is high but it has

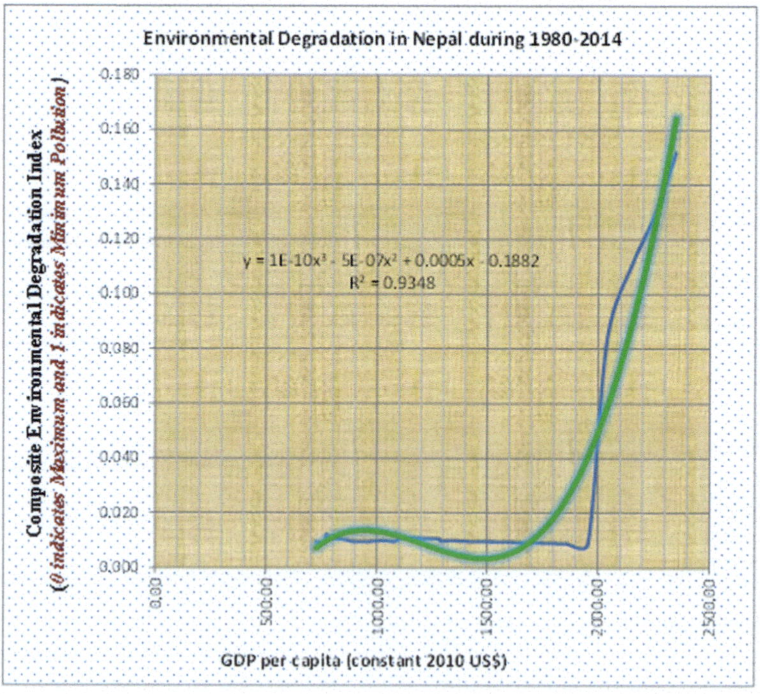


**Fig. 9.6** CEDI in Nepal during 1980–2014

revealed significant improvements showing a rising trend with increase in per capita income.

## Environmental Kuznets Curve for Bangladesh

For Bangladesh, CEDI–GDPC relationship is shown in Fig. 9.7. The trend showed an inverted U shape during our study period in contrast to the EKC hypothesis. Starting with a value of 0.14 with a per capita income of $750, CEDI increased very slowly up to 0.20 associated with per capita income of $2500. This implies that environmental degradation has fallen steadily with increase in per capita income at first. This is the turning point; after this per capita income level, environmental degradation has increased further. From per capita income of $2500–$3100, CEDI has fallen continuously from a value of 0.20 to almost up to 0.17.

**Environmental Kuznets Curve for India**

Figure 9.8 plots EKC defined by us for India. Within the study period, the curve showed an inverted L-shaped relationship between per capita income and CEDI for the country.

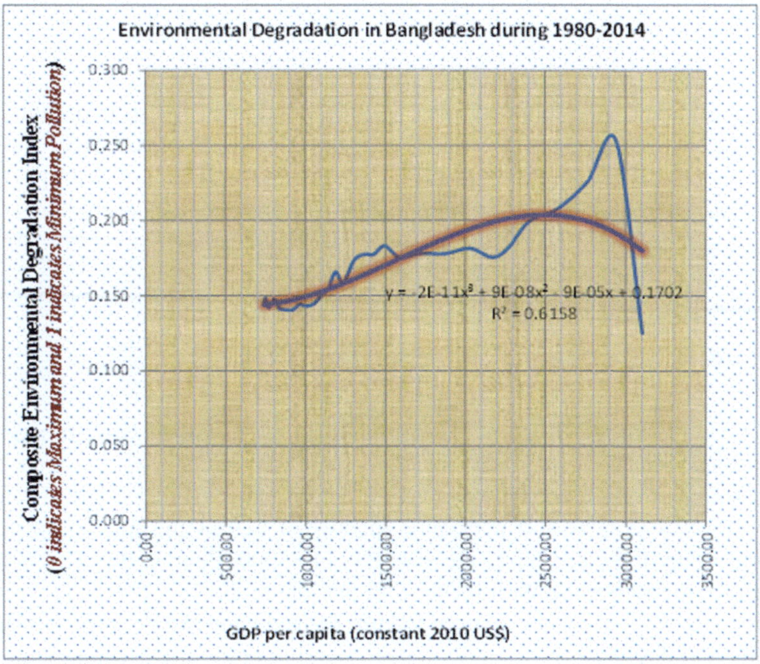


**Fig. 9.7** CEDI in Bangladesh during 1980–2014

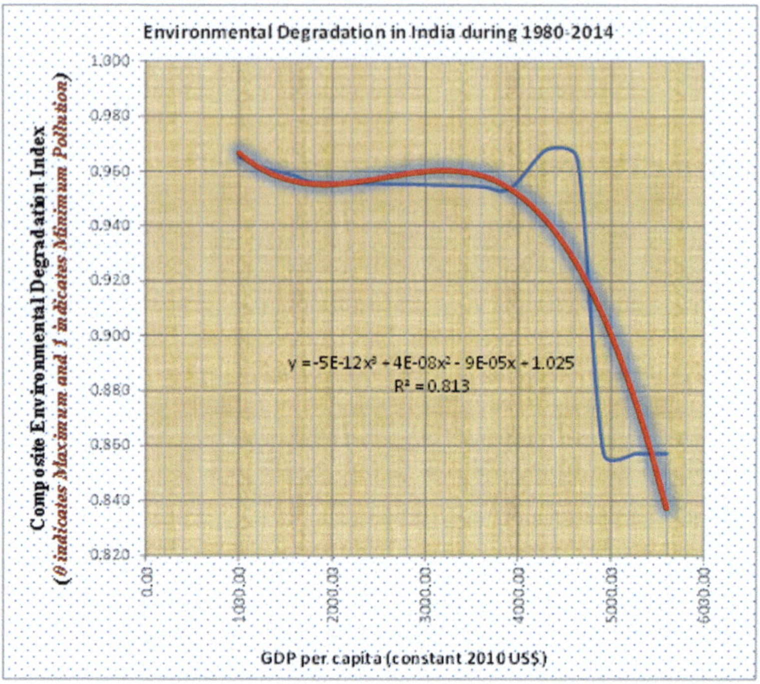


**Fig. 9.8** CEDI in India during 1980–2014

India has shown the highest CEDI values in our study period. It started with a value of 0.97 with corresponding value of per capita income of $1000, and it remained constant up to $4000; after that it started falling at a very high rate up to the per capita income level of $5600, indicating a high level of environmental degradation in the country.

## Environmental Kuznets Curve for Pakistan

Figure 9.9 plots the relationship showing an inverted U-shaped EKC in our study period which is in contrast to the standard theory as defined by us. After India, Pakistan showed the second highest values of CEDI among all other countries studied. CEDI value was 0.26 with a per capita income of $1900 and moved up to 0.28 continuously up to the per capita income level of $3000. Beyond this, CEDI values have shown a falling trend with increase in per capita income. CEDI has fallen to 0.14 with a per capita income level of $4700. Pakistan had shown improvements in environmental degradation first but after the turning point at per capita income of $3000, it has deteriorated further.

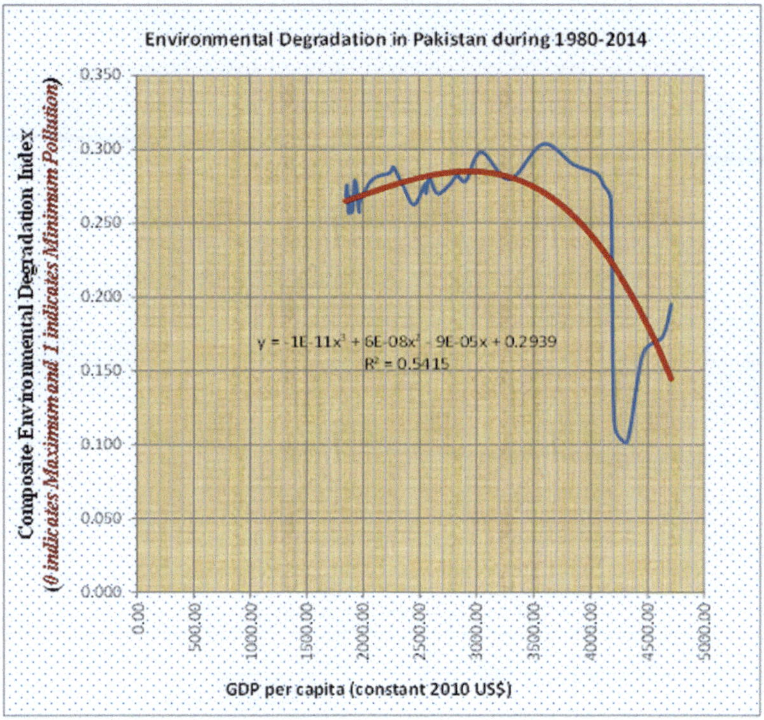


**Fig. 9.9** CEDI in Pakistan during 1980–2014

## Environmental Kuznets Curve for Bhutan

Figure 9.10 has plotted the CEDI–GDPC relationship for Bhutan. Bhutan is the country with lowest CEDI value making it the lowest environmentally degraded country among the selected SAARC countries in the study period. However, CEDI value has increased steadily with an increase in per capita income showing improvements in environmental pollution, but it has shown a falling trend in the later phase indicating increase in environmental pollution again. CEDI value was 0.04 with a per capita income of $1200 and moved up to 0.14 continuously up to per capita income level of $6000. This is the turning point of the country. CEDI has shown a falling trend with further increase in per capita income. The EKC shows an inverted U-shaped trend here, indicating improvements in environmental degradation first with an increase in per capita income, then deterioration in environmental degradation.

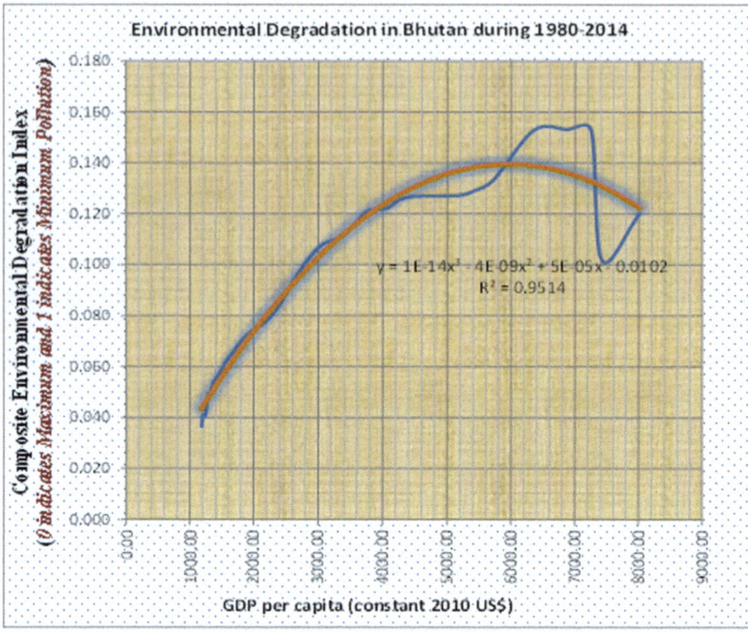


**Fig. 9.10** CEDI in Bhutan during 1980–2014

## Environmental Kuznets Curve for Sri Lanka

The EKC for Sri Lanka is shown in Fig. 9.11. CEDI value has fallen steadily with an increase in per capita income (rise in environmental degradation) but has shown a rising trend in the later phase (fall in environmental degradation). It shows a U-shaped trend for the country supporting the standard EKC hypothesis. CEDI value was 0.11 with a per capita income of $2000 and fell up to 0.02 with rise in per capita income level up to $7500. Beyond this, CEDI value has shown a rising trend with increase in per capita income. CEDI has increased to 0.11 again with a per capita income level of $11,750.

## Summary and Conclusions

During the last quarter of the twentieth century, adoption of economic growth models advocating liberalization, privatization and globalization brought significant improvements to the standard of living in SAARC region. Openness of the economies, volume of international trade and capital flows have increased, employee and capital mobility were facilitated under the ever-increasing concept of globalization. However, that economic growth caused severe consequences to the environment

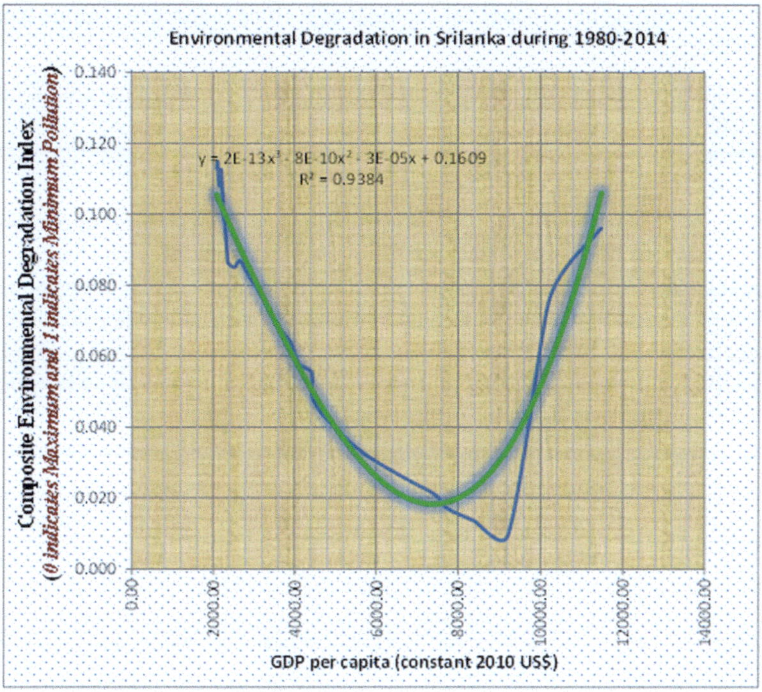


**Fig. 9.11** CEDI in Sri Lanka during 1980–2014

and natural resources, since in the name of prosperity economies undermined the effects of the established economic development model. Environmental concerns have begun to grow a new idea of parallel economic growth with environmental protection.

The present study has focused on the linkages between per capita GDP growth rate and environmental degradation in the SAARC region. Our findings show that Sri Lanka is the most environmentally degraded country followed by Bhutan, Pakistan, Bangladesh and Nepal. An N-shaped EKC (CEDI–GDPC relation) portrays the environmental degradation and per capita income relationship in the SAARC region as a whole the best way. But selected countries showed U-shaped, inverted U-shaped and reverted, and inverted L-shaped EKCs. There is a huge gap in the CEDI scores between highest and lowest polluted countries. Most of the countries showed a rising trend of pollution already except Nepal and Sri Lanka. Nepal and Sri Lanka where environmental degradation is declining significantly should be followed by other countries in taking measures of controlling environmental pollution.

Countries at different development levels are using their resources to boost up respective economic growths, but in this process are making the entire region vulnerable to environmental degradations as shown by the N-shaped EKC for SAARC. As a regional block, SAARC nations need to consider this and take necessary steps to counter it. This study suggests certain future research areas and recommends following issues to be considered for regional as well as individual country-specific policy-makings.

1. Appropriate country-specific model specification to find out the true nature of the EKC.
2. Find out the relative importance of different types of pollutants for different countries as well as the SAARC region as a whole.
3. An acceptable level of pollution of different types is to be set, and economic policies are to be framed accordingly by member countries and violations of that standard are strictly to be compensated through polluters pay principle.
4. Through regional cooperation, financial and technological supports are to be made available for the countries that are in necessity.
5. SAARC-level research institutions are to be made for uniform data keeping and environmental and economic information sharing.

## References

Arrow, K., Bolin, B., Costanza, R., Dasgupta, P., Folke, C., Holling, C. S., Jansson, B. O., Levin, S., Mäler, K. G., Perrings, C., & Pimentel, D. (1995). Economic growth, carrying capacity, and the environment. *Science, 268*, 520–521.

Beckerman, W. (1992). Economic growth and the environment: Whose growth? Whose environment? *World Development, 20*, 481–496.

Coondoo, D., & Dinda, S. (2002). Causality between income and emission: A country group-specific econometric analysis. *Ecological Economics, 40*, 351–367.

Dasgupta, S., Laplante, B., Wang, H., & Wheeler, D. (2002). Confronting the environmental Kuznets curve. *Journal of Economic Perspectives, 16*, 147–168.

Destek, M. A., Ulucak, R., & Dogan, E. (2018). Analyzing the environmental Kuznets curve for the EU countries: The role of ecological footprint. *Environmental Science and Pollution Research, 25*, 29387–29396. https://doi.org/10.1007/s11356-018-2911-4dated13.06.2020

Dinda, S. (2003). *Economic growth with environmental and physical capital: A convergence approach.* Mimeo.

Grossman, G. M., & Krueger, A. B. (1991). *Environmental impacts of a North-American free trade agreement.* National Bureau of Economic Research, working paper 3914, NBER, Cambridge, MA.

Holtz-Eakin, D., & Selden, T. M. (1995). Stoking the fires? $CO_2$ emissions and economic growth. *Journal of Public Economics, 57*, 85–101.

Hsiao, C. (1986). *Analysis of panel data.* Cambridge University Press.

IBRD. (1992). *Development and the environment: World development report 1992.* Oxford University Press.

Lucas, R. E. B., Wheeler, D., & Hettige, H. (1992). Economic development, environmental regulation and the international migration of toxic industrial pollution: 1960–1988. In P. Low (Ed.), International trade and the environment, World Bank discussion paper no. 159. World Bank, Washington, D.C.

Mundlak, Y. (1978). On the pooling of time series and cross section data. *Econometrica, 46*, 69–85.

Ozcan, B., Apergis, N., & Shahbaz, M. (2018). A revisit of the environmental Kuznets curve hypothesis for Turkey: New evidence from bootstrap rolling window causality. *Environmental Science and Pollution Research, 25*, 32381–32394. https://doi.org/10.1007/s11356-018-3165-x,dated13.06.2020

Panayotou, T. (1993). *Empirical tests and policy analysis of environmental degradation at different stages of economic development, working paper WP238*. Technology and Employment Programme, International Labour Office, Geneva.

Panayotou, T. (1997). Demystifying the environmental Kuznets curve: Turning ablack box into a policy tool. *Environment and Development Economics, 2*, 465–484.

Perman, R., & Stern, D. I. (2003). Evidence from panel unit root and cointegration tests that the environmental Kuznets curve does not exist. *Australian Journal of Agricultural and Resource Economics, 47*(3), 325–347.

Schmalensee, R., Stoker, T. M., & Judson, R. A. (1998). World carbon dioxide emissions: 1950–2050. *Review of Economics and Statistics, 80*, 15–27.

Selden, T. M., & Song, D. (1994). Environmental quality and development: Is there a Kuznets curve for air pollution. *Journal of Environmental Economics and Environmental Management, 27*, 147–62.

Shafik, N., & Bandyopadhyay, S. (1992). *Economic growth and environmental quality: Time series and cross-country evidence, background paper for the world development report 1992*. The World Bank, Washington, D.C.

Smulder, S., & Bretschger, L. (2000). Explaining environmental Kuznets curves: How pollution induces policy and new technology, Tilburg University Center Working Paper No. 2000-95. Accessed at https://ssrn.com/abstract=249023 or https://doi.org/10.2139/ssrn.249023. Dated October 12, 2016

Stern, D. I., Common, M. S., & Barbier, E. B. (1996). Economic growth and environmental degradation: The environmental Kuznets curve and sustainable development. *World Development, 24*, 1151–1160.

Suri, V., & Chapman, D. (1998). Economic growth, trade and the energy: Implications for the environmental Kuznets curve. *Ecological Economics, 25*, 195–208.

Torras, M., & Boyce, J. K. (1998). Income, inequality, and pollution: A reassessment of the environmental Kuznets curve. *Ecological Economics, 25*, 147–160.

Usman, O., Iorember, P. T., & Olanipekun, I. O. (2019). Revisiting the environmental Kuznets curve (EKC) hypothesis in India: The effects of energy consumption and democracy. *Environmental Science and Pollution Research, 26*, 13390–13400. https://doi.org/10.1007/s11356-019-04696-z,dated13.6.2020

World Data Bank. (2017). *World development indicators*. Accessed at http://databank.worldbank.org/data/reports.aspx?source=world-development-indicators. Dated January 29, 2017

WRI. (1991). *World resources 1990–91*. World Resources Institute.

# Chapter 10
# Looming Threat of Vehicular Pollution to Human Health

**Alpana Kateja**

**Abstract** Urbanization is typically accompanied with raised productivity, economic growth, enhanced income and living standards of inhabitants. Higher levels of income combined with ever increasing intra-city distances and commuting time inevitably lead to exponential increase in vehicle population, which in turn results in one of the major externality of urban growth—ambient air pollution. Air pollution is one of the leading agents responsible for human health hazards. The present paper briefly discusses trends and patterns of vehicle expansion, explores state of air pollution in India and highlights its effect on human health. In recent years, air pollution in India has emerged as a threat to the public health. The Global Burden of Disease study, 2018 concludes that out of 79 behavioural, environmental and metabolic factors, outdoor $PM_{2.5}$ exposure was the third-most important factor for deaths in India. In 2015, exposure to outdoor $PM_{2.5}$ caused more than 1 million casualties in India. This corresponds to close to one fourth of the global 4.2 million deaths attributable to AAP. Many a time, people inflicted by air-borne serious diseases are left unproductive; the loss is calculated in terms of years of life lived with disability (YLDs), and disability-adjusted life years (DALYs). The mortalities imputable to air pollution have been escalating in India along with growing number of people with ailments arising from exposure to air pollution. India's urbanization is yet to pinnacle, so there is still time to set things right. This paper makes recommendations to facilitate cleaner and healthier urban growth so as to substantially improve the quality of life of people.

**Keywords** Ambient air pollution (AAP) · Disease burden · Life expectancy · Particulate matter (PM) · Vehicular pollution

## Introduction

Urbanization typically accompanies industrialization and modernization which usually move together with increased productivity, higher income levels and better living standards. In fact, rapid productivity gains and improved quality of life are the

---

A. Kateja (✉)
Department of Economics, University of Rajasthan, Jaipur, Rajasthan, India



A. Kateja and R. Jain (eds.), *Urban Growth and Environmental Issues in India*,
https://doi.org/10.1007/978-981-16-4273-9_10

major driving forces for migration from lower productivity rural activities. Resulting demographic changes and multidimensional structural transformation, however, call for elaborate planning, massive investments and meticulous execution to build up orderly urban growth. Historical data reveals that urbanization has always been accompanied with haphazard expansion, congestion, lack of basic amenities like sanitation and environmental degradation. One of the major externality of unsystematic urban expansion is ambient air pollution (AAP). Higher levels of income combined with ever increasing intra-city distances and commuting time inevitably lead to exponential increase in vehicle population and AAP.

Data reveals that the risk associated with outdoor air pollution has been increasing as nations grow, especially the low- and middle-income countries. With economic augmentation, these countries have witnessed rampant industrialized growth along with an increase in urbanization leading to the massive exodus of population to urban areas further aggravating the levels of air pollution due to increase in traffic and unplanned urban and industrial development. Concern for sustainable urban development is a recent phenomenon. The smog episodes in Donora, Pennsylvania, in late October 1948 and in London in December 1952 were among the few crises igniting the environmental dynamism which is quite perceptible in every sphere of life today. Recently, global issues such as climate change, stratospheric ozone depletion and the potential impact of such hazards have also been brought under the ambit of environmental movement along with other issues.

Air pollution is a complex phenomenon. There may be many factors that contribute to pollution across the country. The contribution of particular sources is expected to vary depending on the regions and their climatic conditions. Air pollution can be classified into two categories, namely indoor air pollution and ambient (outdoor) air pollution. Most hazardous gases are found in indoor environment such as carbon monoxide (CO) that is emitted on burning of fossil fuels for purposes like heating and cooking, and smoking tobacco. Vehicular pollution is one of the largest contributors to the AAP since transportation involves burning of fossil fuels to produce energy.

The emission due to incomplete burning of fossil fuels in the vehicles is one of the major causes of vehicular pollution. The resultant pollutants like CO, ozone, sulphur dioxide, nitrogen dioxide, lead; particulate matters, etc. are corollaries of the engine combustion process. Despite the drastic increase in the number of vehicles on the road, only 10–30% of vehicles are chiefly responsible for outdoor air pollution.

The present study attempts to examine the trends and patterns of vehicular pollution in India and relate them with health outcomes of citizens in different states. The paper is organized into four broad sections: After a brief introduction, the trends and patterns in the number of registered vehicles and pollution in major Indian cities have been discussed. In the third part, detrimental health outcomes of air pollution with specific reference to India are elaborated. Finally, concluding remarks and recommendations are presented.

## Trend and Pattern of Registered Motor Vehicles in India

On planet earth, vehicle population exceeds the figure of 1.2 billion; these include cars, trucks, buses, etc. This number was just a quarter of a billion in 1970, a complete 85 years since the inception of vehicles. The number doubled in 1986, a mere 16 years later, and doubled again 24 years later. Following this trend, it is projected that there would be 2.5 billion vehicles in 2050 worldwide. India is no exception to this trend and has been witnessing exponential increase in the number of vehicles.

The trend of motor vehicles registered in India for the period 1991–2016 is shown in Fig. 10.1. Automobiles are generally divided into five categories: (a) two wheelers; (b) cars, jeeps and taxies; (c) buses; (d) good's vehicles; and (e) others. The last category includes vehicles like tractors, trailers, three wheelers/LMV and other miscellaneous vehicles that are not reported in a different category by the State/UT. Until the 1990s, the total number of registered motor vehicles in the country was quite low. However, the numbers have increased sharply since early 1990s. In the post-reform era, there is an alarming growth in the tally of motor vehicles due to increase in income levels, changed consumption patterns and urbanization.

The compound annual growth rate (CAGR) of different types of vehicles (Table 10.1) shows that in three decades, starting from 1961, maximum growth was witnessed in two wheelers but since then the growth rate of cars, jeeps and taxies has almost equalled that of two wheelers. Until 1990s, the growth rate of two wheelers was more than double vis-a-vis personal four wheelers. It is also clear that there has been spectacular rise in the tally for vehicles at a growth rate way up than the population growth rate and the expansion of physical infrastructure for vehicles as is evident from the rising trend of the number of vehicles per thousand persons and per 100 km road length during 2001–16 (Fig. 10.2).

In addition to personal vehicles, the other broad category of vehicles is that of transport vehicles. These are of two types—passenger transport vehicles and goods

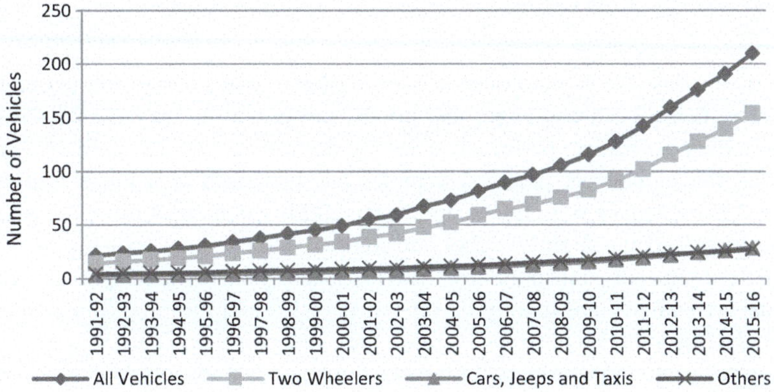


**Fig. 10.1** Trend of total number of registered vehicles in India (1991–2016). *Source* GoI (2017)

**Table 10.1** CAGR of registered motor vehicles in India (1961–2016)

| Period | All vehicles | Personal vehicles | | Transport vehicles | | Others |
|---|---|---|---|---|---|---|
| | | Two wheelers | Cars, Jeeps and Taxies | Buses | Goods' vehicles | |
| 1961–70 | 10.4 | 20.8 | 8.2 | 5.3 | 7.4 | 11.9 |
| 1971–80 | 10.5 | 15.8 | 5.3 | 4.2 | 3.6 | 20.3 |
| 1981–90 | 15.8 | 20.1 | 9.9 | 7.9 | 10.7 | 11.9 |
| 1991–2000 | 9.9 | 10.6 | 8.4 | 6.8 | 8.4 | 8.7 |
| 2001–10 | 10.0 | 10.3 | 10.7 | 11.4 | 9.3 | 6.9 |
| 2010–16 | 10.5 | 11.0 | 10.8 | 5.2 | 7.8 | 7.4 |

*Source* Author's calculation based on data from GoI (2017)

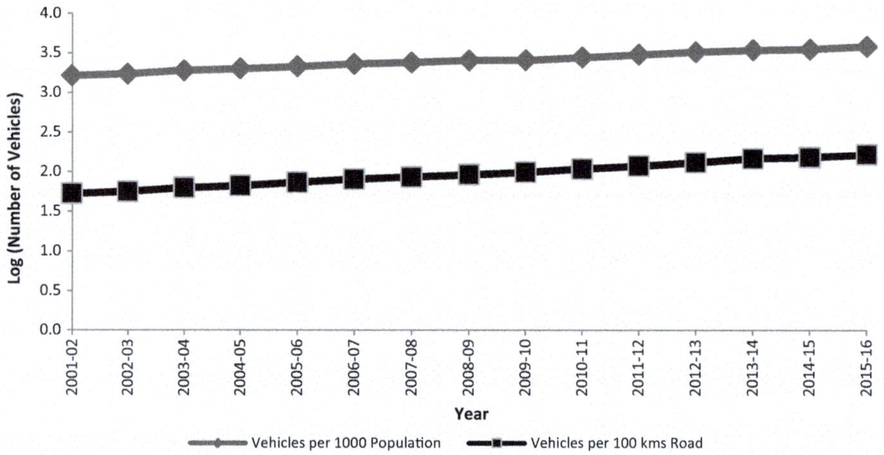


**Fig. 10.2** Total number of registered vehicles per thousand populations and per hundred km of road (2001–16). *Source* GoI (2017)

transport vehicles. While the transport vehicles have grown slowly compared to the personal vehicles, both passenger and goods transport vehicles have shown faster growth in the post-reform period vis-a-vis the pre-reform era. These figures clearly show that the number of personal vehicles has increased at a much higher rate compared to passenger vehicles except for the 2000s when the number of passenger vehicles increased at a slightly higher rate compared to personal vehicles. Moreover, the vehicles for goods' transport have also risen sharply since the start of this century.

Transport vehicles are mainly diesel-run vehicles and more polluting in nature compared to the vehicles powered by other fuels like petrol or CNG. While transport vehicles are by and large more polluting in nature, goods transport vehicles generally pollute more compared to the passenger transport vehicles. The trends in the number of vehicles in different states and the trends in the number of transport (both goods'

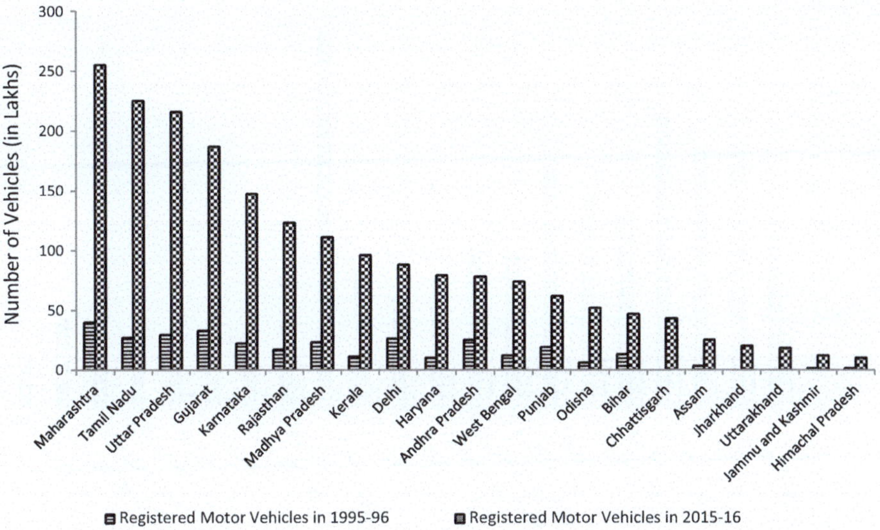


**Fig. 10.3** Registered motor vehicles in major states (1996 vs. 2016). *Source* GoI (2017)

and passenger transport) vehicles exhibit that number of vehicles have increased sharply in last one and half decade. In 2016, the numbers of transport vehicles were the highest in Maharashtra followed by Kerala, Tamil Nadu, Gujarat, Karnataka, Andhra Pradesh, etc.

Figure 10.3 suggests a considerable increase in the motor vehicle tally in the states with greater number of big/mega cities. Maharashtra, Gujarat, Tamil Nadu, Uttar Pradesh are the states where the motor vehicles tally ishigher in comparison to other states in both the years 1995–96 and 2015–16. Moreover, the number of motor vehicles in these states have increased sharply during the period. The largest increase in the vehicle tally has been observed in states irrespective of their level of State Domestic Product (SDP).

Although Fig. 10.3 exhibits the tally of registered vehicles in the states, these large numbers may be due to large size of the states, which may not be a good measure of density of registered motor vehicles and consequent pollution in an area. In order to show density of registered motor vehicles in different states, Fig. 10.4 presents the number of registered vehicles per 1000 population in different states. It suggests that the number of registered motor vehicles per 1000 population is highest in Goa followed by Pondicherry, Delhi, Chandigarh, Gujarat, Haryana, Kerala, Karnataka, Punjab, Maharashtra, etc. The most urbanized states have larger density of vehicles per thousand population. Thus, the chance of vehicular pollution is higher in the more urbanized states.

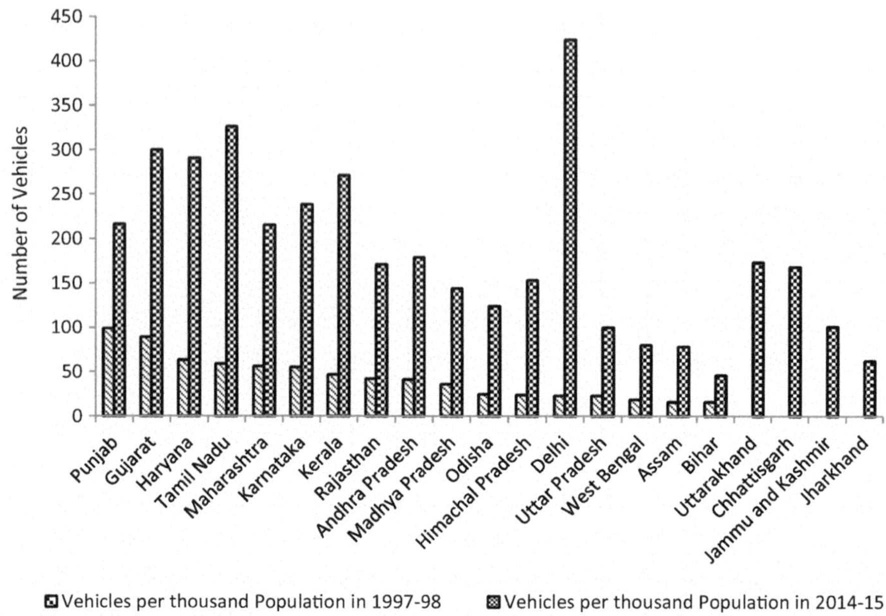


**Fig. 10.4** Motor vehicles per 1000 population. *Source* Based on data from GoI (2017)

## Air Pollution in India

The Environmental Performance Index 2020 (Wendling et al., 2020) ranks 180 countries on 32 performance indicators offering a score card that highlights leaders and laggards in sustainability. It concludes that poor air quality is the leading environmental threat to public health. India ranks 179 in air quality, 174 in $PM_{2.5}$ exposure, 139 in water and sanitation and 174 in heavy metals. India shares 168 rank with Ghana in overall index in the world—down from 141 in 2016 and fares better than only Afghanistan in the region. India's low scores are reflective of degraded air quality.

The WHO guideline for Air Quality is 10 $\mu g/m^3$ annual average $PM_{2.5}$ concentrations. Although the guideline is derived from the evidence on health effects of long-term exposure to $PM_{2.5}$, WHO acknowledges that lower level can also have potential adverse health effects (WHO, 2006). For regions with high air pollution, WHO suggested three interim targets for progressively reduced concentration levels: 35, 25 and 15 $\mu g/m^3$. Dismally, world population residing above these three interim targets is 58, 69 and 85%, respectively (IHME, 2018).

Estimates based on ground-level as well as satellite measurements suggest that 99.9% of Indian population lives in regions with $PM_{2.5}$concentrations exceeding the WHO air quality standards (35 $\mu g/m^3$). Similarly, the populace in majority of Indian states (21) and UTs (6) was exposed to $PM_{2.5}$ levels exceeding the Indian annual standard of 40 $\mu g/m^3$ in 2016. India's annual average $PM_{2.5}$ concentration in 2016 was 54 $\mu g/m^3$. In the last two decades, India has witnessed a 69% increase in $PM_{2.5}$ concentrations. In 2016, Nepal had the highest $PM_{2.5}$ concentration, with India following at the second spot. Nepal's high $PM_{2.5}$ concentration has led to a reduction of 4.4 years in life expectancy (Greenstone & Fan, 2018). Among the Indian states, Delhi has the highest $PM_{2.5}$concentration; next three positions are occupied by Uttar Pradesh, Haryana and Bihar. Based on satellite estimates of the annual $PM_{2.5}$ concentration for 670 districts, an analysis shows upsurge in air pollution over the last 20 years, recording highest levels between 2014 and 2016. For many of the districts, pollution increased by more than 200% since 1998. Of the 30 cities with the highest population worldwide, 21 are in India (WAR, 2019).

While the number of casualties imputable to $\mathbf{PM_{2.5}}$ grew by more than 20% between 1990 and 2016, the rate of deaths decreased by 12% worldwide during the same period. This signals that altering population characteristics may be an explanation of the increase in total deaths during this period. There was a 2.6% increase in deaths world over in the period 2010–2016, reflective of a greater impact of the ageing population and expanding global exposure than of decline in the rates for air pollution borne diseases. In 2016, ambient $\mathbf{PM_{2.5}}$ air pollution accounted for 1.8 million deaths and 22 million Disability-Adjusted Life Years (DALYs) in those older than 70 years. For the 50–69 year age group, it was responsible for 1.3 million deaths and 37 million DALYs. Put together, 4.1 million deaths and a loss of 106 million DALYs worldwide has been attributed to prolonged ambient $\mathbf{PM_{2.5}}$ exposure. Of the total toll, 51% comes from China (26%) and India (25%). The AAP contributes 6.4% to total DALYs in India (GBD, 2016). The AAP, the third largest risk factor in 1990 after malnutrition, and WASH, moved up to second largest place. During this period, the SEV of AAP escalated by 16.6% with increases in all epidemiological transition level (ETL) state groups barring the high ETL group.

Globally, more than 10% of total deaths were due to air pollution and out of those deaths, AAP alone was to be blamed for approximately 4.2 million deaths annually in 2016 (WHO, 2018). The extent of the impact of air pollution in the country is extremely high, and it has already started affecting the economic front along with the health of the citizens. An estimate suggests 3283 deaths per day due to AAP in India (GBD, 2015). Similarly, World Bank estimates that every year India forfeits approximately 3% of its GDP annually because of air pollution. This is due to the loss of productivity due to forced closure of schools, offices, industries, etc.

Table 10.2 enumerates the districts in different ranges across states that have $PM_{10}$ level of pollution. The large metropolitan cities, where the number of vehicles is more, are generally having more pollution compared to small cities where number of vehicles is less. The picture becomes clearer when pollution levels of the cities are compared with aggregate of vehicles they have. In order to show the level of pollution in different cities, we have plotted different cities in Fig. 10.5 by different signs according to their level of pollution. This figure also suggests a clear relationship between the numbers of vehicles in a city with their pollution level. A major portion of the pollution in different cities and in different states is chiefly due to the number of vehicles, especially due to the large number of transport vehicles in those areas which are causing huge amount of environmental pollution there.

**Table 10.2** $PM_{10}$ levels across India (annual average 2015)

| S. No. | State | No. of stations | No. of stations having a normal average in the ranges | | | | |
|---|---|---|---|---|---|---|---|
| | | | Below 60 | 60–120 | 120–180 | 180–240 | 240–300 |
| 1 | Andhra Pradesh | 6 | | 6 | | | |
| 2 | Assam | 5 | | 3 | 2 | | |
| 3 | Bihar | 2 | | | 1 | 1 | |
| 4 | Chhattisgarh | 4 | | 2 | 2 | | |
| 5 | Gujarat | 6 | | 6 | | | |
| 6 | Haryana | 3 | | 1 | 1 | | 1 |
| 7 | Himachal Pradesh | 7 | | 7 | | | |
| 8 | Jammu and Kashmir | 1 | | | 1 | | |
| 9 | Jharkhand | 11 | | 3 | 4 | 4 | |
| 10 | Karnataka | 20 | 11 | 9 | | | |
| 11 | Madhya Pradesh | 8 | | 6 | 2 | | |
| 12 | Maharashtra | 25 | | 20 | 5 | | |
| 13 | Meghalaya | 1 | | | 1 | | |
| 14 | Nagaland | 2 | | 1 | 1 | | |
| 15 | Odisha | 8 | 1 | 6 | 1 | | |
| 16 | Punjab | 14 | | 9 | 4 | 1 | |
| 17 | Rajasthan | 5 | | | 4 | 1 | |
| 18 | Tamil Nadu | 3 | 1 | 2 | | | |
| 19 | Telangana | 7 | 1 | 6 | | | |
| 20 | Uttar Pradesh | 21 | | 2 | 13 | 3 | 3 |
| 21 | Uttrakhand | 5 | | 1 | 3 | 1 | |
| 22 | West Bengal | 1 | | 1 | | | |
| 23 | Chandigarh | 1 | | 1 | | | |
| 24 | Delhi | 1 | | | | | 1 |

*Source* Calculated on the basis of data obtained from Greenpeace (2017)

India is often found lacking on air pollution statistics when compared with other major economies (Greenpeace, 2017):

1. India is the only economy that has seen an increase in satellite-based $PM_{2.5}$ level. While China, USA and EU saw decreases in the range of – 15 to – 20%, India witnessed an increase of + 13%.

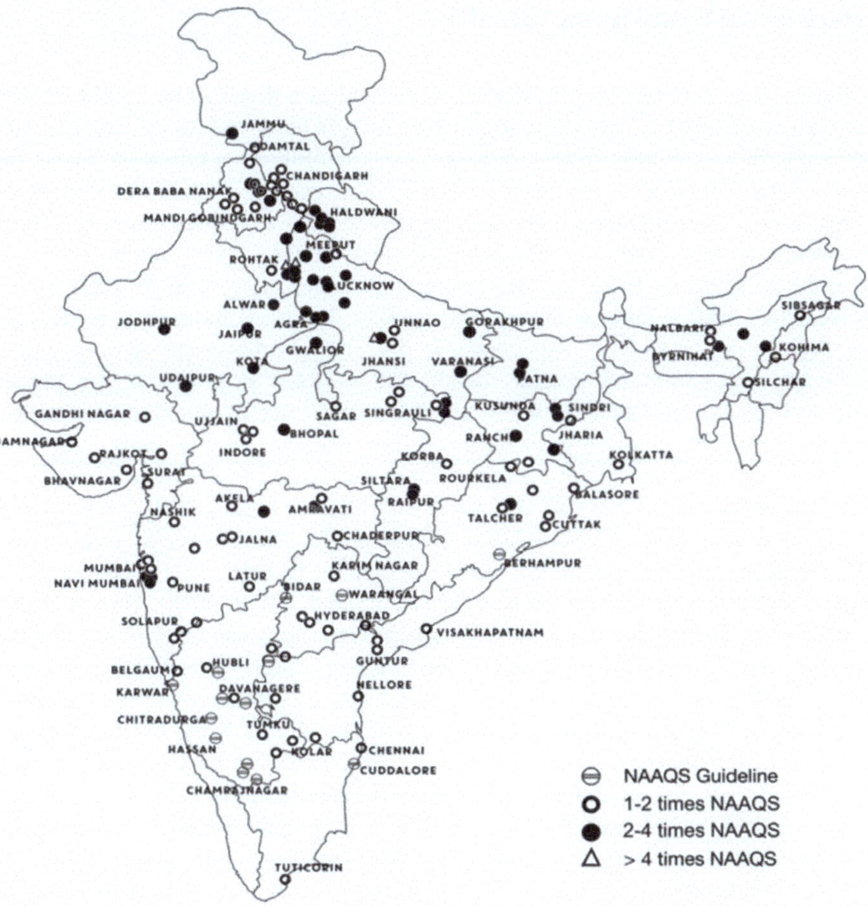


**Fig. 10.5** Cities and $PM_{10}$ levels across India. *Source* Author's presentation based on the data obtained from Greenpeace (2017)

2. India has less than 40 stations in 23 cities to online monitor $PM_{2.5}$ levels. This is order of magnitudes lower as compared to China and USA that have 1500 and 770 stations, respectively.
3. Power generating sources are major source of air pollution. In India, less than 10% of thermal power plants have pollution control measures. The same shares in China, USA and EU are 95, 60 and 75%, respectively.
4. $PM_{2.5}$ levels in India's capital are 10 times higher as compared to USA's capital (128 in India vs. 12 in USA). The same level in China's capital is 81.
5. In addition to poor current state, India neither has targets for national air quality levels, nor has a deadline to meet any targets. Whatever meagre measures individual cities have taken, lack of proper action plan and long-term implementation policy has been quite evident.

## Effect of AAP on Human Health

In recent years, air pollution in India has emerged as a threat to the public health. The GBD study (HEI, 2018) found that out of 79 behavioural, environmental and metabolic factors, outdoor $PM_{2.5}$ exposure was the third-most important factor for deaths in India. In 2015, exposure to outdoor $PM_{2.5}$caused more than 1 million casualties in India. This corresponds to close to one-fourth of the global 4.2 million deaths attributable to AAP. The State of Global Air 2019 (IHME, 2019) report attributed 1.2 million deaths in India in 2017 to air pollution. Lelieveld et al. (2020) also showed that AAP is one of the main causes of increased mortality and loss of life expectancy (LLE). Life expectancy is reduced by an average of three years, amounting to 8.8 million premature deaths due to global air pollution. It is higher than the LLE attributable to all forms of violence and that of smoking. It was also responsible for 29.6 million years of health life lost (DALYs). Disability-adjusted Life Years is a common metric that combines the Years of Life Lost (YLLs) and Years Lived with Disability (YLDs). In 2016, 67.2% of India's DALYs came from YLLs, in contrast to the 82.5% in 1990 (GBD, 2017). The number of air pollution related deaths has been consistently on the rise in India. In 2016, 1892 males and 1272 females died due to acute respiratory infection in India (CBHI, 2018). The economic cost of loss of life is not easy to calculate, yet attempts are made for the same and the costs are enormous. World Bank and the Institute for Health Metrics and Evaluation (IHME, 2016) have estimated that more than US $225 billion is forgone in labour income and US $5.11 trillion in welfare losses globally in 2013. For India alone, these estimates stand at US $55 billion and US $505 billion for lost labour output and for welfare losses respectively. The cost of premature deaths in India accounts to more than US $ 800 billion (OECD, 2016). There are copious indirect costs involved such as costs related to pain and suffering of losing loved ones, possessions, and costs to biodiversity and ecosystems; however, there is no methodology available to estimate it.

Some environmental toxins may cause acute illness leading to reduced life expectancy and even prove fatal (Joshua et al., 2018). Some toxins lead to ailments that may surface years or decades later. Air pollution chiefly impacts the respiratory system. Bronchitis, emphysema, asthma and lung cancer are some of the chronic diseases associated with exposure to polluted air (Kampa & Castanas, 2008). Since illness caused by air pollution, such as asthma attacks, cardiovascular diseases have multiple causes, hence most health issues triggered by air pollution cannot be identified directly. Nevertheless, there is concern that the carcinogens in polluted air are the primary cause of lung cancer. The mortality rate due to air pollution generated diseases is found to be higher in urban areas.

Many pollutants like CO, ozone, sulphur dioxide, nitrogen dioxide, lead, particulate matters, etc. emitted from the vehicles adversely impact the health of the people. Carbon monoxide results from incomplete fuel combustion in vehicles. It has no colour or odour but is poisonous because it reduces the oxygen carrying efficacy of blood. Carbon monoxide's impact is highest on unborn or new-born children, and people with heart disease, although it affects healthy individuals too. Nitric oxide irritates the pulmonary system and is linked with pulmonary haemorrhage (CPCB, 2008).

Sulphur dioxide is a common pollutant which is excessively harmful. Exposure to even small levels of sulphur dioxide can cause smooth muscle spasms in bronchioles, and exposure to higher levels increases mucus (Amato et al., 2010). Sulphur dioxide can also lead to cough, shortness of breath, larynx spasms and acute irritation to the membranes of the eyes. It can cause allergic reactions and can also result in the formation of sulphuric acid (upon reaction with other elements) which can damage lungs.

Vehicular emissions for lead are particularly harmful for children and can potentially damage the brain (WHO, 2015). Lead toxins can negatively impact nervous system, and blood pressure and even intellect. Breathing of even small quantities of lead can be harmful due to its capacity to accumulate over time. This is why it is referred to as a cumulative poison. Oxides of nitrogen ($NO_x$), including nitrogen dioxide, result from fuel combustion and contribute towards ozone formation. As such, exposure to nitrogen oxides has similar adverse effects as that of exposure to ozone and sulphur dioxide.

Toxic compounds like benzene and formaldehyde are by-products of vehicular emissions. They pollute the air, and exposure to even low concentrations may prompt cancer, genetic mutations, birth defects and other harmful illnesses. Small particulates in air (e.g. dust and lint) can carry these chemicals contributing to their inhalation, in addition to direct inhalation.

Other pollutants like asbestos, silica, carbon, beryllium, lead, etc., can exert a noxious (fibrotic) local effect in the interstitial parts of lungs. Radioactive air pollutants harm not only humans, but other living organisms as well. Smog, which also results from air pollution, has a killer effect, and has a death range of few persons to thousands. It is commonly understood that the most dangerous air pollutant is particulate matter. The miniscule particles can reach deep within the lungs and eventually the bloodstream, leading to serious medical conditions like lung diseases, cancer, strokes and heart attacks. $PM_{2.5}$ is more harmful than $PM_{10}$ as it can enter the bloodstream, as opposed to $PM_{10}$ which is confined to nasal cavities and upper airways. More so, sustained exposure to particulate matter and $NO_x$ can prematurely age blood vessels and contribute to a more rapid build-up of calcium in the coronary artery. According to the State of Global Air report (IHME, 2018), $PM_{2.5}$ was responsible for 11% of all global deaths in 2016. An additional 10 ug/m$^3$ of $PM_{2.5}$ leads to reduction of 0.98 years in life expectancy. Worldwide, on average, the life expectancy of a person is 1.8 years less because of particulate air pollutants exceeding the WHO guidelines. This is higher than other dangerous diseases like HIV AIDS and tuberculosis, as well as other behavioural factors like smoking and even war (Greenstone & Fan, 2018).

Children are most vulnerable as the pollutants impact lung development. The under-developed enzyme and immune systems in children are less effective in detoxifying the body and hence more susceptible to air pollution. The problem is exacerbated by a common habit of children breathing via their mouth. Breathing via mouth bypasses the nasal passage filters, providing an easier passage for the pollutants to the lungs. Experts also inform that the surface area of lungs relative to the weight is larger in children and that children also tend to inhale more air. Children also have increased exposure to air pollutants because of their comparatively larger outdoor time. Ignoring early stage symptoms of air pollution in children can lead to more severe problems later.

An infant's mental growth rate is highest in their first thousand days. Any impairment during this crucial period can lead to negative consequences related to their health and wellbeing; children's lungs and brain can be permanently damaged. UNICEF report (UNICEF, 2017) warns that pollution can lower children's IQ and intelligence. It further states that a total of 17 million infants live in risk bearing highly polluted areas, globally, of which 12.2 million live in South Asia. WHO (2018) cites air pollution as the main cause of deaths of children as almost 98% children in less developed countries breath toxic air. Air pollution counteractively impairs children's brains through several mechanisms:

Normally, the blood–brain barrier restricts harmful substances from reaching and affecting the brain. However, particulate matter can break this barrier leading to degenerative memory and memory disorders (Garciduenasa et al., 2015).

A particularly acute risk is posed by ultrafine pollution particles (i.e. particulate matter with diameter equal to or less than 2.5 μm). Their size allows them to enter the blood and reach the brain. On one hand, blood–brain barrier (BBB) breach leads to a high risk of Alzheimer's and Parkinson's disease in elderly people. On the other hand, infants must be protected from ultrafine pollution particles because even small amounts of such toxins can impair an infant's developing brain (Grandjean & Landrigan, 2006).

Specific air pollution particles, like magnetite, are able to get into the body through the olfactory nerve and the gut because of their miniscule size. Magnetite particles are particularly harmful because of their magnetic charge and their role in creating oxidative stress, which can lead to neurodegenerative diseases (Maher et al., 2016).

Polycyclic aromatic hydrocarbons (PAHs) are pollutants resulting from burning of fossil fuels and are found in areas with dense automobile traffic. They are detrimental for the neural connections as they directly diminish white matter essential for different parts of the brain to communicate with each other.

In addition to deleterious effects of air pollution on human health, its impact on environment, such as effect on ozone layer, greenhouse effect, etc. too indirectly affects human health.

If no preventive measures are undertaken, outdoor air pollution could cause 6–9 million premature deaths a year by 2060 (OECD, 2016). Moreover, the resulting sick days, medical bills and reduced agricultural output would incur an annual cost of one percent of the global GDP, i.e. close to US$2.6 trillion. The air pollution would result in an yearly US$176 billion cost of health care, 3.7 billion lost work days, and close to US$330 reduced economic output per capita. The report forecasts double, or even triple, premature deaths in 2060 compared to 2010, disproportionately concentrated among children and elderly. The forecast for India, a 4 × increase in premature deaths, is even worse.

## Conclusion

Healthy lives of humankind and the supporting ecosystem warrants clean air. Unfortunately, anthropogenic emissions of various kinds have led to deterioration of air quality making it the fifth biggest cause of mortality in India. Urbanization levels have swelled from 17% in 1957 to 31% 2011 and are expected to touch a 50% mark by 2050. Growing urbanization has direct linkage with increasing the AAP growth with amplified purchasing power leading to increased mobility and transportation demands. With multifold increase in registered motor vehicles, AAP steadily grew in India over the last 25 years. Indians are exposed to worst degrees of air pollution (GBD, 2016; Wendling et al., 2020) contributing to NCDs and communicable diseases. The GBD estimates indicate that the population weighted mean $PM_{2.5}$ concentration for India as a whole was 74.3 $\mu g/m^3$ in 2015, up from about 60 $\mu g/m^3$ in 1990, registering a 24% increase. Exposure to common air pollutants has been linked to a wide range of adverse health outcomes that include respiratory diseases and heart attack, stroke, asthma exacerbation, reduced lung function and premature deaths.

In recent years, air pollution in many parts of the country has become a threat to the public health. Nonetheless, there is still a lack of consensus regarding air pollution being a national issue which can't be tackled by the region- or city-level measures. The problem of outdoor air pollution in the country is not limited to specific states, regions or cities. However, the intensity of the phenomenon differs depending on the climatic and other geographical conditions of the regions. To win the battle against air pollution, we need to act as a nation transcending city and regional boundaries. The moot point is how to minimize the health risks caused by exposure to vehicle emissions. Policies must be framed which encourage production and usage of newer and more efficient technologies with controlled emissions. Conscientious monitoring of $PM_{2.5}$ by government agencies is called upon for better understanding of air pollution trends.

If timely action is not taken, the posterity will have to bear the brunt of air pollution generated diseases. The flip side, however, assures avoidance of nearly 1.2 million deaths if aggressive action is taken. At the direction of the apex court, though the Government of India has issued a slew of measures ranging from graded response system to emergency response measures and issuance of advisories, yet significant challenges remain to be addressed. Good governance has emerged as the crucial factor required for balancing the distinct dimensions of sustainable development, viz. economic growth, urbanization, industrialization and environmental health. There is an urgent need for national sustainability effort for reducing air pollution and improving environmental health. The lesser vehicles on the roads, the better the quality of the air. Individual contribution as responsible citizens through adoption of measures like carpool, using mass transit, biking or walking whenever possible are suggested for a healthier and happier future.

## References

Amato, G. D., Cecchi, L., Amato, M. D., & Liccardi, G. (2010). Urbanair pollution and climate change as environmental risk factors of respiratory allergy: An update. *Journal of Investigational Allergology and Clinical Immunology, 20*(2), 95–102.

CBHI. (2018). *National health profile 2018.* Central Bureau of Health Intelligence, Directorate General of Health Services, Ministry of Health and Welfare, Government of India.

CBHI. (2019). *National Health Profile 2019.* Directorate General of Health Services, Ministry of Health & Family Welfare, Government of India, New Delhi. http://www.cbhidghs.nic.in/showfile.php?lid=1147.

CPCB. (2008). *Epidemiological study on effect of air pollution on human health (adults) in Delhi.* Central Pollution Control Board Ministry of Environment and Forests.

Garciduenasa, C. L., Melo-Sanchezb, G., Vargas-Martineza, J. E., Macias-Escobedoc, O. V. L., Ramirezf, J. A., & Torres-Jardong, R. (2015). Air pollution and children: Barrier breakdown, inflammation, brain immunity and neurodegeneration. *Journal of the Neurological Sciences, 357*, 1039–1058.

GBD. (2015). Global, regional, and national comparative risk assessment of 79 behavioural, environmental and occupational, and metabolic risks or clusters of risks, 1990–2015: A systematic analysis for the global burden of disease study 2015. *Lancet, 388*, 1659–1724.

GBD. (2016). Measuring progress and projecting attainment on the basis of past trends of the health-related sustainable development goals in 188 countries: An analysis from the global burden of disease study. *Lancet, 390*, 1423–1459.

GBD. (2017). Nations within a nation: Variations in epidemiological transition across the states of India, 1990–2016 in the global burden of disease study. *Lancet, 390*, 2437–2460.

GoI. (2017). *Road Transport Year Book.* Ministry of road transport and highways transport research wing, Government of India.

Grandjean, P., & Landrigan, P. J. (2006). Developmental neurotoxicity of industrial chemicals. *Lancet, 368*(9553), 2167–2178.

Greenpeace. (2017). *Airpocalypse: Assessment of air pollution in Indian cities.* Green Peace, GPET, pp. 1–33, Bangalore, India. https://storage.googleapis.com/planet4-india-stateless/2018/06/Airpoclypse-Not-just-Delhi-Air-in-most-Indian-cities-hazardous-Greenpeace-report.pdf. Accessed on February 12, 2020.

Greenstone, M., & Fan, C. Q. (2018). *Introducing the air quality index.* University of Chicago.

HEI. (2018). *Burden of disease attributable to major air pollution sources in India.* Health Effects Institute.

IHME. (2016). *The cost of air pollution: Strengthening the economic case for action.* Institute for Health Metrics and Evaluation, World Bank Group.

IHME. (2018). *State of Global Air, 2018: A special report on global exposure to air pollution and its disease burden.* Health Effects Institute and Institute for Health Metrics and Evaluation's Global Burden of Disease Project. Health Effects Institute.

IHME. (2019). *State of Global Air, 2019: A special report on global exposure to air pollution and its disease burden.* Health Effects Institute and Institute for Health Metrics and Evaluation's Global Burden of Disease Project. Health Effects Institute.

Joshua, S. A., Michael, B., Aaron, J. C., Majid, E., & Pope, C. A. (2018). Ambient $PM_{2.5}$ reduces global and regional life expectancy. *Environmental Science and Technology Letters, 5*, 546–551.

Kampa, M., & Castanas, E. (2008). Human health effects of air pollution. *Environmental Pollution, 151*(2), 362–367.

Lelieveld, J., Pozzer, A., Poschl, U., Fnais, M., Haines, A., Munzel, T. (2020). *Loss of life expectancy from air pollution compared to other risk factors: A worldwide perspective. Cardiovascular Research, 16*(7).

Maher, B. A. M., Ahmed, I. A. M., Mac Laren, V. K., Foulds, P. G., David, A., David, M. A., Ricardo, T.-J., & Lilian, C. G. (2016). Magnetite pollution nanoparticles in the human brain. *Proceedings of the National Academy of Sciences of the United States of America, 113*(39), 10797–10801.

OECD. (2016). *The economic consequences of outdoor air pollution.* Environment and Economy Integration Division OCED Environment Directorate, Task Force on Integrated Assessment Modelling.

UNICEF. (2017). *Danger in the air: How air pollution can affect brain development in young children.* United Nations Children's Fund.

WAR. (2019). *World air quality report.* Retrieved March 18, 2020 from iqair.com/world.most.polluted.countries.

Wendling, Z. A., Emerson, J. W., de Sherbinin, A., Esty, D. C., et al. (2020). *2020 environmental performance index.* Yale Center for Environmental Law & Policy. epi.yale.edu

WHO. (2006). *WHO air quality guidelines for particulate matter, ozone, nitrogen dioxide and sulphur dioxide.* World Health Organisation.

WHO. (2015). *Global health observatory data repository.* World Health Organization.

WHO. (2018). *Ambient air pollution: A global assessment of exposure and burden of disease.* World Health Organization.

# Chapter 11
# Climate Change, Urbanization and Impact on Natural Environment: The Indian Scenario

**M. K. Pandit**

**Abstract** 'Climate change', the term interchangeably used to denote 'global warming', is of serious concern. A correlation between global warming and the rise in the concentration of atmospheric carbon dioxide (greenhouse gas) during the last two centuries points towards fossil fuel consumption as a significant anthropogenic contribution to its increase. Industrialization has also resulted in the mass migration of rural populations to the cities to fulfil the workforce needs. This has caused unbearable stress on the limited land and water resources of cities/metros, and further removal of the vegetative cover has led to the enrichment of $CO_2$ in the atmosphere. All these anthropogenic factors have contributed in accelerating the pace of 'climate change' that has been occurring simultaneously. Indian urbanization can be described as an involute phenomenon, unlike the evolute types, as seen in several other countries, implying the transfer of inherent misery and poverty with the population shift. The urbanization pattern in India is skewed and complex, compounded by the growth of slum areas and the development of residential zones at the city periphery with inadequate infrastructure and poor sanitation. The Indian urbanization has a unique characteristic of being a product of the population explosion and poverty-induced rural to urban shift.

**Keywords** Global warming and climate change · Milenkovitch Cycles · Urbanization · Land and water resources

## Introduction

One of the most significant global environmental issues of contemporary times is the unprecedented rise in the average temperature of the Earth's surface during the past decades, called 'global warming' (a term used interchangeably with climate change). Its consequences are a big threat to the survival of the human population if it continues unabated. Unlike several similar issues, environmental degradation is not restricted to the source region of the activity only. The primary reason for

M. K. Pandit (✉)
Department of Geology, University of Rajasthan, Jaipur, Rajasthan, India

 
A. Kateja and R. Jain (eds.), *Urban Growth and Environmental Issues in India*,
https://doi.org/10.1007/978-981-16-4273-9_11

stratospheric ozone depletion is the chlorofluorocarbon (CFC) bearing industrial exhaust into the atmosphere that disrupts the ozone cycle by accelerating the pace of ozone molecule break-up. It is a paradox that Antarctica, despite being the only exception of an industry free continent, has recorded the most severe ozone depletion. The inherent message is that the environmental issues should not be seen in isolation and warrant a concerted global effort. Therefore, environmental degradation, on a larger scale and global warming as of immediate consequence, have been identified as the top priority global issues. The United Nations (UN) summit, held in Rio de Janeiro in June 1992, represents a landmark event, and participation of 172 (including Indian) heads of states underlines the seriousness of the issue and global concern. Since then, the environment has prominently figured as an area of major concern for the United Nations, and several such summits mark a positive global move in the right direction. The Intergovernmental Panel on Climate Change (IPCC), constituted by the UN, has been assigned the responsibility of technical assessment of the problem and to suggest remedial/mitigation measures.

The 'climate change' in the present-day context has several dimensions and what we see is the result of the complex interplay of multitude of factors operative over the past more than two centuries. The impact became perceptible and grew exponentially during the last few decades. It was in the 1950s that the impact of environmental degradation on human life and the need to halt the trend was felt all over the world. The Centre for Biodiversity, Philippines, foresees a great loss in biodiversity in the Southeast Asia by the end of the current century, apprehending erosion of 70–90% of the species in the region (Shamasundari, 2017).

It was not too difficult to correlate the phenomenon of global warming with rapid industrialization that also induced increased natural resources utilization. Phenomenal rise in energy consumption is a significant and direct impact of urbanization, so much so that the per capita energy and natural resource consumption became the yardsticks to discriminate 'developed' and 'underdeveloped' nations, while a third category, 'developing', was also introduced to denote nations where such trends were showing a rise.

Several human activities in the past two centuries or so have affected the climate change; the urbanization being one of them. In most cases, the urbanization is associated with a search for employment and economic opportunities, better educational facilities and living for self and family. It can also be a compulsive movement to survive natural calamities and to avoid ethnic/religious or political conflicts. Whether by choice or a forced activity, the mass migration of population from rural to urban regions has put severe pressure on the urban land and water resources, further adding to the environmental degradation and consequent global warming.

Urban growth would mean an increase in commercial and industrial activity, consequently a rise in the use of fossil fuels and energy. Their higher consumption would result in additional heat. In the rural scenario, a high density of vegetation would normally absorb the heat, while the severity gets compounded in 'urban areas' due to the removal of an already scanty vegetation, to pave way for infrastructure, such as roads, commercial and community and residential buildings. The cement concrete buildings and road asphalt would absorb the heat, thus creating several 'high-temperature sinks'. These would rob the atmosphere off its moisture, more relevant in the Indian scenario, which is a predominantly arid to semi-arid terrain. These phenomena trigger a series of reactions, cumulatively contributing to the trend of rise in surface and atmospheric temperature, i.e. 'global warming'. The uniqueness of the Indian urbanization is underlined in one of the largest infrastructure growth in coming years, coupled with poor compliance of climate change mitigation and implication in the Indian cities. The Indian scenario is also distinctive in terms of an additional dimension of 'high population density', contributing to climate change, in addition to urbanization (Ram Avatar et al., 2019). In the Indian urbanization context, Khosla and Bhardwaj (2018) have posited a change from climate vulnerabilities and risks in the early stages to climate mitigation strategies during later stages of urbanization. They have suggested a multidimensional strategy, underlining the application of local development priorities at the initial level, followed by climate mitigation and adaptation, emphasizing the role of non-governmental agencies in achieving the goals. In an urbanization impact study covering > 1 million population spread over 43 Indian cities, Singh et al. (2021) observed adaptation actions in only half the cities, implemented typically through sectoral projects with some synergy interventions of co-benefit. They have compared the approaches in Ahmedabad, Surat, Coimbatore and Mumbai and highlighted variation in implications on risk management. These examples necessitate a relook into the closely and complexly intertwined issue of urbanization and environmental concerns in India. The recent Disaster Risk Reduction Status Report (UNDRR, 2020) has also apprehended that climate change will exacerbate the impacts and frequency of hydrometeorological hazards in the country. The effects would translate into worsened healthcare facilities for the poor and damages and losses to the infrastructure.

In this article, the interdependence of urbanization and resource exploitation is discussed in terms of impact on the Indian urban environment.

## Climate Change and Global Warming

Although the climate is generally referred to in a rather static sense, it is dynamic and has been changing. The Thar Desert of western India, for example, bears evidence of marine conditions in the past, and the rocks at the summit of Mount Everest have a marine origin. The present-day oceans are not older than 200 million years. Various continental fragments have been wandering, in the geologic past, from one climate zone to the other, and the climate has been influenced by tectonic (internal) and external forces. If the climate has been naturally 'changing', why do we make such a hue and cry about 'climate change' now? In a normal process, climate change occurs over a period of hundreds, thousands or millions of years. The cause of concern is the 'rapid climate change', loosely referred to as the 'climate change' in the present context. The United States Climate Change Science Program defines 'abrupt climate' change as 'a change in the climate (for example, in temperature or precipitation) that takes place over a few decades or less, persists for at least a few decades, and causes substantial disruptions in human and natural systems'. The effect of 'abrupt climate change' may be limited to a small region or to the entire globe. Significantly different from 'climate change', the 'global warming' has several connotations. It refers to an increase in the average temperature of the surface of the Earth, hydrosphere and atmosphere over some time. It has also been described as a process that leads to the rise in the lower atmospheric air temperature due to the heat trapped by 'greenhouse gases' (GHG). The term 'global warming' is used interchangeably with 'climate change', to denote planet-wide changes in weather patterns to suggest a significant rise in mean temperatures.

'Climate change' has several socio-scientific aspects. While its adverse effect on human life and humankind's survival is more than obvious, the cause and process continue to be debated among scientists. Some argue in favour of the temperature rise being a natural process with evidence of 'ice ages' and intervening warmer 'interglacial' periods in the geological past. On the other hand, a perfect correlation between temperature increase and rise in atmospheric carbon dioxide levels in the post-industrialization era links global warming to a significant anthropogenic contribution. The discernable anthropogenic influence on global climate is manifested mainly in the form of increased emission of carbon dioxide, a major GHG and destruction of forests that would have otherwise absorbed the harmful carbon dioxide. It has also been influenced by several other factors, such as population rise, industrialization and urbanization, energy and natural resource consumption and changing land-use patterns. The projected implications of global warming include sea-level rise and submergence of coastal cities, extreme weather fluctuations, more frequent cyclonic storms, hotter summers, flooding and droughts, etc. Changes in oceanic water current patterns would affect wildlife even in the Polar regions. Yet another group of scientists believes that global warming may eventually lead to reduced sea levels since seawater evaporation would result in more rains and snowfall. However, the majority of concerns are placed on the opposite.

The Earth originated about 4.56 billion years ago as an extremely hot gaseous mass that eventually cooled down externally to the level to sustain water and support life. The initial atmosphere was carbon dioxide dominant, and oxygen was introduced much later into it, after the appearance of elementary forms of plant life that absorbed carbon dioxide and released oxygen through photosynthesis. The history of the Earth is replete with several 'glaciation' events when almost entire Earth got covered by ice and each glaciation event is followed by an intervening warmer or interglacial period. Secular variations in Earth's temperature have also been attributed to the Milankovitch Cycles that postulate changes in the Earth's orbital eccentricity every 1,00,000 years, angle of inclination of the axis of rotation every 41,000 years and wobble in spin axis every 23,000 years.

The surface temperature record for the last 800,000 years of the Earth's history indicates cyclic variations in the surface temperature. During the past 8000 years, the favourable and relatively constant climatic conditions during the interglacial period facilitated the advancement of the human civilization. Since the end of the latest glaciation (the Pleistocene Glaciation) event about 18,000 years ago, a gradual rise in temperature is discernible. Global warming, resulting entirely from the anthropogenic contribution, is not fully supported by the historical account as one can see that the hottest period in the human history occurred approximately 7500–4000 years ago, much before the industrial revolution. At the same time, a perfect correlation between atmospheric carbon dioxide levels and Earth's surface temperature attributes the temperature rise to increased carbon dioxide emissions into the atmosphere.

## Urbanization Vis-À-Vis Air, Water and Land Resources

The industrialization has resulted in the mass migration of rural population to cities to fulfil the manpower requirements for running the industries. Although the issue is of global relevance, it needs to be discussed in terms of local impact. The discussion here is confined primarily to the Indian context. Besides industrial factors, several other causes have also provided an impetus to the urbanization process. Due to the skewed developmental priorities and prevalence of strong traditions, especially in the rural areas, a huge gap in terms of basic infrastructural facilities exists in India between urban (metro/city/town) and rural areas. Traditionally, basic hygiene has not been an important component of the social life in the majority of the rural sector, irrespective of the economic background of persons concerned. However, concerted efforts by the government to generate awareness through campaigns and incentives, active involvement of NGOs and other agencies and educational institutions have resulted in improved sanitation in rural India in the recent years. The predominantly agrarian, rural economy in India is heavily dependent on monsoon rains. Due to the absence of perennial surface water bodies, a major part of peninsular India utilizes groundwater, which in turn is augmented/recharged through soil absorption of monsoon rains. Low and erratic rainfall, reduced number of rain days during monsoon seasons, frequent monsoon failures and successive years of drought during the last decades have all contributed to rural exodus and urban influx. The pattern is quite different from industrial/commercial activity triggered migration as in this case the 'undeveloped' urban marginal areas, with no basic infrastructure, get occupied and habituated, commonly called the 'slum areas'.

Urban development planning must consider an environment conducive to family life, child development and social interaction. The executive part may deviate from place to place, depending upon prevailing cultural and climatic factors. Any urban development plan must include open public spaces (parks) close to the residential areas with provision to play sports, socialize and organize recreational events and meetings, accessible to local citizens. Trees and plants alongside roads and pathways and any other appropriate places not only provide respite from the heat by creating a micro-environment but also add to the aesthetics of the area. The city master plans always have the provision of parks and community centres nearby residential areas and reflected in the residential/commercial areas developed by the governmental agencies and by the organized sector. In the latter case, it is usually to meet the obligatory requirements for environmental clearance norms.

## Urbanization Scenario in India

Urbanization can be described as the concentration of population in the spatial sense due to mass movement of population (Geruson & McGrath, 1977). The relocation of the human population is an outcome of simultaneous economic and city growth. However, 'urbanization and migration' have diverse implications in developed and developing (including the so-called underdeveloped) nations. In case of developed nations, the movement from rural to urban setting is more an 'urban pull', while in the Indian context, it is mainly the 'rural push' on account of several socio-economic factors including the natural ones, such as droughts due to consecutive monsoon failures and epidemic breakouts, etc. Mukherji (1995) has described Indian urbanization as an involute and not evolute phenomenon. As a result, population shift also transfers inherent misery and poverty. Such a poverty-induced migration to megacities adds only to the urban population size but not in the prosperity and culture of urban society (Nayak, 1962). Therefore, such a scenario can be best described as an urbanization process without the urban functional characteristics. The Indian urban population has risen from about 11% in 1901 to 17% in 1951 and currently stands at > 35%, registering a steady rise during the last decades. However, the pattern is skewed, and the major movement is towards the large (mega) cities that are already saturated and technically 'overurbanized'. Unable to cope with the extra pressure, these megacities have become the dysfunctional urbanization. The net result of inadequate infrastructure and insufficient resources to cope with the population growth is a virtual collapse of functional structure, more so, without any related increase in the economic base. As a result, the cities are subject to the growth of unhygienic slum areas that lack access to housing, potable water and electric supply, civic amenities, etc., to a large population of extremely poor, rural migrants (Kundu et al., 1999). There is already huge pressure on the urban land resources in cities and megacities, coupled with poor economic conditions, and prohibitive prices the land is unaffordable to the migrating population. This results in the growth of slums and poorly developed residential areas at the city periphery, without basic facilities such as electricity, water and sanitation. Poor quality of the migration shift from rural to urban setup would lead to poor quality of urbanization (Bhagat, 1992). Jaysawal and Saha (2014) have described the Indian urbanization as a poverty-triggered shift of rural population, a consequence of demographic explosion. Some unique features of the Indian urbanization scenario, highlighted by these authors, are listed below:

1. Unplanned urbanization in India has induced haphazard growth of Category-I cities.
2. Urbanization occurs without any strong economic base; it is distress migration and detrimental to cities.
3. Rapid and unplanned urbanization leads to the development and substantial growth of slums that eventually become epicentres of socio-economic regression, manifested in desolation, scarcity, redundancy, mistreatment, discrimination, etc. The net result is the degradation in the quality of urban life.

Besides, there are several other direct and indirect impacts, but here we focus on the land and water resources.

The problems of urbanization in India stem from lopsided urbanization, lack of planning or rather execution failure, poor economic base and socially and economically heterogeneous societal fabric without functional categories. These are manifested in some basic problems in urban India, especially in Category-I cities that can be recognized in the fields of housing, public transport, water supply, sewerage system, urban waste disposal, roads, insufficient social infrastructure (schools, hospitals, recreation centres), unscientific land use and environmental pollution (air, water and soil). Such urbanization, with stark socio-economic disparities would trigger community conflicts, criminal and anti-social activities (Kundu & Gupta, 1996). Since these cities had reached saturation long back in terms of employment generation, they cannot absorb the (largely) poor, landless, illiterate and unskilled, stress-driven migrants (Kundu, 2009). This snowball effect reasons further urban poverty, housing shortage, unemployment and poor health conditions, jointly causing the 'urban decay'.

## Environmental Impact of Urbanization

The physical environment in the context of urbanization primarily refers to the land (soil), water and air. There is a clear interdependence between the population and the environment and both influence each other. One of the major environmental impacts of urbanization is the degradation of air quality. Called 'air pollution', it is a cause of serious concern on account of its direct impact on human health. Air pollution in the recent years has recorded a significant increase as compared to the past decades. The number of deaths resulting from air pollution in India has been estimated by the World Bank at an alarming high of around 40,000 persons per year. Several studies have revealed a quantum increase in respiratory diseases and related allergies in a large population and a significant growth in such incidences since the 1990s. The Covid-19 pandemic has also caused a spurt in the recent times. Several Indian cities have reached the danger mark in terms of air pollution due to high vehicular exhausts and industrial emissions. Felling of trees and removal of vegetative cover for urban expansion is further adding to the air pollution as these would have otherwise absorbed the toxic carbon dioxide from the vehicular exhaust and industrial and domestic emissions. Encroachment and illegal constructions on earmarked open lands for public utilities and parks are also a factor, though not a significant contributor directly. Engulfing of peripheral green cover (rural pastures and agriculture fields) by city expansion is yet another dimension of the problem.

The air composition for pollution purposes can be discriminated into physical and chemical criteria. All the minute solid or liquid suspended particles in the air are physical components. The atmospheric aerosol particles are called atmospheric particulate matter, particulate matter (PM) or particulates. However, the term suspended particulate matter (SPM) more specifically refers to the organic or inorganic solid particles that can be further discriminated into two size categories (USEPA, 1982, 1996). The > 10 μm size particles are non-inhalable, while < 10 μm thoracic and inhalable particles are further divided into coarse respirable (10–2.5 μm) and fine respirable (<2.5 μm) categories. Recent WHO Global Urban Ambient Air Pollution Database (May 2018) on 2.5 μm SPM in the air underlines severe air pollution in major Indian cities. The particulate matter may have a natural or anthropogenic source. The International Agency for Research on Cancer (IARC) and World Health Organization (WHO) have designated respirable SPM as carcinogenic (cancer-causing) due to their ability to penetrate deep into lungs and bloodstreams, the PM-2.5 μm being more deadly (WHO, 2000). Very high frequency and long hours of smog in winter months are a serious consequence of urban air pollution. Fog can be described as a natural phenomenon of the condensation of vapour into minute water droplets, close to the cold ground surface, during winter. Fine particles of smoke, soot or sand get attached to these droplets due to pollution, causing 'smog' conditions with very poor visibility that disturbs the routine activities and causes disruption in rail, road and air transport.

Chemical pollution of the air is more detrimental to human health because it cannot be filtered. This relates to the higher levels of toxic gases, mainly carbon dioxide (including carbon monoxide) and oxides of sulphur and nitrogen. These can be directly attributed to the vehicular exhaust and industrial emissions. These emissions, in some specific cases, also contain minute quantities of heavy metals that are toxic for human beings and also degrade the soil where they finally settle down. Sulphur bearing industrial exhausts can react with atmospheric water to cause 'acid rain' that is harmful to soil fertility. Besides several direct consequences, urbanization has also indirectly contributed to air pollution. Das et al. (1998) noted high values of cadmium (a toxic heavy metal) in the air samples at several places in Jaipur city. The dispersion pattern analyses of cadmium values in the air could trace the cadmium source to the quarrying activity in the Jhalana Hills, located to the east of the city, from where cadmium (locked in the mineral composition of rock) was released into the air during blasting. The quartzite rocks in these hills (quarrying activity stopped since then) were being quarried for building construction purposes.

## Land-Use Changes

The term 'land use' refers to the use of a geographic domain for economic, residential, recreational, conservational and governmental purposes, a concept that is closely related to societal development. In an ideal scenario, land use is decided on the basis of land/soil type(s) and its suitability for the specific community activity. While land use for commercial, residential and industrial purposes is interchangeable, land suitability for agriculture needs special attention. Conversion of agriculture land for other usage is an irreversible process, and soil fertility is lost forever. Long-term urban 'master plans' take into account the soil types to propose future development of cities. Horizontal expansion is the basic aspect and the most dominant factor during the initial stages of urban growth. City expansions are usually need (and often greed) driven and many times, do not follow the land use approved and stipulated in the master plan(s). A major impact of urban expansion is its extension into fringe rural areas and the conversion of agricultural land for industrial, residential and/or commercial use. This results in permanent loss of agricultural land, besides the removal of green cover. Inappropriate land use may result in rocky areas left as free spaces that are unsuitable for plantation and parks. The zone of urban–rural interface needs special attention to identify the current land usage to manage the growth of developed areas, developmental activities, population distribution and the prevailing infrastructure for the community. Urbanization-induced changes in the Land-Use and Land-Cover (LULC) are a significant cause of climate change. The satellite data has helped in quantifying such effects and has shown a significant rise of 3–5 °C in the near-surface temperatures, accompanied by decreased humidity (up to 10%) over a number of newly formed urban zones in north India (Kedia et al., 2021).

## Effect on Water Resources

Water is the most essential resource to support human life, next only to the air. The increasing urbanization and land-use deviations, and climate change have affected availability of water and its quality during the last decades. Many parts of India have witnessed conflicts over water resources in the past, and future projections paint a gloomy picture, showing further deterioration in the water scenario. Most of the rivers in India are polluted by the sewage and industrial waste disposal into them, and traditional anthropogenic activities. Several legislations and norms, enacted during the recent years and massive governmental efforts for cleaning rivers, have met with little success. Similarly, most of the aquatic ecosystems and the living beings therein have either been lost completely or reduced significantly due to non-point source contamination of surface and groundwater from agricultural and urban lands. Because of the severe shortage of water, the quality issues are usually ignored as the entire emphasis is shifted to the water availability. As a result, groundwater has become a dwindling (finite) rather than a 'renewable' infinite resource. The seriousness of the crisis can be seen in interstate disputes and conflicts over water sharing. Historical and archaeological accounts underline a close link between surface water and habitation as the ancient civilizations are always associated with major river systems. This basic concept continued, and several historic cities were developed at riverbanks. Later city development activities were also around surface water bodies. Urban dwellings away from the surface water bodies were provided with adequate infrastructure for receiving water supply from surface water sources, such as natural ponds and dams. The city of Jaipur can be cited as an example for which Ramgarh reservoir, built about 30 km from the city, catered to the water requirements of the citizens. Increasing population pressure and new urban centres rendered the surface water bodies inadequate, and groundwater was tapped as the only available alternative. Since water is a replenishable resource, it was viable with the annual monsoonal recharge to balance the water withdrawal. Water is a renewable but finite resource, and its availability is further hampered by low and erratic rainfall, dwindling land surface for recharge and increased draft. Therefore, the groundwater levels have gone down to alarming depths, resulting in substantial rise in the number of designated 'dark zones', implying no further groundwater development. As per the State and Central Groundwater agencies, only ten per cent area in the State of Rajasthan remained under the 'safe' category in 2011 for groundwater extraction (25 out of 248 blocks). The situation had marginally improved to 44 blocks in 2013 on account of above-average rainfall during the preceding years and efforts for artificial groundwater recharge. Besides the reduced availability of groundwater, the extraction rates in excess of recharge have cumulatively effected water quality deterioration. In the coastal regions, the groundwater draft would result in the formation of 'cone of depression' and saline water incursion from the sea, rendering the water unsuitable for human consumption and causing permanent damage to the aquifer (natural rock/soil formation that holds and transmits groundwater). Higher levels of total dissolved solids (TDS) in water is a common feature of deeper inland aquifers because of water–rock interaction and long residence time of groundwater in direct contact with bedrocks. Deepening of the water table (the upper level of groundwater in unconfined aquifers) means longer passage for groundwater recharge and increased evaporation losses in the transit. Pandit et al. (2009) reported higher nitrate levels in the groundwater in Jaipur city that could be directly associated with numerous septic tanks in the 'walled city'. Higher fluoride levels in groundwater can be traced to the release of fluoride from the fluorine bearing minerals in the granitic bedrock in deeper aquifers. Although there are several geogenic factors, anthropogenic contributions, detrimental to ground and surface water, exist beyond any doubts.

The United States Geological Survey (USGS) has evaluated a stage-wise impact of urbanization on groundwater. During the early stages of urbanization (levelling of ground, filling up of ponds, and construction of roads and other infrastructure), the water system is affected by increased land erosion, storm flooding due to blockage of small drains and reduced groundwater recharge due to increase in paved or constructed areas. Continued urbanization means additional roads, houses, sewage disposal systems, additional bore wells and deepening of existing ones. The net effect is additional land erosion, increased water impounding, lower recharge and high surface run-off over paved areas and roads, and additional sewage into local streams.

## India and Global Climate Negotiations

A major outcome of the 1992 Rio Earth Summit was agreeing to collective measures to tackle the issue of environmental degradation. The summit paved way for creation of the UN Commission on Sustainable Development and formation of the United Nations Framework Conventions on Climate Change (UNFCCC), the United Nations

Convention to Combat Desertification (UNCCD) and the Convention on Biological Diversity. Several international or UN-centred agreements have paved way for the present-day international cooperation on environmental concerns and mitigation efforts. Prior to Rio, the 1989 Montreal Protocol had led to an agreement on phasing out of ozone depleting substances, and the positive results are seen in the reduction of 'ozone hole' over Antarctica, in the recent years.

The current international efforts on environmental governance can be attributed to the climate change. It has a significant political component with disagreement between developed and developing nations, due to conflicting national economic priorities and resources to handle global environmental challenges. The matter regarding reduction in carbon footprint is of special mention here. Being a developing nation, India has followed a cautious approach, highlighted by a transformation from a passive role as a protest voice, peripheral to the global environmental policy during the early years, to an active role in outlining the global environmental efforts (Michaelowa & Michaelowa, 2012). Although the active role of India in the success of the Paris Agreement negotiations was hailed by a number of countries, some local sections found it contrary to the founding principles of the Indian Climate Policy.

Mohan (2017) has provided a historical account of the Indian role in global climate politics, dating back to the establishment of the UNFCCC in Rio in 1992 to the Conference of Parties (COP) 21 at Paris in 2015. In this document, the author has highlighted the main narratives of the Indian Climate Policy and has critically assessed the shift in India's negotiation position during the preceding two and a half decades. The first formal Indian stand on environment can be traced to the 1972 United Nations Conference on the Human Environment, underlined in the conflict between socio-economic development and environmental protection, further, holding the developed countries responsible for the global environmental problems (Vihma, 2011). A report by the Centre for Science and Environment (CSE) called for 'Global Warming in an Unequal World' and underlined the role of developed countries in 'carbon colonialism' (Agarwal & Narain, 1991). This report provided the basic framework for the formal Indian Climate Policy. Based on the historical and prevailing per capita emission trends of the developed nations, the report commended that the developed nations should accept the major responsibility for climate change and urged them to shoulder major responsibility for climate mitigation (ibid). India's per capita emissions are still about 33% of the global average, electricity consumption only about 25% of the global average and just 10% of that of the Organisation for Economic Co-operation and Development (OECD) (World Bank, 2015; World Resources Institute, 2014). On the other hand, India is the third largest in the world, in terms of GHG emissions. Continued rise in the emissions makes India a significant constituent of any climate action. India holds a unique position in the global climate diplomacy, as a nation with current high levels, but with insignificant emissions in the past (Dubash, 2013).

Strong economic growth of some developing nations (India, China, Brazil and South Africa together called BASICS) during the last two decades or so has put an increasing burden of expectations on these countries to take the lead in influencing the outcomes of global environmental governance (Hallding, 2013). The developed countries call these nations as 'emerging economies', therefore, urging them to contribute to mitigation action (ibid) and argued that Kyoto exemptions should not be applicable to these 'advanced' developing countries such as India (Stern & Antholis, 2008). India accepted at least voluntary participation by the developing nations in global mitigation efforts, at COP 13 at Bali in 2007, hailed as a major deviation from its traditional stand (Michaelowa & Michaelowa, 2012). The BASICS countries (including India) announced in 2009 voluntary targets to reduce the emissions intensity of the GDP by 20–25% against 2005 levels, by 2020. The 2009 Copenhagen summit could not reach a climate policy agreement to replace the Kyoto Protocol. However, the summit did succeed in initiating the process of inverting the climate governance architecture from a top-down approach to bottom-up commitments (Hurrell & Sengupta, 2012).

Prior to the Paris summit in 2015, negotiations were held at Durban in 2011, Warsaw in 2013, and the final version of Intended Nationally Determined Contributions (INDCs) was adopted by several countries at COP 20 at Limain in 2014, asking all the countries to submit their plans for climate action up to 2030. At the Paris 2015 negotiations, India accepted the 1.5 degrees goal for climate policy. The Paris 2015 Agreement is an International Treaty on Climate Change with a legal binding. India is one of the 196 signatories of this agreement, signed during the Paris convention on 12 December 2015 and enforced on 4 November 2016. As compared to the previous agreements, this is a landmark agreement due to inherent binding on all the member nations to undertake determined efforts in climate change mitigation. The most important provision of the treaty is to bring down the global warming to below 2 degrees Celsius, preferably to 1.5 degrees, compared to the pre-industrial levels. It envisages achieving greenhouse gases peaking levels to global neutral by the year 2050. India ratified the Paris Agreement to help bring it into force, despite concerns about developed countries first fulfilling their pre-2020 commitments, and launched the Global Solar Alliance, a part of an ambitious expansion of its renewable energy program. These actions were praised during the COP 21 negotiations (PTI, 2016), and India's proactive role in global climate policy was duly acknowledged. Notwithstanding the debate over the magnitude of shift in the Indian climate policy, the early narrative of economic development as a competing interest with environmental protection is no longer a dominant perception (Stevenson, 2011).

## Summary and Recommendations

Although the urbanization has affected the environment globally, the impact depends upon several local natural and anthropogenic factors including the biota. The impact on regional environmental security in southeast Asia, including India, is of real concern on account of lower levels of societal awareness, poverty, climate vagaries, lack of access to safe drinking water and poor sanitation and infrastructure. All these factors make the best of the policies ineffective, leading to inconsistencies in achieving the goals of environmental security and sustainable development. The climate change issues can plausibly be managed through innovative technologies with more efficient use of fossil fuel, blending with alternative fuel, reduced traffic through efficient public transport systems and improved roads and infrastructure. The greenohuse gas emissions are a major concern; despite several international efforts, the GHG emissions in the developing countries are set to rise in coming decades.

The adverse impact of unplanned urbanization on the surrounding land, air and water environment has been well documented. High population density and increasing population growth have further compounded the problem, highlighting the uniqueness of the Indian scenario. Indian megacities have received major population migration on account of better infrastructure and economy to offer jobs, although mostly petty or in the unorganized sector. These cities are oversaturated, and there is a need to develop relatively smaller cities to avoid further congestion in already overcongested slums in these cities. There should be incentives for investment in smaller cities. Strengthening of the functional base in such cities would accelerate the local economy and may lead to more uniform urbanization. Urban planning should emphasize on operational, developmental and restorative activities with particular attention on improvement of urban infrastructure, for example, roads, transport, etc. Development of the city into nearby rural interface should be carefully monitored, keeping in view the land quality and land capability. Yet another important aspect of development is to preserve and restore the structure and status of historic buildings and monuments. In general, urban planning must aim at the following components: (a) balanced regional and urban planning (Mukherjee, 2001), (b) development of strong economic base for the urban economy, and (c) integration of rural and urban sectors (Kundu et al., 2003). The policy of judicious use of water resources needs to be integrated into urban planning with an emphasis on reuse and augmentation. Good water and bad water must be segregated, and dual plumbing can be adopted to preserve good quality water for drinking purposes. Water cannot be created but can be recycled, and therefore, each drop of rainwater should be utilized to replenish the groundwater through an appropriate rainwater harvesting structure. The location and design of the structure must be finalized by considering technical suitability. There are several existing laws promulgated for protection and preservation of environment and water resources that need to be enforced, but the most important thing is community participation to ensure sustainability. Human dependence on the nature is a fact, and there is no doubt that human race needs to live in harmony with the biological and physical environment. India is endowed with the biodiversity and a wealth of natural resources, and can show the way to reverse the trend of human-induced adverse effects on the environment. This would require a combined effort by policymakers, administrators, executing agencies, NGOs and the people; otherwise, the livelihood and survival of the humankind would continue to be threatened.

## References

Agarwal, A., & Narain, S. (1991). *Global warming in an unequal world: A case of environmental colonialism*, New Delhi.

Avatar, R., Tripathi, S., & Aggarwal, A. K. (2019). Assessment of energy—pollution—urbanization nexus with changing energy industry scenario in India. *Land, 8*(8), 124. https://doi.org/10.3390/land8080124

Bhagat, R. B. (1992). Components of urban growth in India concerning Haryana: Findings from recent censuses. *Nagarlok, 25*(3), 10–14.

Das, D. B., Bhargava, A., Kushwaha, R., Gupta, A. B., & Pandit, M. K. (1998). Quarrying induced air pollution in Jaipur city, western India. *Journal of Nepal Geological Society, 18*, 369–378.

Dubash, N. K. (2013). The politics of climate change in India: Narratives of equity and cobenefits. *Wiley Interdisciplinary Reviews: Climate Change, 4*(3), 191–201.

Geruson, R. T., & McGrath, D. (1977). Cities, and urbanization. United States Agency for International Development (USAID) Document No.210251.

Hallding, K. (2013). Rising powers: The evolving role of BASIC countries. *Climate Policy, 13*(5), 608–631.

Hurrell, A., & Sengupta, S. (2012). Emerging powers, North-South relations and global climate politics. *International Affairs, 88*(3), 463–484.

Jaysawal, N., & Saha, S. (2014). Urbanization in India: An impact assessment. *International Journal of Applied Sociology, 4*(2), 60–65.

Kedia, S., Bhakare, S. P., Dwivedi, A. K., Islam, S., & Kaginalkar, A. (2021) Estimates of change in surface meteorology and urban heat island over northwest India: Impacts of urbanization. *Urban Climate* 36 Article number 100782.

Khosla, R., & Bhardwaj, A. (2018). Urbanization in the time of climate change: Examining the response of Indian cities. *Wires Climate Change, Focus Article, 9*, 2020. https://doi.org/10.1002/wcc.560AccessedonMarch10

Kundu, A. (2009). *Urbanization and migration: An analysis of trend, pattern and policies in Asia*. United Nations Development Program, Human Development Reports, Research Paper 2009/16, MPRA Paper No 19197.

Kundu, A., & Gupta, S. (1996). Migration, urbanization and regional inequality. *Economic & Political Weekly, 31*(52), 3391–3398.

Kundu, A., Bagchi, S., & Kundu, D. (1999). Regional distribution of infrastructure and basic amenities in urban India. *Economic & Political Weekly, 38*(28), 1893–1906.

Kundu, A., Sarangi, N., & Dash, B. P. (2003). *Rural non-farm employment: An analysis of rural urban interdependencies* (p. 36.), working paper no. 196. Overseas Development Institute, London.

Michaelowa, K., & Michaelowa, A. (2012). India as an emerging power in international climate negotiations. *Climate Policy, 12*, 575–590.

Mohan, A. (2017). *From Rio to Paris: India in global climate politics*. India and Global Governance. Observer Research Foundation Occasional Paper #130 38.

Mukherjee, S. (2001). The nature of migration and urbanization in India: The search for alternative planning strategies. In K. Srinivasan & K. B. Pathak (Eds.), *Dynamics of population and family welfare* (pp. 204–249). Himalayan Publishing.

Mukherji, S. (1995). *Poverty induced migration and urban involution in ESCAP countries*. Paper presented at UNESCAP, Expert Group Meeting on Poverty and Population in ESCAP Region, Bangkok, pp. 1–45.

Nayak, P. R. (1962). The challenges of urban growth to Indian local government. In: R. Turner (Ed.) *India's Urban Future*. University of California Press, California.

Pandit, M. K., Bhardwaj, V., & Pareek, N. (2009). Urbanization impact on hydrogeological regime in Jaipur urban block: A rapidly growing urban center in NW India. *The Environmentalist, 29*, 341–347.

PTI. (2016). *Modi demonstrated Indian leadership on climate change: White House*. Times of India. Available at: http://timesofindia.indiatimes.com/india/Modi-demonstrated-Indian-leadership-on-climate-change-White-House/articleshow/52634055.cms Accessed on April 5, 2020.

Shamasundari, R. (2017). *Climate change dilemma driven by urbanization*. The ASEAN Post (daily), 26 August 2017.

Singh, C., Madhavan, M., Arvind, J., & Bazaz, A. (2021). Climate change adaptation in Indian cities: A review of existing actions and spaces for triple wins. *Urban Climate* 36 Article number 100783.

Stern, T., & Antholis, W. (2008). A changing climate: The road ahead for the United States. *The Washington Quarterly, 31*(1), 175–188.

Stevenson, H. (2011). India and international norms of climate governance: A constructivist analysis of normative congruence building. *Review of International Studies, 37*(3), 997–1019.

UNDRR. (2020). *Disaster risk reduction in the republic of India: Status Report 2020* (p. 36). United Nations Office for Disaster Risk Reduction (UNDRR), Regional Office for Asia and the Pacific, Bangkok, Thailand.

USEPA. (1982). *Air quality criteria for particulate matter and sulphur oxides, vols. I, II, and III*. The United States Environmental Protection Agency, Research Triangle Park, NC.

USEPA. (1996). *Air quality criteria for particulate matter* (Vol. I). United States Environmental Protection Agency.

Vihma, A. (2011). India and the global climate governance: Between principles and pragmatism. *The Journal of Environment & Development, 20*(1), 69–94.

WHO. (2000). *Air quality guidelines* (2nd Ed.). World Health Organization Regional Office for Europe, Copenhagen, Denmark.

World Bank. (2015). *Electric power consumption (kWh per capita) | Data*. Available at: http://data.worldbank.org/indicator/EG.USE.ELEC.KH.PC. Accessed on March 30, 2020.

World Resources Institute. (2014). *6 Graphs Explain the World's Top 10 Emitters*. Available at: https://wri.org/blog/2014/11/6-graphs-explain-world's-top-10-emitters. Accessed on March 17, 2020.

# Chapter 12
# An Analysis of Climate Change Related Vulnerability Among the Forced Migrants in India

**Sunetra Ghatak**

**Abstract** One response to climate change and natural disasters is a human migration to relatively safe places where the effect of climate change is less likely. While this climate-induced migration has an impact on human well-being and political status, research on this issue is limited and often fragmented. The objective is to find out the facets of the changing patterns of the forced migration, especially, in the context of India. In Indian migration data, we can observe a various range of migration flows, comprising with internal and international migrations as well as permanent and temporary migrations. In all these migration flows, a considerable number of forced migrants have been reported due to different environmental conditions. On this viewpoint, this paper focuses on migrated human due to climate-induced in India and wants to recognize their characteristic along with social and economic impacts. The paper also draws attention to the probable vulnerability of environmental-induced migration in the context of natural disaster to identify the most vulnerable classes related to these migrations. The paper has used 55th (1999–2000) and 64th (2007–2008) NSS rounds unit level data and compares the forced migration in India over time. The paper was set to review overall scenario of forced migration in India and found that upper-caste Hindus and Muslims, women, aged and poor people are more vulnerable to the effect of climate hazard and migrated more in response to environment disasters. Natural disasters become one of the growing causes of migration in some of the Indian states along with the reasons noted as social and political problems.

**Keywords** Climate change · Forced migration · Vulnerability · Natural disasters

## Introduction

Migration is a complex dynamics of humanitarian movement from their native to a new destination either for work or for social reasons. It involves multiple streams of movements and includes different patterns, like short term, iterative, permanent and

---

S. Ghatak (✉)
National Institute of Public Finance and Policy, New Delhi, India

 
A. Kateja and R. Jain (eds.), *Urban Growth and Environmental Issues in India*,
https://doi.org/10.1007/978-981-16-4273-9_12

return migration, and happens due to various reasons (Chandrasekhar & Sharma, 2014; Kundu & Gupta, 2001; Lusome & Bhagat, 2006). Available secondary data sources (Census of India, 2001; NSSO, 2010) record different reasons for leaving the last usual place of residence. According to the reasons for migration, the migrants are mainly classified into two broad categories: (a) economic migrants, those migrate for work, and (b) non-economic migrants who migrate for other than economic reasons (e.g. marriage, studies, natural disaster, etc.). In the literature, migration in pursuit of work is not new, and labour migration appears to be a growing phenomenon in almost all the developing countries. Besides labour migration, scientific indications specify that a big number of persons might be displacing due to environment degradation. Scientists expect climate-induced migration or climate migration to be one of the key humanitarian consequences of climate change (Boas, 2012; Panda, 2010). Therefore, a whispering view is gaining ground in the literature that environmental disaster can relocate people from their homes and vigour them to flee by making people migrants. These disasters displace a big number of inhabitants from their roots, leaving them houseless and jobless (such as Odisha cyclone in 1999, Indian Ocean earthquake and consequent tsunami in 2004, etc.). One obvious response to these natural disasters is human migration since vulnerable people change their place of residence in search of safer and more stable living conditions. Consequently, it is expected that a large population in India will migrate as an impact or consequence of natural hazard. Since Indian population mostly depends on climate-sensitive sectors such as agriculture, water and forestry for their livelihood, a change in climatic condition and the consequent natural disasters have an impact on human well-being by changing their job pattern, living style and of course displacing them from their roots. In India, a diverse range of migration flows can be observed, in which a considerable number of migrations have been reported due to different environmental conditions. However, research on the effect of these changes in the environment on internal migration in India is limited and often fragmented. This paper is set to provide a methodical approach on the overall scenario of forced migration with special emphasis to environment-induced migration. The paper also tries to identify the vulnerable classes associated with these migrations using secondary unit level data from National Sample Survey (NSS) during the period 1999–2008 (55th round and 64th round).

It is now broadly recognized that climate disaster and change have adverse effect on the vulnerable population in India. As a result, recent migration data (64th round of NSS) has included 'natural disaster' as one of the reasons for migration. This paper is based to provide a summary of the forced migration in India using the NSS rounds to examine the association of climate-induced migration with different socio-economic factors in India. Moreover, the paper draws consideration to the vulnerability faced by the climate-induced migration in the framework of ecological, political, and developmental foundations. Therefore, the specific objectives of the study are listed as follows:

1. To compare the environment-induced migration with other types of forced migration, like sociopolitical and development-induced migration.

2. To provide an insight for the observed regional variation of environment-induced migration in India.
3. To examine the classed differentials by different demographic, social and economic factors in the context of environment-induced migration in India.
4. To observe the vulnerable socio-economic classes, which are mostly affected by environment-induced migration in comparison with other types of migrants and other forced migrants in India.

In the subsequent section, we review the concept of climate-induced migration with the available literature. Third section deals with the discussion on data and methodology used in the paper. Fourth section presents the discussion and findings of the paper. Finally, conclusions of the study are presented.

## Review of Literature

Semple (1911) supported and argued that many people change their location in search for better land, slighter climate and easier conditions of living to settle down. However, until the late nineties environment was absent as an explanatory factor in the neoclassical migration theories (Harris & Todaro, 1970; Olsson, 1965). The concept of ecological refugees was presented by Lester Brown from World Watch Institute in the early 1970s (Black, 2001). This became known in the literature by forced migrants with the reasons environmental degradation and natural disasters (El-Hinnawi, 1985; Jacobson, 1988). Later, the works of Myers (1993, 1997, 2002, 2005) and Myers and Kent (1995) became the best-known works on environmental refugees in the existing literature. Myers in his study suggested that climate-induced migration is going to be the foremost crisis for human beings in coming years and millions of natives are going to escape their homes and may look for shelter in the neighbouring countries (Castles, 2002). Further, climate-induced migration largely took place at intra- and international and/or regional, international scales, and it is continuing under the different climate change conditions (Massey et al., 2010). However, these studies are based on international refugees and are yet to reach consensus on the question of whether this migrants are as a significance of climate change should be considered as climate refugees.

Another recent branch of the literature suggests that in developing countries, which are mainly reliant on climate-dependent economic activities such as agriculture and forestry, extreme climate and changing climatic conditions may accelerate growing levels of human migration (McLeman & Hunter, 2010). Both permanent and temporary migrations are the survival strategy of people dealing with the opportunity and also the impact or consequences of disasters (Hugo, 1996). Cyclical migration for smaller time duration in reply to droughts may last or even grow due to climate alteration (Deshingkar & Start, 2003). Although it is now widely documented by the environmental specialists that climate change will adversely affect India, there are little studies existing on how weather change is going to affect the migrant people

(Boas, 2012). According to the existing literature, climate transform might result in two types of migration. First, this can increase internal migration as a result of climate change especially due to natural disaster like drought, desertification, rise in sea level, water shortage, low agriculture output, melting glaciers, etc. Climate change can cause international migration by flowing migrants to the neighboring countries (Myers, 2002; Panda, 2010). In the Indian context, very recently Dallman and Millock (2012) have tried to identify the impact of climatic factors on interstate migration and found that drought frequency in the origin state acts as push factor on interstate migration in this country.

While the effect of climate change on migration is clear, the mechanisms through which climate change induces migration are yet to be identified clearly. Moreover, in the Indian context, none of the existing studies have identified the groups which are most vulnerable due to climate change. In this study, we try to identify such groups of migrants using recently introduced climatic reasons for migration in the NSS 64th round and compare these migrants with other forced migrants available from the 55th round of NSS.

## Data and Methodology

The paper is relied on secondary data sources. Most widely used secondary data of migration in India is the Census of India which provides information on the volume of migrants and the migration streams in the decadal intervals. The data set typically imitates mobility of the population and not essentially of workers. But this data set fails to capture interpersonal variation on migrants which is an important factor while judging migrants' characteristics and their behaviour. In contrast, NSS conducts a number of surveys to collect information on migration. NSS rounds on 'Employment, Unemployment, and Migration' as a part of employment and unemployment inquiries provide various information on migration particular of the migrants. The current paper is based on unit level data from 55 and 66th rounds of NSS. The 55th round was conducted during July 1999–June 2000 as a part of the 'Employment–Unemployment' schedule of which we have taken data from Schedule 10. Meanwhile, the 64th round of NSS provides more recent data based on the survey carried out during July 2007–June 2008. We used Schedule 10.2 of NSS 64th round. In both rounds, information is being collected on migration type and various other dimensions as an extension of the employment–unemployment situation.

The 55th and 64th rounds of NSS define 'migrants' as those whose current place of enumeration of each household member differs from the last 'usual place of residence'. Usual place of residence of a person is defined as a place (village/town) where the person had stayed continuously for a period of six months or more. Then, NSS asks a question regarding the reason for leaving the last 'usual place of residence' which helps us to classify the reasons for migration into three broad categories: economic, non-economic (marriage, associational, etc.) and forced migration. However, in this study the reason 'marriage' has been excepted as the allocation of migrants is highly distorted towards marriage migration of women. Forced migrants are those migrants who have reported natural disaster, communal or political troubles, and displacement by development projects as their causes for migration. The persons who migrated in search of employment, in search of better employment, business, to take up employment/better employment, transfer of services/contact, proximity to place of work have been categorized as 'economic migrants', while the others reporting any other reasons for migration have been named as 'other/non-economic migrants' to examine the various comparisons with the category of forced migrants. In this study, we are dealing with forced migration, that is, the persons who have migrated due to either natural disasters or sociopolitical problems or are displaced due to development projects. Further, our focus is on the natural disaster-related migrants, that is, those who have migrated due to extreme climatic conditions such as drought, flood and tsunami.

One of the purposes of this study is to analyse the patterns of forced migration in India, that is, whether the pattern is changing over time. Later, we focus on natural disaster-related migrants—a relatively new dimension in the migration literature. The analysis begins with simple cross-classification, that is, bivariate analysis followed by an empirical modelling. The relations between migration type and different socio-economic classes have been observed by applying bivariate distribution analysis with Pearson chi-square tests. In order to identify the most exposed cluster that is affected by forced migration, binary logistic regression has been carried out. Therefore, the study assumes migration occurred as a function of a set of variables expressed as the following:

$$\text{Migrant}_i\,(M_i) = f(\text{FORCED}_i, \text{ECONOMIC}_i, \text{NON-ECONOMIC}_i)$$

Further, the study concentrates on forced migration in terms of their reasons as expressed in the following:

$$\text{Forced Migrant}_i\,(\text{FM}_i) = f(\text{NATURAL DISASTER}_i, \text{SOCIAL/POLITICAL PROBLEMS}_i, \text{DISPLACEMENT BY DEVELOPMENT PROJECT}_i)$$

## Distribution of Migrants According to the Reasons for Migration

Table 12.1 analyses the reasons for migration that have been broadly categorized into three components—forced, economic and non-economic factors of migration. The analysis is done with and without 'marriage' to get the significant causes of the migratory movements of people. From the table, it is seen that forced and non-economic migrations have emerged as insignificant if we consider all types of reasons, particularly, when marriage has been considered as a reason for migration. Marriage

**Table 12.1** Distribution of migrants by reasons for migration (in per cent)

| Reasons for migration | | With 'marriage' migration | | Without 'marriage' migration | |
|---|---|---|---|---|---|
| | | 1999–2000 | 2007–2008 | 1999–2000 | 2007–2008 |
| Forced migration | Natural disaster | – | 0.19 | – | 0.59 |
| | Social/political problems | 1.04 | 0.45 | 2.89 | 1.44 |
| | Displacement by development project | – | 0.14 | – | 0.45 |
| | Total | 1.04 | 0.78 | 2.89 | 2.48 |
| Economic migration | | 10.92 | 9.91 | 30.25 | 31.37 |
| Non-economic migration | | 88.04 | 89.31 | 66.86 | 66.15 |

*Sources* Computed from unit level data of NSS 55th and 64th rounds

migration happens in India mainly due to sociocultural set-up where women change their parental residence after marriage. Hence, our calculations are based on the percentage of migrants by different reasons of migration without taking into account the 'reasons for marriage'.

Table 12.1 provides the distribution of migrants according to the reasons for migration using the two rounds of NSS mentioned earlier. This enables us to compare the percentage of migrants and provide a picture on whether the percentages are changing. While the percentage of migrants under forced migration has been found to be 2.5% (2,243,280) during 2007–2008, the number was slightly higher (2.9%) during 1999–2000. The absolute numbers of migrants under forced migration have marginally increased from 2,243,280 in 1999–2000 to 2,545,226 in 2007–2008. During 2007–2008, extra two million migrants have dropped under the forced migrant category. The numbers of natural disasters-related migrants come around 537,336 in 2007–2008. During this time frame, social and political troubles have occurred as the central reasons for forced migration (1.44% of total migration) followed by the reasons connected to natural disasters (0.59%).

Table 12.2 shows all the three major reasons for migration recorded in the 64th round along with their sub-categories. The table presents percentage supply of migrants by comprehensive causes under each group of migration. 'Marriage' migration has been deducted from all the figures of migration presented in the table. It can be interpreted as that non-economic migrants are 7.55% people who migrated for studies followed by 3.99% migrated for the attainment of own house/flat, and 2.44% migrated due to reasons related to housing problems. Healthcare-related migration has accounted for 0.71%. Among economic reasons of migration are search for better employment (9.40) and taking up employment/better employment (7.94) as the most important migration reasons in India in 2007–2008.

**Table 12.2** Distribution of migrants by detailed reasons for migration

| Reasons for migration | | 2007–2008 | |
|---|---|---|---|
| Forced migration | Natural disaster | 0.59 | 2.48 |
| | Social/political problems | 1.44 | |
| | Displacement by development project | 0.45 | |
| Economic migration | In search of employment | 7.65 | 31.37 |
| | In search of better employment | 9.40 | |
| | Business | 1.69 | |
| | To take up employment/better employment | 7.94 | |
| | Transfer of services/contact | 3.81 | |
| | Proximity to the place of work | 0.88 | |
| Non-economic migration | Studies | 7.55 | 66.15 |
| | Acquisition of own house/flat | 3.39 | |
| | Housing problems | 2.44 | |
| | Health care | 0.71 | |
| | Post-retirement | 0.88 | |
| | Others | 51.19 | |

*Sources* Computed from unit level data of NSS 64th rounds

## *Regional Differential in Forced Migration*

Since different regions and states have a different degree of incidence of climatic disasters, social conflicts and progression of development projects, we can expect a large variation of the numbers of forced migrants among the states. While forced migration in India is the lowest among the three broad categories of migrants recorded in the data, vulnerability and uncertainty associated with them are far greater than the economic migrants who mainly moved for a better employment opportunity. In this section, we have discussed the spatial allocation of different forms of forced migrants across states during 1999–2000 and 2007–2008. The Indian map presented in Fig. 12.1 is based on the regional differential of migration among Indian states during this period. We have placed two bars on each state representing the percentage of forced migrants, respectively, in the years 1999–2000 and 2007–2008. It shows that while the highest percentage of forced migrants has been observed in Mizoram in 1999–2000, the largest percentage of forced migrants has been observed in Tripura in 2007–2008. The most notable phenomenon is that both the states are situated in the north-east region which is one of the most politically disturbed regions in the country.

Figure 12.1 shows that in the northern region, Jammu and Kashmir recorded the highest migrants under the category of forced migration. In the central region, the number of forced migrants is created as highest in Uttar Pradesh. In the eastern region, West Bengal and Bihar have more enforced migrants. Assam and Tripura

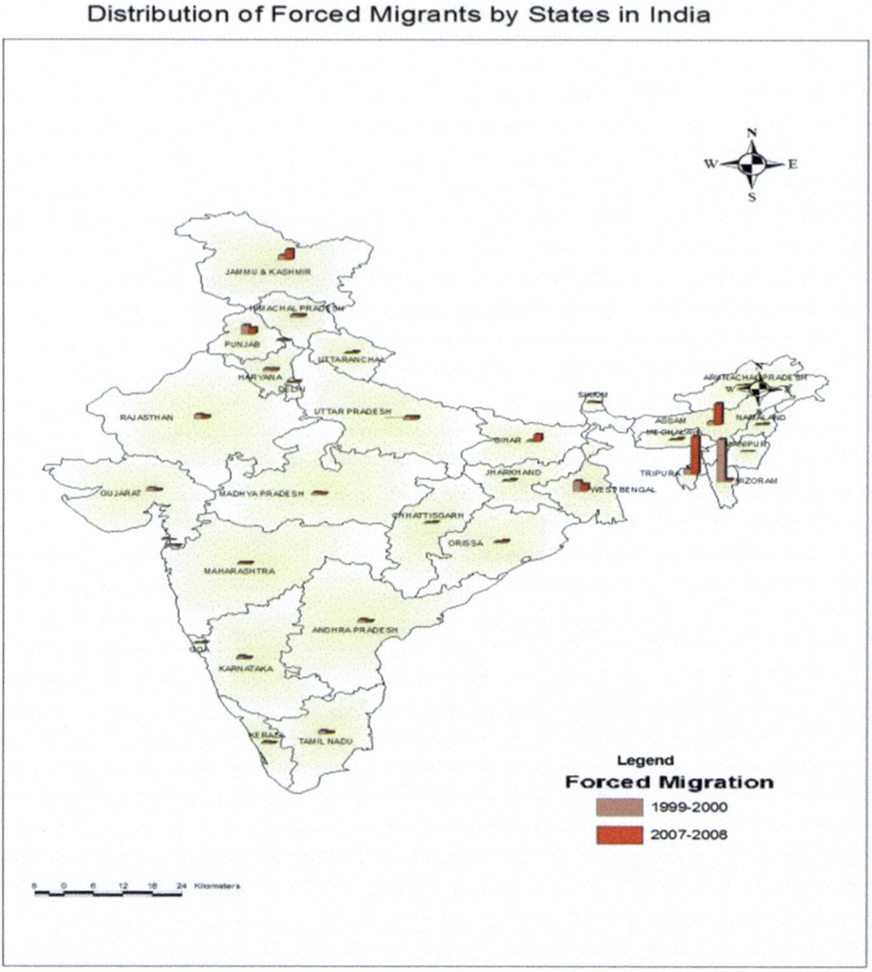


**Fig. 12.1** Geographical spread of forced migration. *Sources* Computed from unit level data of NSS 55th and 64th rounds

in north-east region, Maharashtra in the west region and Andaman and Nicobar in South region represent the highest percentage of forced migrants in the respective regions. It is interesting to note that the highest percentage of forced migrants is found in Tripura (37.68%), followed by Assam (30.9%)—both situated in the north-east. Forced migration is reported to be least in the states of central, western and southern regions in comparison with other regions during 2007–2008.

In the 55th round, since the reason behind forced migration was only social/political reasons, the percentage of forced migration between 1999–2000 and 2007–2008 could not be compared. Region-wise, Punjab (northern), West Bengal (eastern), Gujarat (western) and Lakshadweep (southern) are at the top according to the percentage of forced migrants in these states during 1999–2000. However, the composition of higher forced migrating states changes when we consider two more reasons of forced migration available in the latest round of NSS (64th). In the northern region, Punjab was on the top in 1999–2000 while Jammu and Kashmir recorded more forced migrants in 2007–2008. In the eastern region, while the percentage of forced migration in Bihar had increased sharply during the period from 1999–2000 to 2007–2008, the percentage of forced migration in West Bengal had declined sharply during the same period. Finally, in the north-east region, two more states had recorded a sharp increase in forced migration in 2007–2008, that is, Tripura and Assam. The overall observation is that inclusion of two more components in the reasons for migration in the 64th round captures more specific reasons for displacement of people from their places of residence and as a consequence of this, the distribution of forced migrants had changed during the periods 1999–2000 and 2007–2008.

In order to get a clearer picture of different types of forced migration, we analyse the distribution of types of forced migration by regions and states in Table 12.3. To analyse the regional variation of the number of forced migrants, we divide the states into six regions: east, west, north, south, north-east and central.

In the cause of the natural disaster, Rajasthan (23.1%) in the northern region is found to have the highest percentage of forced migrants. The desert state Rajasthan has the maximum occurrence of natural disasters, followed by Uttaranchal (12.3%) which is again hilly and mountainous region. While the highest percentage of forced migrants have been found in Assam due to natural calamities (90.0 %) in the north-eastern region. Assam is a relatively developed north-eastern region and has less mountainous topography. Therefore, it is expected that when a natural disaster happens in neighbouring states in the north-east, people tend to move to Assam for resettlement. In the eastern region, the highest number of climate-induced migrants has been identified in Bihar (72.9%) followed by Orissa. This can be due to the fact that the population in Bihar is very poor and vulnerable due to natural disasters. Poor people preferred to migrate within the state instead of going out of the state. Orissa is a coastal state which records a considerable amount of climate-induced migrants. In the western region, Maharashtra has attracted the highest climate-induced migrants. Similar to the coastal state of Orissa, in the south region, Tamil Nadu and Andaman and Nicobar also have a considerable percentage of natural disaster-induced forced migrants.

Another major cause of forced migration of people is sociopolitical troubles. States like Sikkim, Nagaland, Mizoram and Meghalaya in the north-eastern region have 100% forced migrants due to sociopolitical problems. The present political position in the north-eastern region validates this finding; 100% forced migrants due to sociopolitical reasons have been identified in Chandigarh in north zone and Dadra and Nagar Haveli in the west.

Lastly, in case of migration due to developmental projects, Madhya Pradesh (56.2%), Jharkhand (93.6%) and Arunachal Pradesh (83.6%) reflect highest percentages of migrants. It is interesting to note that these states have a large number of tribal

**Table 12.3** Distribution of forced migrants by regions and states in India (in per cent)

| Region | State | Natural disaster | Social/political problems | Displacement by development project |
|---|---|---|---|---|
| North | Jammu and Kashmir | 0.00 | 91.71 | 8.29 |
| | Himachal Pradesh | 8.60 | 75.25 | 16.15 |
| | Punjab | 3.35 | 95.06 | 1.60 |
| | Chandigarh | 0.00 | 100.00 | 0.00 |
| | Uttaranchal | 12.28 | 62.62 | 25.11 |
| | Haryana | 0.64 | 81.70 | 17.66 |
| | Delhi | 3.86 | 54.54 | 41.60 |
| | Rajasthan | 23.10 | 52.69 | 24.20 |
| Central | Uttar Pradesh | 23.87 | 74.82 | 1.32 |
| | Chhattisgarh | 1.65 | 94.60 | 3.75 |
| | Madhya Pradesh | 0.13 | 43.70 | 56.17 |
| East | Bihar | 72.96 | 23.62 | 3.42 |
| | West Bengal | 16.99 | 75.98 | 7.04 |
| | Jharkhand | 0.00 | 6.40 | 93.60 |
| | Orissa | 61.25 | 9.29 | 29.46 |
| North-east | Sikkim | 0.00 | 100.00 | 0.00 |
| | Arunachal Pradesh | 0.00 | 16.36 | 83.64 |
| | Nagaland | 0.00 | 100.00 | 0.00 |
| | Manipur | | | |
| | Mizoram | 0.00 | 100.00 | 0.00 |
| | Tripura | 1.99 | 98.01 | 0.00 |
| | Meghalaya | 0.00 | 100.00 | 0.00 |
| | Assam | 90.70 | 5.39 | 3.91 |
| West | Gujarat | 0.76 | 49.01 | 50.23 |
| | Daman and Diu | 0.00 | 0.00 | 0.00 |
| | Dadra and Nagar Haveli | 0.00 | 100.00 | 0.00 |
| | Maharashtra | 11.77 | 73.86 | 14.36 |
| | Goa | 0.00 | 15.11 | 84.89 |
| South | Andhra Pradesh | 0.00 | 21.89 | 78.11 |
| | Karnataka | 0.00 | 64.99 | 35.01 |
| | Lakshadweep | 0.00 | 0.00 | 0.00 |
| | Kerala | 19.70 | 51.99 | 28.31 |
| | Tamil Nadu | 24.85 | 75.15 | 0.00 |
| | Pondicherry | 1.61 | 98.39 | 0.00 |

(continued)

**Table 12.3** (continued)

| Region | State | Natural disaster | Social/political problems | Displacement by development project |
|---|---|---|---|---|
| | Andaman and Nicobar | 77.59 | 22.41 | 0.00 |

*Sources* Computed from unit level data of NSS 64th rounds
*Note* Pearson chi-square is significant at one per cent level

population and large mines. Many recent incidents of displacement of the tribal population due to development projects reported in the popular discourses in recent years provide support for these figures.

## *Socio-economic Profile of the Forced Migrants*

As mentioned earlier, the forced migrants are captured in three different categories in the 64th round which were not available in the previous rounds of NSS; therefore, this segment analyses the socio-economic characteristics of forced migrants only for the 64th round of NSS (2007–2008). In demographic classes, we have age cohort and gender variables. In the social classes, we have taken religion and caste indicators. Finally, in the economic parameters, we have discussed occupations and different MPCE quartiles. Table 12.4 analyses the socio-economic profile of forced migrants.

In the case of the age cohort, we have five classifications (0–14, 15–24, 25–39, 40–59 and 60 and above). While social or political problems are the main reasons for forced migration among elderly migrants (75.46%), children and teens are found to migrate largely due to natural disasters. We see that the younger population took migration decision due to development projects. Among both the genders, sociopolitical grounds are the main reasons for forced migration. However, the share of migrants as a result of sociopolitical troubles is high across females compared to male migrants. Climate-induced migration has the same impact on both the genders. In our analysis, sociopolitical troubles are the key reason for all social classes for migration except Muslim migrants. Muslim migrants are recorded the highest percentage of climate-induced migrants (51.24%). Nearly, 21.0% of Hindu migrants are being displaced for development activities. Among different castes, sociopolitical difficulties have appeared as the foremost reason for forced migration among Scheduled Caste and upper-caste population. Again 38.3% of Scheduled Tribe population has been forced to migrate for developmental assignments. In the economic class, the data of different occupations of the forced migrants narrates that people who are involved in the agriculture and fishery activities are more prone to get affected by natural disasters. Other than agriculture or fishery activities, craft or allied works are another sector to be affected by the natural disaster. On the other hand, clerks and professionals are being affected by sociopolitical troubles. Finally, developmental projects hamper highest percentage of people who are engaged in basic occupation.

**Table 12.4** Distribution of forced migrants by socio-economic classes in India (in per cent)

| Class | Variable | Forced migration | | |
|---|---|---|---|---|
| | | Natural disaster | Social/political problems | Displacement by development projects |
| *Demographic indicator* | | | | |
| Age cohort | 0–14 | 56.42 | 15.02 | 28.56 |
| | 15–24 | 34.06 | 49.07 | 16.87 |
| | 25–39 | 22.38 | 57.73 | 19.89 |
| | 40–59 | 21.02 | 57.20 | 21.79 |
| | 60 and above | 16.18 | 75.46 | 8.36 |
| Gender | Female | 23.01 | 60.88 | 16.11 |
| | Male | 24.62 | 55.89 | 19.48 |
| *Social indicator* | | | | |
| Religion | Hindu | 21.94 | 57.18 | 20.88 |
| | Muslim | 51.24 | 40.52 | 8.24 |
| | Christian | 4.60 | 89.72 | 5.68 |
| | Others | 3.50 | 96.50 | 0.00 |
| Caste | Scheduled Tribe (ST) | 17.54 | 44.16 | 38.30 |
| | Scheduled Caste (SC) | 19.51 | 64.79 | 15.70 |
| | Other Backward Classes (OBCs) | 26.32 | 51.12 | 22.56 |
| | Others | 25.93 | 64.80 | 9.27 |
| *Economic indicator* | | | | |
| Occupation | Professionals | 17.15 | 76.33 | 6.52 |
| | Technicians and associate professionals | 21.98 | 54.85 | 23.17 |
| | Clerks | 2.86 | 97.14 | 0.00 |
| | Service workers and shop and market sales workers | 13.39 | 67.93 | 18.68 |
| | Skilled agricultural and fishery workers | 29.69 | 55.87 | 14.44 |
| | Craft and related trade workers | 30.15 | 54.03 | 15.82 |

(continued)

**Table 12.4** (continued)

| Class | Variable | Forced migration | | |
|---|---|---|---|---|
| | | Natural disaster | Social/political problems | Displacement by development projects |
| | Plant and machine operators and assemblers | 22.69 | 71.54 | 5.76 |
| | Elementary occupations | 19.20 | 55.61 | 25.20 |
| MPCE quartile | Highest | 13.83 | 72.42 | 13.75 |
| | Higher | 22.11 | 58.15 | 19.74 |
| | Medium | 33.55 | 48.51 | 17.94 |
| | Lowest | 24.51 | 55.28 | 20.21 |

*Sources* Computed from unit level data of NSS 64th rounds
*Note* Pearson chi-square is significant at one per cent level

## *Perspectives of Climate-Induced Migration*

We have found that among the forced migrants in India sociopolitical problems-related migrants were the highest in number during 2007–2008. However, in some states, the numbers of natural disaster-related migrants are also seen to dominate. In this section, we have focused only on climate-induced forced migration because these types of migrations have larger impacts on human beings. According to Human Development Report (UNDP, 2009), 'those who are moving in the face of local economic crisis, drought or other causes of desperate poverty, may not know what capabilities they will have; they only know that they cannot remain. Even migrants who end up well off after a move often start out with very restricted capabilities and high uncertainty'.

In this section, we have employed the binary regression models to identify the vulnerability of climate-induced force migrants using the demographic, social and economic variables as independent variables. In the empirical models, 'types of migrants' have been taken as dependent variable. We have applied two different regression models to identify the vulnerability of different socio-economic groups due to forced migration. In the first model, dependent variable takes '0' if the reasons for migration are other than forced migration and '1' for forced migration (natural disaster, sociopolitical problems and displacement by developmental projects). The second binary logistic model attaches '0' if the reason for migration is either sociopolitical problems or displacement by developmental projects and takes value '1' to climate-induced migrants. The likelihood of fitting a forced migrant has been studied by controlling different independent variables. These are demographic (age and sex), social (religion and caste) and economic (monthly household consumption expenditure class) parameters. The likelihood of fitting a forced migrant in the first model and climate-induced migrant in the second model by different socio-economic classes has been identified with the help of odds ratio resulting from the logistic regression model. The model specification is as follows:

$$M_i = \alpha + \beta_1 \text{age cohorts} + \beta_2 \text{gender} + \beta_3 \text{religion} \\ + \beta_4 \text{caste} + \beta_5 \text{economic status} + \mu_i$$

where $M_i$ is the migrant, $\alpha$ is constant term, $\beta_1$ to $\beta_5$ are the regression coefficients associated with the independent variables and $\mu_i$ is the error term.

The results of the two economic models are presented, respectively, in two different columns in Table 12.5. The table gives the result of odds ratios derived from the two logistic regression models. In the third column, the results of the first regression model have been presented, where we have compared forced migration with the other types of migrations in India. It has been recorded that person aged more than 60 is 2.91 more prone to be forced migrants than a child below aged 14. Females are prone to be associational migrants with their spouse with the likelihood of 2.4. Muslims and other religious believers are more probable to become forced migrants than Hindus. The probability of becoming a forced migrant is more among Scheduled Tribes than any other caste. The likelihood of attractive forced migrant reduces among higher MPCE classes.

The fourth column presents the probabilities of being a forced migrant by natural disaster conditioned on different factors discussed in the second regression model. The results show that migration due to natural disaster increases with growing age. Moreover, females are less likely to migrate due to climate-induced factors. Among the religious groups, Muslims are the highest affected due to a natural disaster in India. Among the different Hindu castes, upper-caste Hindus are more likely to be affected by natural disasters compared to different backward classes. Among backward class Hindus, Scheduled Tribes are least probable of being affected by natural disasters. Most interestingly, the individuals included in the medium MPCE quartile are more likely to be migrated due to a natural disaster in India during 2007–2008. All categories in each variable are found to be highly statistically significant.

In addition to the aforementioned results, we have estimated the marginal effects to confirm the understanding completed on the basis of the estimate of the parameters. The marginal effects are described in Table 12.6. The marginal effects also support the results presented in Table 12.5. Findings suggest that upper-caste Hindus and Muslims, women, aged and poor people are more vulnerable to the effect of climate hazard and migrate more in response to environmental disasters.

**Table 12.5** Results of logistic regression model

| Class | Variable | Odds ratio[a] | Odds ratio[b] |
|---|---|---|---|
| *Demographic indicator* | | | |
| Age cohort | 0–14® | 1.00 | 1.00 |
| | 15–24 | 0.67 | 1.06 |
| | 25–39 | 1.13 | 1.85 |
| | 40–59 | 1.84 | 2.93 |
| | 60 and above | 2.91 | 4.33 |
| Gender | Male® | 1.00 | 1.00 |
| | Female | 2.38 | 0.69 |
| *Social indicator* | | | |
| Religion | Hindu® | 1.00 | 1.00 |
| | Muslim | 1.45 | 2.06 |
| | Christian | 0.23 | 0.13 |
| | Others | 1.87 | 0.37 |
| Caste | Scheduled Tribe (ST)® | 1.00 | 1.00 |
| | Scheduled Caste (SC) | 0.92 | 1.16 |
| | Other Backward Classes (OBCs) | 0.39 | 1.16 |
| | Others | 0.62 | 1.54 |
| *Economic indicator* | | | |
| MPCE quartile | Highest® | 1.00 | 1.00 |
| | Higher | 0.34 | 1.57 |
| | Medium | 1.89 | 3.14 |
| | Lowest | 2.14 | 2.73 |
| Constant | | 0.08 | 0.01 |
| Number of observations | | 22,43,279 | 5,72,254 |
| −2 log likelihood | | 5,62,801.74 | 88,49,004.88 |
| Chi-square | | 25,143.39 | 3,26,365.50 |
| Pseudo $R^2$ | | 0.201 | 0.036 |

*Sources* Computed from unit level data of NSS 64th rounds
*Note*: ®: Reference category
[a]Dependent variable: types of migrants '0' = non-forced migrants and '1' = forced migrants
[b]Dependent variable: types of forced migrants '0' = other types of forced migrants and '1' = natural disaster-induced forced migrants

## Conclusion

Given the complexity of the association between environmental change and migration, it is difficult to capture climate-induced migrants from secondary data sources. This is mainly due to two reasons: (a) non-availability of comparable data for earlier periods and (b) since migration is an adaptation strategy in itself; therefore, rational

**Table 12.6** Results of marginal effects

| Class | Variable | Marginal effect[a] | Marginal effect[b] |
|---|---|---|---|
| *Demographic indicator* | | | |
| Age cohort | 15–24 | 0.036 | 0.374 |
| | 25–39 | 0.011 | 0.364 |
| | 40–59 | 0.254 | 0.432 |
| | 60 and above | 0.021 | 0.762 |
| Gender | Female | 0.002 | 0.216 |
| Social indicator | | | |
| Religion | Muslim | 0.024 | 0.472 |
| | Christian | 0.043 | 0.198 |
| | Others | 0.002 | 0.035 |
| Caste | Scheduled Caste (SC) | 0.042 | 0.016 |
| | Other Backward Classes (OBCs) | 0.054 | 0.516 |
| | Others | 0.032 | 0.094 |
| *Economic indicator* | | | |
| MPCE quartile | Higher | 0.009 | 0.087 |
| | Medium | 0.039 | 0.914 |
| | Lowest | 0.214 | 0.373 |

a Dependent variable: types of migrants '0' = non-forced migrants and '1'= forced migrants and b Dependent variable: types of forced migrants '0' = other types forced migrants and '1'= natural disaster induced forced migrants. All categories in each variable are found to be statistically highly significant

*Sources* Computed from unit level data of NSS 64th rounds

people want to migrate towards the places where employment opportunities are better. Despite these difficulties, the paper tried to analyse the facets of forced migration on the light of climate-induced forced migration. Also, the analysis checked if there any vulnerability exists with this type of migration using recently available secondary data. We see that the percentage of forced migration in India is negligible in terms of total migration but it is growing and has importance in the development perspective. An attempt has been made to identify the population groups which are the most vulnerable to migrate due to climate-induced factors.

In this respect, first, we discussed the distribution of migrants according to the reasons for migration in India. Then, we presented regional variation in forced migration and socio-economic profiles of the forced migrants. Finally, we presented the results of the binary regression models which identify the vulnerability of different forced migrant groups due to climatic factors by controlling for the different migrant characteristics.

Regional analysis of forced migrants shows that the highest percentage of forced migration has been observed in the north-eastern region mainly due to political or social troubles. We have seen that forced migrants are found in the north-eastern region and other states as marginalized populations like Scheduled Tribes and Scheduled Castes populations are higher there. Also, these people have been forced to migrate due to building up developmental projects. This may have an effect on the social cohesion of these regions by enhancing ongoing sociopolitical tensions in these states in the form of Maoist conflicts. Therefore, there is a need to study region-specific forced migration policy. As per expectation the distribution of forced migrants across different socioeconomic groups clearly recognizes its patterns and hence raises the issue of vulnerability associated with lower MPCE classes people. Policies should be intended at these population groups seeing different regional extents of forced migrations, especially migration due to development-related projects. We have more strong evidence of this class selectivity when we employ our regression model.

Regarding environmental disaster-induced forced migration, the current study shows that population living in hilly, coastal and desert states are mostly affected by climate change. The policies to counter the impact of climate change need to focus on these states to reduce climate-induced vulnerabilities of people. The high percentage of climate-induced forced migrants found in the coastal regions may raise the issue of future vulnerability of the population in the coastal states due to increase in sea level.

The findings of empirical model clearly show that there is a clear class-based differentiation among forced migrants in India. Specific classes are significantly associated with forced migration. Male and elderly are found to be mostly migrating as a result of climate-induced forced migration. Muslims are more prone to be victims of climate change compared to other religious groups in India. However, while marginalized social classes like Scheduled Caste and Scheduled Tribes are most likely to be migrated due to developmental projects, they are less likely to become sufferers of forced migrations due to climate change. This may be due to the fact that these social groups use some indigenous techniques that provide them warning before such natural calamities. While poor people are also most likely to be affected by developmental projects in the country, people belonging to the medium MPCE quartile are most affected due to climate change. It may be concluded that policies should be aimed at these population groups who are suffering the most due to different dimensions of forced migration, including climate-induced forced migration, which has recently been included as the reason for migration in the secondary data sources of migration in India. Further research in this direction is necessary to form a concrete policy to tackle the future problems most likely to be faced by the country as climate change intensifies.

## References

Black, R. (2001). *Environmental refugees: Myth or reality*? Working paper no 34, New issues in refugee research. United Nations High Commissioner for Refugees.

Boas, I. (2012). *Climate change migration (India)*. Berkshire Publishing Group. pp 61–64. The Berkshire encyclopaedia of sustainability: china, India, and east and southeast Asia: assessing sustainability, US.

Castles, S. (2002). *Environmental change and forced migration: Making sense of the debate*. New issues in refugee research working paper 70. United Nations High Commissioner for Refugees, Geneva.

Census of India. (2001). *India D-series, migration tables*. Registrar General and Census Commissioner, India.

Chandrasekhar, S., & Sharma, A. J. (2014). *Urbanization and spatial patterns of internal migration in India*. Indira Gandhi Institute of Development Research, Mumbai.

Dallman, I., & Millock, K. (2012). *Climate variability and internal migration: A test on Indian inter-state migration*. Paper presented at the ENMRDTE Pre-conference on Migration and Environment, Clermont-Ferrand, France. October 17, 2012.

Deshingkar, P., & Start, D. (2003). *Seasonal migration for livelihoods in India: Coping, accumulation and exclusion*. Overseas Development Institute.

El-Hinnawi, E. (1985). *Environmental refugees*. United Nations Environment Programme, Nairobi.

Harris, J., & Todaro, M. P. (1970). Migration, unemployment and development: A two-sector analysis. *American Economic Review, 60*(1), 126–142.

Hugo, G. J. (1996). Environmental concerns and international migration. *International Migration Review, 30*(1), 105–131.

Jacobson, J. (1988). *Environmental refugees: A yardstick of habitability, World Watch paper no 86*. World Watch Institute.

Kundu, A., & Gupta, S. (2001). Declining population mobility, liberalisation and growing regional imbalances. In Kundu, A. (Ed.), *Inequality, mobility and urbanization*. ICSSR, Manak.

Lusome, R., & Bhagat, R. B. (2006). *Trends and patterns of internal migration in India, 1971–2001*. Paper presented at the Annual Conference of Indian Association for the Study of Population (IASP), Thiruvananthapuram. June 7–9, 2006.

Massey, D. S., Axinn, W. G., & Ghimire, D. J. (2010). Environmental change and out-migration: Evidence from Nepal. *Population and Environment, 32*(2–3), 109–136.

McLeman, R., & Hunter, L. M. (2010). Migration in the context of vulnerability and adaptation to climate change: Insights from analogues. *Wiley Interdisciplinary Reviews: Climate Change, 1*(3), 450–461.

Myers, N. (1993). Environmental refugees in a globally warmed world. *BioScience, 43*, 752–761.

Myers, N. (1997). Environmental refugees. *Population and Environment, 19*(2), 167–182.

Myers, N. (2002). Environmental refugees: A growing phenomenon of the 21st century. *Philosophical Transactions of the Royal Society, 357*(1420), 609–613.

Myers, N. (2005). *Environmental refugees: An emergent security issue*. 13th Economic Forum, Prague, 23–27 May. Accessed on July 10, 2009. http://www.osce.org/documents/eea/2005/05/14488_en.pdf.

Myers, N., & Kent, J. (1995). *Environmental exodus: An emergent crisis in the global arena*. The Climate Institute.

NSSO. (2001). *Migration in India 1999–2000*. National Sample Survey Office, Ministry of Statistics and Programme Implementation, Government of India, New Delhi.

NSSO. (2010). *Migration in India 2007–2008*. National Sample Survey Office, Ministry of Statistics and Programme Implementation, Government of India, New Delhi.

Olsson, G. (1965). Distance and human interaction: A migration study. *Geografiska Annaler Series B Human Geography, 47*(1), 3–43.

Panda, A. (2010). Climate refugees: Implications for India. *Economic and Political Weekly, 45*(20), 76–79.

Semple, E. C. (1911). *Influences of geographic environment*. Henry Holt.

UNDP. (2009). *Human development report 2009: Overcoming barriers: Human mobility and development*. United Nations Development Programme.

# Chapter 13
# Vertical Farming: Resilience Towards Climate Change

**Sonia Maheshwari**

**Abstract** Climate change is nearly irreversible aspect of futuristic progression. Research undertaken globally to understand the implications of greenhouse effect in past few decades suggests that the accumulation of greenhouse gases (GHGs) including major concentration of carbon dioxide ($CO_2$) is responsible for global warming and other significant climatic changes. Climate change is responsible for changing weather patterns and the resultant impact in the form of rising surface temperatures, worsening droughts and other forms of natural disasters. The social and economic impacts of increase in frequency of natural disasters are adding to the existing set of challenges that includes distressed population, wretched infrastructure and appalling agricultural and food systems. In addition, climate change impacts are increasingly adding to the risk of Global Food Insecurity that in turn may require early warning systems and effective development programs for increased resilience of urban settlements. United Nations' Report released in 2015 (UN in Department of Economic and Social Affairs, 2015) projected the world's population count to touch 9.7 billion by the year 2050. With worldwide acceptance for increasing population, changes in climate patterns and urbanization trends are already leading to alarming situations. Referring to basic human needs, food security is under threat due to series of interrelated concerns including but not limited to depleting water aquifers, top soil erosion and increasing urban sprawl on arable land. As much as is the need to provide shelter to increasing population, so is the need to arrange for their nutrition. As the scarcity of land is creating challenge to provide for the essential requirements of humans, therefore, it is desirable to develop innovative, sustainable and integrated solution finding approaches for built environment and farming activities. The focus area is to develop consistent, high-quality production alternatives for farming in built environment, with optimized usage of resources. In the defined scenario, the model of vertical farming has evolved as a multidimensional approach to support urban farming as well as other global challenges, viz. urban heat island effects, carbon emissions, depletion of non-renewable resources and waste management, to name a few. A newer concept in India, but well adopted in countries like Japan, Korea and Netherlands, vertical farming has opened up new avenue to address many challenges

S. Maheshwari (✉)
Aayojan School of Architecture, Jaipur, Rajasthan, India
e-mail: som@aayojan.edu.in



A. Kateja and R. Jain (eds.), *Urban Growth and Environmental Issues in India*,
https://doi.org/10.1007/978-981-16-4273-9_13

through one solution (Maheshwari and Garg in Sustainable Cities: A New Outlook; Reflections from Past to Reinforce the Future. Roorkee, 2015). The Vertical Farm System developed by Dickson Despommier in 1999 was based on conventional techniques of hydroponics and aeroponics to produce farm yields at a consistent rate (Bourne in Nat Geogr 215(6):26–59, 2009). The paper intends to review and discuss the impacts of climate change and usefulness of vertical farming as a suitable technique to meet the global food requirements with energy efficiency in indoor environments and suggestive waste management approaches for built environment.

**Keywords** Climate change · Food security · Vertical farming · Urban agriculture

## Introduction: Climate Change and Impacts

Anthropogenic activities have created the alarming situation, changing climate patterns being most visible consequence. Climate change is a challenge, being directly responsible for increase in frequency of natural disaster, for developed as well as developing nations. The rising diversity of challenges due to global climate change calls for proactive initiatives to develop and implement curative measures for building up resilience in future developments.

Research studies support the urban population projection around the globe till 2050 with about 60% of total increase to be concentrated only in Asia. Further suggesting the trend, it is estimated that the half of the projected growth will occur in Asian cities that have fewer than 500,000 inhabitants today (2012). In addition, Asian cities house larger population of vulnerable urban poor, surviving on illegal encroachment of vacant but hazard-prone, ecologically sensitive land parcels within built environment, mostly left undefined in planning schemes (Banerjee & Adenaeuer, 2014). Subsequently, this concentration of urban population faces scarcity of physical and social amenities, seasonal and perennial water scarcity being predominantly pressurizing today, as well as in future based on projected water shortage due to climate change by 2050 (Despommier, 2010). Above mentioned, when put together, is a complex set of issues to address that calls for a practical model to implement whether it is land-use planning, water management or the food security, to mention a few.

Global climate change is adversely affecting the environment and thereby the well-being of humans. The uncertainties and threat due to increased disaster frequencies is causing concern for safety and security of resources inclusive of food security. Scientists, researchers and urban professionals are exploring systemic and secure solutions to combat the challenges as a result of fast pace of climate change patterns along with urbanization trends. Environmental concerns focusing on land as a resource in rapidly urbanizing areas support the concept of vertical farms in such urban areas. Acceptance for this new food production model over traditional method will give way to solutions for food security problem as well as will develop the interrelationship of diverse professionals for developing sensitive agro-ecosystems in new kinds of vertical building structures. Vertical farming offers the opportunity to rework the resources required for food production thereby reducing the demand pressure on land for agricultural activities. The shift in farming approaches supports the probability of defining new usage of farmlands for multiple ecological functions that would be beneficial for environment and humanity (Despommier, 2011).

At this point, it is essential to bring to notice that the deliberations at Rio Summit led to the formation of United Nations Convention to Combat Desertification (UNCCD). The agenda for the newly formulated UN Convention was inclusive of issues pertaining land degradation in the most vulnerable ecosystems and affected populations, across the world. UN Secretary-General Ban Ki-moon stated in his speech at concluding ceremony for ninth session of the Conference of the Parties (COP 9), UNCCD, organized at Buenos Aires, Argentina, 'These disasters are exacerbated by desertification, land degradation and drought, which, are among the most pressing global environmental challenges of our time, threatening to reverse the gains of sustainable development that have emerged in many parts of the world over the past few years'.

## Building up Urban Resilience

According to World Bank Report, resilience may be defined as the resisting capacity of a system, community or society developed through considerate approaches for nullifying the ill effects of disasters through absorption, accommodation and recovery in a timely and efficient manner, including the preservation and restoration of its essential basic structures. The global challenges worldwide are consequences of uncontrolled anthropogenic activities resulting in alarming increase in natural disasters. The extreme weather conditions are resulting into continuous threat of hazards and disasters. The scenario being realized world over, nations are now building up strategies to build up urban resilience in urban development strategies for the settlements.

## Urban Agriculture: Vertical Farming as a Resilient Model

Further, research studies bring out severe impacts of climatic changes on agricultural production leading to the preparedness for food security (Kamonpatna & Anuntaboranich, 2013). Regarding climate change impacts, Lobell et al. state that in the next two decades owing to increasing temperatures and thereby declining precipitation, especially at semi-arid regions, there is a risk of yield reduction for corn, wheat, rice and other primary crops (Smit, Nasr & Ratta, 2001).

To address the challenge of food security, urban agriculture has come up as a sustainable approach with manifold benefits for the future cities. Apart from food security, the approach is helpful in disaster risk reduction also. It has been observed that the ecosensitive sites within the cities are left unbuilt and thereby unattended. As a result, in most of the cases, these vulnerable land pieces are quite often occupied by the urban poor, especially in India. The occupancy of vulnerable sites always keeps alive the probability of disaster risk, and it is after impacts, both social and economic. For other segment of the citizens, apart from infrastructure and life losses, the disaster risk brings in the increased probability of depletion of natural resources, food and water being most essential. Urban agriculture as a resilient prototype is suggestive for these unoccupied, ecosensitive land masses. The farming activities will maintain the ecosystem and restrict the occupancy of land by urban poor.

The concept of urban farming is based on the fundamental concept of reducing distance between product and consumer. Instead of trucking the farm produce from distant areas, grow it in nearest vicinity to use the readily available crop produce. With limited availability of arable land within the urban limits, the alternate approach of vertical farming that uses vertical built surfaces, walls and floors for growing produce has come out. Vertical farming is an emerging option of future for the urban resource management in cities. Dickson Despommier, an American ecologist, revived the idea of vertical farming in 1999. Despommier proposed the technology-driven vertical farming model over the conventional farming to combat climate change effects along with the high costs and resource requirement of the conventional farming. The supporting projections highlighted the possibility of producing half of the world's food requirement through vertical farms in the time span of next 50 years (Maheshwari & Garg, 2015).

Food, being one of the three essential necessities to sustain life, needs immediate action to meet the present demand as well as ensure surplus to secure the future. The impact of climate change and usefulness of vertical farming as a resilient approach towards climate change impacts needs to be understood by policymakers. The projected increase in urban population till 2050 is envisaged of creating pressure on agricultural land to meet the food requirements for the urban areas. Vertical farming technique assures to promulgate as a novel method of meeting the demands of food security in an urbanizing world.

## Vertical Farming

The increasing urban population is bringing continuous transformation in urban morphology of settlements. The pace of increase is separating apart the functional compatibility of development approaches to achieve sustenance in reference to limited natural resources available on planet earth (Maheshwari & Garg, 2015). Till the beginning of twenty-first century, modern urban settlements were following the explicitly linear resource metabolisms to meet the resource demands. Importation of food produce, freshwater, energy and other resources in demand from great distances, high consumption and then swift disposal as sewage or solid waste (that the natural world cannot easily process) was the fundamental principle for urban management. Likewise, the conventional, high-yield farming methods adopted to support food demands for increasing population are also derived upon their insatiable consumption of limited reserves of freshwater, fossil-fuel energy and soil.

Vertical farming offers sustainable approach to address food security for the ever-increasing urban population world over (Al-Kodmany, 2018a, b; Corvalan et al., 2005; Healy et al., 2013; Thomaier et al., 2015). In principle, the concept is simple; rise up than sprawling unlimited (Despommier, ). The land scarcity for meeting the requirements of growing urban population is being addressed with vertical urban growth model. Similarly, model of vertical farming is apparently favourable for scarcity of arable land and supporting resources.

A Vertical Farm System is configured by stacking farm trays on top of one another, within a skyscraper greenhouse or on its vertical surfaces, over horizontal branching. Vertical Farm System runs on modern techniques like hydroponics and aeroponics for consistent yields production (Bourne, 2009). Energy requirement for functional operations in vertical farms is principally similar to glass house effects, even with artificial lighting over natural sunlight. Optimum range of temperature and humidity levels is essential requisite for each crop being grown. Additional quantities of carbon dioxide are desirable to further increase the crop growth and yield rates. Water used for irrigation of Vertical Farm System installations is treated beforehand to remove traces of particulate, fluoride and heavy metal contaminant (Maheshwari & Garg, 2015).

The literature review about vertical farming refers to three categories (Despommier, 2014; Touliatos et al., 2016, Muller, 2017). The first category includes tall structures, provided with several levels of growing beds, often lined with artificial lights. These modestly sized structures can be created as urban farm in new as well as old buildings, being reused for agricultural activities (Despommier, 2013). The second category includes vertical farms created at a level using the rooftops of old and new buildings supporting commercial, residential and other functions as well (Despommier, 2014; Touliatos et al., 2016). The third category of vertical farm is created on the vertical façade of the multi-story building. In the past few decades, visionary proposals for vertical farms have been considerably increasing. However, it is important to note the connection between the three types of vertical farms and the success of their technologies that will pave out the way for the skyscraper farming in future (Despommier, 2014).

## Vertical Farming: Chronological Review

Although Despommier proposed the concept of vertical farming at the end of twentieth century (Despommier, 2010), the conceptual model for vertical farming dates back to the period of River Valley Civilizations, in reference to available records. Hanging Gardens of Babylon, built around 600 BC, are worth an example to quote. After a major gap of centuries, the references are available for 1915 when Gilbert Ellis Bailey coined the term 'vertical farming' and wrote a book '*Vertical Farming*'. His arguments about vertical farming are somewhat similar to hydroponic farming in a controlled vertical environment. In addition, Ellis also pointed out the benefits, economic and environmental, for the farming model suggested by him. Further, in the early 1930s, William Frederick Gericke from the University of California at Berkley pioneered hydroponic farming in 1930. Later in 1980s, Åke Olsson, a Swedish ecological farmer, proposed vertical farming in cities for producing vegetables. Researchers refer to a spiral-shaped rail system for vegetation, being invented by Olsson (Al-Kodmany, 2018a, b; Corvalan et al., 2005; Healy et al., 2013; Thomaier et al., 2015).

## Relevance of Vertical Farming

The relevance of vertical farming in future cities is on account of its manifold benefits in reference to global challenges like climate change, food security and increasing urban density. The physical rationale of vertical farms is similar to vertical growth of cities, the reason being common: land scarcity and land values. The fundamental of vertical farm is simple: in spite of less land being available, produce more food with appropriate alternatives (Muller et al., 2017; Touliatos et al., 2016).

Vertical farming model ensures efficient and sustainable production of healthy food. The future speculation of cities with vertical farms assures restoration of ecosystem with opportunities to access fresh air, besides employment opportunities and enhanced economy (UN, 2017). The vertical farm designs lead to compact and self-sustained ecosystems that cover multiple functions, from food production to utilization of treated grey water, in case of occupied buildings as site for the same (Al-Kodmany, 2018a, b). The change in approach as compared to traditional farming allows high probability of greater yields with judicious usage of resources. The carefully designed mechanisms to regulate the crop environment support the vertical farming crops, keeping the same less effected due to impact of adverse weather conditions (Katz, 2013). Furthermore, the locational benefits of vertical urban farms result in significant reduction of carbon emissions as well as the travel costs, by bringing the farm produce and consumers close enough (Mukherji, 2010).

The technological advancements have been able to address diverse urban issues holistically with hopeful assurance for future cities to achieve development sustenance. Controlled environment agriculture similar to well-established Greenhouse technologies guarantees safe food production year-round in urban centres. By utilizing the high-rise building stacks in an integrated and well-engineered manner, agricultural footprints can be reduced significantly, and the vertical farm concept can be made applicable to every urban centre, regardless of its geographical location and climate type (Depsommier, 2011).

## *Climate Change*

The global research on climate change and resultant issues is suggestive that the climate change is going to adversely affect the agricultural produce. It is being estimated that for every 01 degree of increase in atmospheric temperature, 10% of the land, now being used for growing food crops, will become unusable as a consequence (Despommier, 2011). Soil erosion due to floods and loss of nutrients of soil due to the droughts contributes as consequences of climate change that significantly reduces the potentiality of farm lands being used to grow food, world over.

Climate change impacts across the globe account for decrease of arable land, which in turn have affected the world economy severely. The decrease in arable land is a consequential damage due to increased frequency of natural disasters like flooding, hurricane, storms and drought (Despommier, 2010; Muller et al., 2017; UN, 2017; Kalantari et al., 2017a, b).

World over, climate change impacts on arable land are leading to increasing shortage of cultivable lands for farming (Al-Kodmany, 2018a, b). In addition to the above, traditional farming practices are dependent on fossil fuels. Further, the food miles are proportionate to the increasing distance between farms and cities due to global urbanization. The greenhouse gas emissions from food transport and agricultural activities significantly contribute to climate change.

## *Food Security*

Urban Scientists have declared the twenty-first century to be the 'Urban Century' based on demographic projections about increasing urban population in the coming decades. Food security, as a result, has become an increasingly important issue for nations across the world. At the same time, climate change impacts being studied by land specialists (e.g. agronomists, ecologists and geologists) suggest the rise in deterioration of arable land (Corvalan et al., 2005; Healy, 2013; Thomaier, 2015). For these reasons, the speculations suggest that food demand could exponentially surpass supply, adding to global issue of hunger and poverty.

The United Nation (UN) predicts that by 2050, the food requirement will increase by 70% to meet the demands of expected 3 billion more inhabitants worldwide (UN, 2017). In addition to the increase in demand of food produce, the food prices may further increase due to increasing prices of draining agricultural resources like oil, water and arable land energy (Despommier, 2010, 2013, 2014; Touliatos, 2016). The increasing urban sprawl continues to eat up more and more farmland. Besides, conceptual framework for urban agriculture has never been thought of due to lack of agricultural land as well as high values of urbanizable land parcels available in cities. In addition to above, traditional farming is also prone to attacks by plant pathogens and similar, that leads to lower levels of yield for food grains and vegetable crops, worldwide (Despommier, 2011). Urbanization is accepted as a global phenomenon which is inevitable. Concept of Vertical Farming offers a futuristic model that may combat multiple global challenges. The model is apparently a promising concept for the benefit of mankind and needs to be implemented at urban scale.

## *Urban Density*

Increasing urban populations are dependent on urban land availability for procuring their basic need of shelter and infrastructure, both physical and social. The increasing urban sprawls are taking away agricultural and forest lands to meet the high demands generated by increasing urban density. In this scenario, where traditional farming on horizontal land parcels is evidently facing threat from urbanization, vertical farming if reviewed over traditional farming supports the availability of more free land for essential urban provisions like housing, infrastructure services and social amenities (Despommier, 2010). Research revealed that designated urban farm lands lead to decreased population density with longer commutes and higher energy usage which generates more air and water pollution.

Referring to space efficiency of vertical farms, Despommier suggested that to produce a crop yield equivalent to 971.2 ha (2400 ac) of conventional horizontal farming, a 30-story building (about 100 m high) with a basal area of 2.02 ha (5 ac) would be required. The proposed configuration suggests that the one high-rise farm would be equivalent to 480 conventional horizontal farms (Cho, 2011; FAO, 2013).

## *Health and Hygiene*

Traditional farming practices, especially in developing nations, often adapt to chemical additives like fertilizers and pesticides for assuring high quantity of produce and thereby commercial gain. The adaptation neglects the harmful consequences from usage of chemical compounds, both in respect of humans and other animals along with natural environment and its features like soil and water (Despommier, 2010, 2013; Touliatos, 2016). The practices being adapted and followed to meet the pressure of increasing (urban) population are not consciously analysed for side effects being caused like depletion of qualitative nutrients and contamination of top soil and use of excessive freshwater in societies where drinking water is deficient. The World Health Organization referred to risks involved in agricultural practices being followed, especially in developing countries. The regular usage of unprocessed animal waste as crop booster increases the threat of attracting flies and thereby transmission of disease vectors to plants (Al-Kodmany, 2018a, b). This affects people's health adversely. Besides, the uncontrolled usage of chemical fertilizers and pesticides has adversely affected the human bodies as the traces of unwanted chemicals are getting deposited in human bodies due to consumption of artificially supported food produce. The alarming increase of chronic diseases, irrespective of age and lifestyle being followed, is a passive but visible impact of techniques in traditional farming. Contrary to the above, farming for food crops in a controlled environment requires negligible quantity of chemical compounds like pesticide and herbicide, thereby reducing the probability of generating polluting agricultural run-off (Cho, 2011). According to Renee Cho, controlled environment considerably reduces the possibility of crops getting affected by pests, pathogens and weeds (Cho, 2011).

The excessive use of fertilizers and pesticides affects the soil quality gradually and moreover creates a threat of continuous contamination of surface and underground water resources. Rainfall washes away the traces of excess fertilizer into on surface water bodies (e.g. rivers, streams and oceans) and simultaneously supports the percolation of these chemicals in dissolved form. A biochemical process called eutrophication occurs in surface water bodies due to high concentration of chemical nutrients which could disturb the ecological equilibrium (Cho, 2011). Further, vertical farming techniques require less water, about 1/10th of the quantity required for traditional farming. The optimized water usage is owed to high-tech irrigation model offering precision and efficient scheduling for irrigation (Cho, 2011; Wood, 2001). The high-tech model of vertical farming also utilizes treated grey water, further reducing the need of freshwater for crop production unlike traditional farming. This can significantly affect the dire situation faced by urban areas across the globe since potable water demands are consistently increasing as the urban population grows. Traditional farming techniques supposedly depend on two-third of the available world's freshwater world over. However, consistently expanding urban areas are creating excessive water demands, and the alarming rate of consumption thereby affects the water requirement for traditional farming. In addition, water scarcity is likely to become severer owing to global temperature increase leading to proliferation of droughts (Cho, 2011). Passive impact on population and their health due to shortage of freshwater may also be addressed well by vertical farming model.

## *The Ecosystem*

The uncontrolled urban development and excessive usage of natural resources have created a severe threat for the well-being of humans as well as the ecosystem on planet earth. The researchers have identified that anthropogenic activities like industrialization, transportation, construction and surprisingly the farming activities are collectively responsible for the degradation of earth's ecosystem. The nations across the globe are working together to attain the compliance of Sustainable Development Goals for prosperous and progressive future. Apart from other domains, the traditional agricultural practices have also been instrumental in adversely affecting the natural ecosystems. The increasing world population generates the demand for food supply which in turn creates the requirement for more arable land.

According to Dickson Despommier, the hardwood forests are being encroached like anything by agricultural activities. This encroachment on ancient ecosystems is further adding to climate change. The limited resources available are losing out on meeting the demands of increasing population. The vertical farms in urban areas may combat the adversity and help reduce the local temperatures as well. In addition, the reduced activities of cutting, collecting and transporting the harvest will further reduce the agricultural footprints. The new ecosystems in the form of vertical urban farms may aid the restoration of natural ecosystems on planet earth.

Despommier suggested that if cities may employ vertical farms to produce food crops, the effort might in turn help to minimize the impact of carbon emissions. The vertical farming promises to be one of the best technological innovations capable for bringing multifaceted improvement in the condition of biosphere in coming years (Despommier, 2011). Wood, et al. added to above stating that 'The best reason to consider converting most food production to vertical farming is the promise of restoring the services and functions of ecosystems' (Wood, 2001).

## Hi-Tech Methods of Vertical Farming

Sustainable Development Guidelines have generated hope amidst nations about achieving the development goals while upgrading the urban resilience. The advanced researches in various domains with special mention of futuristic agricultural techniques are being carried out successfully. Vertical urban farming is being accepted and being upgraded consistently as sustainable approach for food production. The advanced techniques devised for vertical farming could provide good yields with optimized usage of land, water and other resources (Kalantari, 2017a, b). These vertical farms are designed with optimum space layout, getting exposure to required quantity of sunlight and are supported with precisely devised mechanism for nutrient and temperature monitoring.

Researchers have been working to deploy these techniques world over, customization being required according to geographical location and subsequent climate type. In addition to above, local food crops need to be experimented upon for developing the seeds and saplings to be used in farming. The concept of vertical farming is revolutionary and will change the urban ecosystems if implemented in totality. The issues like global warming, extreme weather conditions and food security could be addressed by local communities in urban regions bringing a participatory approach in growth and development. The ecomodel would generate opportunities for employment and entrepreneurship to add to the economic benefits of vertical farming, besides reduction in transportation and storage costs for food yields. New technological systems for hydroponics, aeroponics and aquaponics along with associated technologies are being consistently upgraded, to offer diverse range of food production opportunities (Table 13.1).

**Table 13.1** Technological systems for vertical farming

| Technique | Key attributes | Advantages |
|---|---|---|
| Hydroponics | Farming bed is soil-less, and water is used as growing medium | Water medium supports nutrition-based plant growth; no traces of soil-borne cultivation problems; suppresses the use of fertilizers or pesticides |
| Aeroponics | Controlled variant for hydroponics; plant roots are kept hydrated by spraying nutrient solution or mist | In addition to benefits of hydroponic technique, aeroponics requires less water |
| Aquaponics | Aquaponics integrates aquaculture (fish farming) with soil-less growing beds similar to that used in hydroponics | Principally configured on interdependent relationships between plant and animal kingdom; nutrient-rich waste from fish habitat created in hydroponic grow beds is used as nutrient for plants growing in the hydroponics beds which in turn cleans water for fish habitat |

*Source* Al-Kodmany (2018b)

## Benefits of Vertical Farming

The benefits of vertical farming are inclusive of direct and indirect impact as mentioned below:

- Integration of vertical farms in buildings has led to design and technological innovations that may pave way to ecotourism.
- Vertical farms developed as the innovative technological marvels may act as incubators for skill development and research studies.
- Vertical farming in and on the buildings offers the advantage of minimising heating and cooling loads for indoor thermal comfort.
- Judicious usage of waste generated by building operations such as heat, grey water and solid waste in vertical farming bridges the gap in desired symbiotic relationship between nature and built environment.
- Controlled environment in vertical farms allows year-round crop production bringing an equivalence of 1 indoor acre to 4–6 outdoor acres or more, depending upon the crop (e.g. strawberries: 1 indoor acre = 30 outdoor acres).
- Vertical farming remains unaffected with extreme weather conditions and disaster impacts caused due to droughts, floods and pests.
- Vertical farming is used for organic crop yield along with reduced dependency on fossil fuels too.
- Vertical farming model is progressive to holistically meet the requirements of three pillars of sustainability, i.e. environment, society and economy.

Vertical farming supports food cultivation in close proximity to city centres. The location in cities brings two-fold benefits: firstly, reduction in use of fossil fuels for growing the crops in comparison with traditional farming; secondly, reduction in need of transportation for food produce from farm to fork. This would enhance the sustenance levels for vertical farming. Apart from being an added feature in new constructions, vertical farming is sprouting in urban areas across the international lands, as a retrofit measure for some of the old and unused buildings.

Vertical farming is still far from being commercialized, and it will take time to develop the level for growing enough to feed increasing urban population, may be in the next few years. With advancements in technology, vertical farming will become more cost effective and feasible. It is observed that around 30% of farmers in developing countries are facing food scarcity owing to extreme weather conditions, insect and pest attacks, poor soil quality and unavailability of consistent irrigation facilities, manhandling and accidents like fire, etc. Scholars suggest that added ill effects of climate change may adversely affect the impoverished communities by decreasing small-scale crop production. Simultaneously, the adverse impacts of climate change will affect large-scale production via traditional farming too bringing a reduction in crop produce and thereby contributing to a global increase in food prices. The significant reduction worldwide, in the production of corn, wheat and rice, the staple food crops, supports the projections for global food market scenarios (Smit et al., 2001).

## Limitations at Present

The main challenge till day is the limited technical know-how. The expertise to propagate vertical farming model further would need to develop skilled labour for various activities like (FAO, 2008).

- harvesting the crop.
- maintaining the indoor operations.
- managing the defined farming strategies.
- addressing uncertainties associated with design and detailing.
- construction and occupancy stage of vertical built masses.

As a developing concept, vertical farming may get critical reviews for being expensive to experiment; useful but not mandatory to adapt, with reference to the establishment cost, in particular. However, the labour cost will still be manageable as implementation cost for vertical farm is still disproportionately on the higher side. The primary contributor to high cost of implementation of vertical farms is the construction process of building to be used as Green House System. The technological inputs for building and finishing material along with operational systems supported by fittings and fixtures tremendously add up to the cost. In addition, artificial lighting is used 24 × 7, throughout the year, along with technology-driven Monitoring and Management Systems, leading to higher rates of power consumption and therefore the high cost factor. As a cost cutting measure, the existing buildings may be substantially retrofitted to run vertical farming, thereby saving upon the initial construction cost. Further, the buildings must be equipped to generate their own energy with reliance on resources like solar or wind power, what may be appropriate to site selected (Maheshwari & Garg, 2015).

## Discussion and Results

Since advent of civilization, the apparent victory of man on nature has slowly but steadily generated alarming global concerns like dying ecosystems, extinction of species, depletion of fossil reserves, polluted urban environment, global warming and climate change along with anticipated crisis of food and freshwater (Maheshwari & Garg, 2015). Climate change impacts on agricultural produce and thereby the food security may be varied from one region to another, as a resultant of technology being used. Technological excellence may contribute in the productivity of a farm, but the very fact that arable land is decreasing both as an impact of disaster as well as ills of urbanization is an existing challenge, for future generations (Kamanopatana, & Anuntavoranich, 2013; FAO, 2012).

Referring to development in Asia and other parts of the developing world, rapid pace of urbanization is unable to meet the infrastructure demands, especially referring to second- and third-tier cities. In addition to high pace of urbanization, the climate change impacts are responsible for increased frequency of natural disasters too. The available development models being limited, city authorities face the pressure of defining effective development strategies for resilience and sustenance. Furthermore, availability of natural resources like land, water and clean air, being imperative for development, needs careful utilization for future scenarios. The survival of humanity essentially calls for resilient development approaches to achieve development sustenance.

## Conclusion

The research studied the effectiveness of conceptualizing vertical farming as a resilient development approach and concludes that vertical farming concept can be adopted in urban planning. The preference is for a system that could serve various income groups in urban population, ensuring better living indices in urban settlements, leading to healthy environment and convenient access to natural products, especially food items.

With the advent of twenty-first century, nations across the world have realized the importance of redefining urban development approaches based on self-sustenance and resilience as achievable targets. In continuation to working models for resilient and sustainable urban development, urban farming through vertical farms has been well acknowledged and is gaining popularity all over the world. As a part of urban farming, rooftop gardens have been in place for quite some time, in many countries, across the globe. However, the efforts and consequent numbers of rooftop gardens for growing food items in the urban areas which are capable of reducing food transportation costs and promoting self-reliance are still very limited. The concept of vertical farming in buildings is still in experimental mode, especially in developing countries due to its high cost of implementation as well as defined need for highly skilled technical expertise. Future research endeavours must address the details of built form design and its holistic integration with vertical farming techniques through research initiatives.

## References


Al-Kodmany, K. (2018a). *The vertical city: A sustainable development model*. WIT Press.

Al-Kodmany, K. (2018b). The vertical farm: A review of developments and implications for the vertical city. *Buildings, Open Access Journal of MDPI*.

Banerjee, C., & Adenaeuer, L. (2014). Up, up and away! The economics of vertical farming.*Journal of Agricultural Studies, 2*(1).

Bourne, Jr., J. (2009). The global food crisis—The end of plenty. *National Geographic, 215*(6), 26–59.

Corvalan, C., Hales, S., & McMichael, A. J. (2005).*Ecosystems and human well-being: Health synthesis*. World Health Organization.

Cho, R. (2011). *Vertical farms: From vision to reality*. State of the Planet, Blogs from the Earth Institute, 13 October 2011. Available online http://blogs.ei.columbia.edu/2011/10/13/vertical-farms-from-visionto-reality/comment-page-1/. Accessed on February 04, 2021.

Despommier, D. (2010). *The vertical farm: Feeding the world in the 21st century*. St. Martin's Press.

Despommier, D. (2011). The vertical farm: Controlled environment agriculture carried out in tall buildings would create greater food safety and security for large urban populations. *Journal of Consumer Protection and Food Safety*. https://doi.org/10.1007/s00003-010-0654-3.

Despommier, D. (2013). Farming up the city: The rise of urban vertical farms. *Trends Biotechnology, 31*, 388–389.

Despommier, D. (2014). *Encyclopedia of food and agricultural ethics (vertical farms in horticulture)*. Springer.

FAO (2008). *The state of food and hunger*. Food and Agriculture Organization of the United Nations.

FAO (2012). *The state of food insecurity in the world*. Food and Agriculture Organization of the United Nations.

Food and Agriculture Organization (FAO) (2013). *Good agricultural practices for greenhouse vegetable crops: Principles for mediterranean climate areas*. FAO; Chapter 15.

Healy, R. G., Rosenberg, J. S. (2013). *Land use and the states*. Routledge.

Kalantari, F., Tahir, O. M., Lahijani, A., Kalantari, S. A. (2017a). Review of vertical farming technology: A guide for implementation of building integrated agriculture in cities. *Advanced Engineering Forum, 24*, 76–91. ISSN 2234-991X.

Kalantari, F., Tahir, O. M., Joni, R. A., Fatemi, E. (2017b). Opportunities and challenges in sustainability of vertical farming: A review. *Journal of Landscape Ecology, 2*, 2.

Kamonpatana, K., & Anuntavoranich, P. (2013). Vertical farming concept in Thailand: Important decision variables. *International Journal of Innovative Research in Science, Engineering and Technology, 2*(12), 7851–7860.

Katz, R., Bradley, J. (2013). *The metropolitan revolution. How cities and metropolitan areas are fixing broken politics and fragile economy*. The Brookings Institution.

Maheshwari, S., & Garg, N. K. (2015). *Sustainable cities: A new outlook; reflections from past to reinforce the future," sustainable built environment 2015*. In National conference organized by Department of Architecture and Planning at IIT Roorkee, Roorkee.

Mukherji, N., & Morales, A. (2010). *Zoning for urban agriculture. Zoning practice 3.* American Planning Association.

Muller, A., Ferré, M., Engel, S., Gattinger, A., Holzkämper, A., Huber, R., Müller, M., Six, J. (2017). Can soil-less crop production be a sustainable option for soil conservation and future agriculture? *Land Use Policy, 69*, 102–105.

Smit, J., Nasr, J., & Ratta, A. (2001). *Urban agriculture food*. The Urban Agriculture Network Inc.

The science of vertical farming: Why farm vertically? (n.d.). Retrieved from Ri Aus Austrailia's Science Channel. http://riaus.org.au/articles/the-science-of-vertical-farming-why-farm-vertically/

Thomaier, S., Specht, K., Henckel, D., Dierich, A., Siebert, R., Freisinger, U. B., & Sawicka, M. (2015). Farming in and on urban buildings: Present practice and specific novelties of zero-acreage farming (Z farming). *Renewable Agriculture and Food System, 30*, 43–54.

Touliatos, D., Dodd, I. C., McAinsh, M. (2016). Vertical farming increases lettuce yield per unit area compared to conventional horizontal hydroponics. *Food Energy Security, 5*, 184–191.

United Nations Department of Economic and Social Affairs (2015). Retrieved from United Nations: https://esa.un.org/unpd/wpp/publications/files/key_findings_wpp_2015.pdf.

United Nations (2017). *World population prospects: The 2017 revision*. United Nations.

Wood, S., Sebastian, K., Scherr, S. J. (2001). *Pilot Analysis of global ecosystems: Agroecosystems* (p. 110). International Food Policy Research Institute and World Resources Institute. Available online http://www.wri.org/publication/pilot-analysis-global-ecosystems-agroecosystems. Accessed on February 5, 2021.

# Chapter 14
# Ecology of Lakes and Reservoirs in Semiarid Regions of Rajasthan

**P. D. Charan and K. C. Sharma**

**Abstract** Agriculture has been the cornerstone of all economic activities in India for centuries which is mainly driven by monsoon. The isohyet map of India shows that the distribution of rainfall pattern is significantly variable for different regions. It is highly influenced by the topography of the regions. Geographically, Rajasthan is the largest state of India having most of its area under arid and semiarid climate. More than 60% population of Rajasthan is inhabited in these regions. The semiarid and arid regions experience extreme disparity in the temperature, high rate of evaporation, scanty and unpredictable rainfall. The lakes and reservoirs of the region are facing severe threats of existence due to multifold pressure. Droughts are very frequent in the state which has shrunken the water level in these water bodies. The restoration and conservation of the lakes and reservoirs of the region through sustainable practices are the need of the time. The present study focuses on the problems of the lakes and reservoirs of semiarid regions of Rajasthan and to draw attention of policymakers and common residents for conservation of these water bodies.

**Keywords** Lakes and reservoirs · Semiarid regions · Restoration · Conservation · Sustainable practices

## Introduction

Clean environment is the basic need for healthy life of living beings. The rapid urbanization has adversely affected the natural resources including water, air and land in the urban region and its periphery. Excessive stress on the natural resources deteriorates their quality. The population of urban region in developed countries is almost at saturation stage, and it is stabilized there, but in developing and underdeveloped countries, the rate of urbanization is very fast. It is reported that the global population will touch the 9 billion mark by the year 2050, with the majority of

P. D. Charan (✉)
Department of Environmental Science, MGS University, Bikaner, Rajasthan, India

K. C. Sharma
Department of Environmental Science, Maharishi Dayanand University, Ajmer, Rajasthan, India

 
A. Kateja and R. Jain (eds.), *Urban Growth and Environmental Issues in India*,
https://doi.org/10.1007/978-981-16-4273-9_14

the residents in the metropolitan regions (Tucci, 2017). It has been predicted that about one billion urban people will have to live in 'water-vulnerable conditions' but if the water potential stress from poor water quality and climate change is added, another one billion people, especially in developing countries, will come under such conditions (McDonald, 2011). The quality as well as quantity of water is directly associated with health, agriculture, energy and biodiversity (Mwanza, 2003). The healthy ecosystem is repercussion of its regular maintenance through ecological services including nutrient cycling, biological diversity, fertile soil, clean water and air and energy transfer at varying trophic levels (Cairns & Pratt, 1995; Costanza, 1997; Costanza et al., 2011; Xu, 2001). It is observed that the water bodies have been severely degenerated due to encroachment of catchment areas and human interferences during last few decades. Water bodies located near urban and semi-urban settlements are critically degraded due to input of human waste and soil erosion, which results in excessive siltation and eutrophication of lakes.

## Water Scenario in Rajasthan

Rajasthan, being a desert State of India, typically evinces high rate of evaporation, extreme disparity in the temperature, scanty and unpredictable rainfall. The average rainfall of the state is 574 mm compared to the all-India average of 1100 mm, and a significant variation is seen across different regions (TERI, 2010). The climatic conditions of the State are significantly different in different regions.

Lakes and reservoirs in arid and semiarid regions of Rajasthan play an important role in maintaining the microclimatic conditions of the region. Under the Ramsar Convention, total 42 wetlands are identified as Ramsar sites in India, out of which, two sites, namely Sambhar Lake and Keoladeo National Park, are pertaining to the state of Rajasthan. According to a report on (National Wetland Atlas, 2011) published by *Space Application Centre* under *National Wetland Inventory and Assessment project*, the Bhilwara district had the highest 6.94% of its total geographic area occupied by wetlands, while Churu had the lowest 0.08% of its total geographical area under wetlands in Rajasthan state. As per the Vision-2045 document published by the Department of Planning, Government of Rajasthan, the status of availability and utilization of water resources of the state is at alarming situations (Table 14.1). It is very serious matter that 104% of groundwater is explored and utilized annually, which means that the rate of recharge of groundwater is slower than the rate of exploration.

The comparative analysis of category-wise status of groundwater depicts that the number of safe blocks of groundwater in Rajasthan has reduced from 49 in Year 2001 to 31 in 2009, while the number of overexploited blocks have drastically increased from 86 to 166 during the same period (Table 14.2).

The multifold pressure on water resources of Rajasthan for last few decades has attracted the concern of governments, scientists and common public of the state. The present study is an attempt to focus on issues, problems, status and management

**Table 14.1** Status of water resources in Rajasthan

| Category | Availability in BCM | Utilization in BCM | Per cent utilized |
|---|---|---|---|
| Surface water<br>Economically utilizable<br>Economically non-utilizable | 21.71 (42.6%)<br>16. 05 (35.6%)<br>05.66 | –<br>11.29 (31.6%)<br>– | 52<br>70<br>– |
| Groundwater | 11.36 (22.3/25.2%) | 11.77 (39.3%) | 104 |
| External water | 17.88 (35.1/39.2%) | 12.66 (35.4%) | 71 |
| Total<br>(a) Economically utilizable water | 50.96 (100.0)<br>45.0 | –<br>35.72 (100.0) | 70<br>79 |

*Source* Government of Rajasthan (2003)

**Table 14.2** Status of 'groundwater blocks' in Rajasthan

| S. No. | Category of blocks | 2001 | 2004 | 2009 |
|---|---|---|---|---|
| 1 | Safe | 49 | 32 | 31 |
| 2 | Semi-critical | 21 | 14 | 16 |
| 3 | Critical | 80 | 50 | 25 |
| 4 | Overexploited | 86 | 140 | 166 |
| 5 | Saline | 1 | 1 | 1 |
| Total | | 237 | 237 | 239 |

*Source* Central Ground Water Board (2017)

of lakes and reservoirs of Rajasthan and to sketch out the future possibilities for conservation of these vulnerable but valuable resources.

## Characteristics of Water Bodies

Lakes are natural water bodies where the flow from one or several rivers is impounded by a natural obstacle, while a reservoir is an artificial lake created by constructing a dam on river. Physical properties of reservoirs and lakes depend on various features including inflow characteristics and insolation, water level and temperature, dissolved oxygen as a percentage of saturation, etc. These features play a significant role in wetland ecology (Fischer et al., 1979). In India, except the rivers and canals, the total water body cover is about seven million hectare (Ramachandra & Solanki, 2007). Due to limited water resources in the arid and semiarid areas, it is very important to be familiar with the quality of groundwater, which is an important source of drinking water. The potable water is the basic need for healthy human life. Therefore, regular monitoring of the water quality is an important issue for effective management of water resources (Sharma et al., 2010; Sharma et al., 2011a).The investigations of

**Table 14.3** Major researches associated with wetlands of Rajasthan

| S. No. | District | Water body |
|---|---|---|
| 1 | Ajmer | Ana Sagar (Charan & Sharma, 2014; Charan et al., 2008, 2010; Gupta & Sharma, 1998; Mathur, 2010; Sharma & Sharma, 1993), Foysagar (Sukhija, 2010), Budha Pushkar (Chouhan & Sharma, 2009), Pushkar (Thomases, 2017; Sharma et al., 2011b) |
| 2 | Alwar | Siliserh Lake (Jain, 2000; Khinchi, 2015) |
| 3 | Bharatpur | Keoladev Ghana Lake (Reddy et al., 2010; Sharma, 2016) |
| 4 | Bhilwara | Gandhi Sagar (Ojha & Sharma, 1994) |
| 5 | Bikaner | Kolayat Lake (,Barupal & Gehlot, 2015) |
| 6 | Bundi | Jaitsagar Lake (Rathore, 2014; Salodia, 1995), Nawalsagar Lake(RUIDP, 2011) |
| 7 | Jaipur | Jalmahal Lake/Man Sagar Lake (Raina, 2008), JamwaRamgarh Lake(Sharma, 1978; Srivastava, 2003), Mewtha Lake Amer (Sinha & Rajora, 2014), Newta (Srivastava et al., 2003) |
| 8 | Jodhpur | Kaylana (Kumar & Yadav, 2014; Mishra, 2012), TakhtSagar Lake (Kumar & Yadav, 2014; Mishra, 2012), Balsamand Lake (Kumar & Yadav, 2014) |
| 9 | Kota | Kisoresagar Lake (Kumar, 2016; Sarang & Sharma, 2009) |
| 10 | Rajsamand | Rajsamand Lake (Naruka & Sharma, 2017), Amarchand reservoir (Sharma & Gupta, 1994) |
| 11 | Udaipur | Fatehsagar (Mangal & Pathania, 2015; Paliwal, 1984), Udaisagar (Vijayvergia & Tiagi, 2007; Vijavergia, 2008), Jaisamand (Balai, 2015; Rao & Durve, 1987), Pichhola (Das & Singh, 1996; Rawal, 2014), Doodhtalai (Kumar, 2008) |

various reservoirs of Rajasthan for bathymetric and sedimentation studies, physico-chemical analyses of water and sediments, quantitative analyses of bacteria and phytoplankton, etc., were carried out by many researchers (Table 14.3).

The satellite-based techniques like remote sensing and geographical information system have become an effective tool for assessment of large lakes and reservoirs (Gao, 2012). Assessment of spatiotemporal changes in Ana Sagar Lake of Ajmer shows that it has lost its 55.8% area between year 1972 and 2005 (Charan & Sharma, 2014). As per the Indian Meteorological department, the cumulative rainfall in India from 1 June 2015 to 30 September 2015 was 14% less than the long period average of 89 cm. (Rohilla et al., 2017). The reduction in rainfall pattern has shrunken the area of water bodies drastically. Therefore, afforestation, reforestation, recycling, reuse of wastewater, rainwater harvesting, sustainable industrialization, etc., are some of the practices which must be implemented to save the water bodies and reservoirs of the state.

## Pollution of Lakes and Reservoirs

Rapid urbanization and increased human interferences with the nature have eventually culminated in the loss of biodiversity and habitat degradation (Arunpandiyan et al., 2015). Both the natural processes and as well as the anthropogenic disturbances have adversely affected the quality of the water (Mokaya et al., 2004; Melina, 2005; Singh et al., 2005; Charan et al., 2008; Lodhi et al., 2014; Lal & Sharma, 2017). Fast economic growth, population blast and urbanization have resulted in the significantly increased environmental degradation. The point and non-point sources of water pollution play a key role in deterioration of the quality of water, especially in urban areas. The point sources are identifiable individual locations such as an industrial or municipal effluent discharge directly into the water body. The nonpoint or diffused sources of water pollution are non-identifiable sources covering wider area, and they are not pertaining to a single source. Such sources are imparting pathogens, organic wastes, toxic chemicals, nutrients, hazardous sediments and heavy metals into water resources (Yohannes & Elias, 2017). The water bodies of Rajasthan are also affected due to such activities. Unjustified use of pesticides, fertilizers and other agrochemicals has severely deteriorated the aquatic ecosystems of many lakes and reservoirs, especially located near urban and suburban areas (Charanet al., 2010). The limnological studies and physico-chemical assessment of the various lakes of Rajasthan show that the quality of water is not fit for drinking and other purposes due to excessive pollution load (Bhatnagar & Jani, 2007; Saxena, 2007; Sharma, 2009; Sharma & Durve, 1991).

## Nutrient Release Potential

Nutrients released from sediments of water bodies can enhance the nutrient level and productivity of phytoplankton. The rate of release of such nutrients may be directly proportional to the phytoplankton density in lakes and reservoirs. Many studies have been done on the discharge of nitrogen and phosphorus from sediments of lakes (Andersen & Jensen, 1992; Gardner, 2001; Weston, 2005). It is also reported that low level of dissolved oxygen in the water bodies enhances the release of nutrients from the sediments (Walker et al., 2007). There may be different mechanisms for releasing of phosphorus and nitrates to the water column from underlying sediments, including mineralization of organic matter by microbes (Gachter & Meyer, 1993), desorption and dissolution of precipitates and inorganic material (Moore & Reddy, 1994) and the diffusion of dissolved phosphorous from sediment pore waters (Moore, 1998; Pettersson, 1998). It is reported that phosphate concentration in holy lakes, including Pushkar and Budha Pushkar in Rajasthan, is very high due to offering of religious materials and addition of funeral ash to be the source of pollution (Chouhan & Sharma, 2009). The insolation and fluctuation in temperature are important factors responsible for proper mixing of lower and upper waters of any reservoir. The stratification and mixing process vary among lakes depending on their depth and location. The mixing and stratification in shallow lakes occur very rapidly as compared to the deeper ones (Horne & Goldman, 1994). Most of the lakes of Rajasthan are shallow, and therefore, the mixing of water within the lakes is quite frequent, which allows rapid circulation of nutrients throughout the column. The excessive release of such nutrients may lead to eutrophication problems to the reservoirs and lakes. Thus, regular monitoring of such physico-chemical properties of water of such water bodies may help in their conservation and implementation of effective management practices.

## Macrophytes

Macrophytes in an aquatic system have deep impact on the environment (Sculthorpe, 1967). The invertebrates and algal colonization are occurred due to the presence of macrophytes, as they provide a substrate for them (Morin, 1986; Schram et al., 1987; Miller et al., 1989). The biogeochemical cycles and productivity of any water body are influenced by the presence of macrophytes (Wetzel, 1990; Wetzel & Sondergaard, 1998). The macrophytes play a significant role in nutrients dynamism, oxygen level, pH and dissolved organic and inorganic carbon in the lakes and reservoirs. Fewer research studies have examined the potential of submerged macrophytes to modulate the complex and dynamics of pelagic and benthic food webs (Jeppesen, 1998). It is reported that biotic interactions are more vigorous in those lakes and reservoirs where the diversity and density of macrophytes are rich (Gasith & Hoyer, 1998; Heck & Crowder, 1991). The macrophytes attract the thronging of animals and support the biological diversity by providing shelters and offering them feeding sites (Thomaz & Cunha, 2010). The roles of macrophytes in the influence of habitat complexity in the littoral zones have been assessed by few researchers. The compaction of macrophytes bed, vegetation types and structural patterns of roots and stems are the deciding factors of the degree of complexity of an aquatic habitat (Petry, 2003). Similarly, the habitat complexity estimation by measuring the densities of leaf and stem is one of the accurate indices to assess the importance of macrophytes in littoral zones of water bodies (Theel et al., 2008). The growth of macrophytes is limited by various factors including water depth, availability of nutrients, substrate type, turbidity and physical disturbances from waves, tides, water currents, animals and humans (Kimber & Barco, 1994). The presence of macrophytes in peripheral areas of any water body may help in checking infiltration of phosphorus and nitrates from surrounding region, which is useful in mitigation of eutrophication of the water body (Pandey et al., 2012). So far as the wetlands of Rajasthan are concerned, few studies on macrophytes are carried out by different researchers. FatehSagar Lake, Udaipur, was studied based on ecological study with special reference to macrophytic vegetation (Paliwal, 1984). Similarly, the assessment of primary productivity of phytoplankton and macrophytes of Ana Sagar Lake, Ajmer, was also carried out (Dutta & Sharma, 2001; Sharma &

Sharma, 1993). The aquatic biodiversity of Rana Pratap Sagar Lake of Rajasthan was reported, in which 19 species of macrophytes were found in the lake (Verma et al., 2008). It is noticed that various lakes of Rajasthan are invaded by exotic macrophyte species, like *Nelumbonucifera* in Kolayat Lake (Sidhu, 1991), *Eichhorniacrassipes* in Udaisagar (Vijayvergia & Tiagi, 2007) and *Trapabispinosa* in Ana Sagar Lake (Charan et al., 2010). Such invasive macrophytes may alter the structure of habitat and influence the organisms associated with them. Therefore, management strategies should be focused on effective restoration and conservation of these aquatic ecosystems.

## Restoration Practices

Restoration of water bodies means reinstating original/natural aquatic functions and the related physico-chemical and biological features with the purpose of reproducing natural and automated system that is amalgamated ecologically with the landscape and the system of the wetlands (Cairns, 1988; Ramachandra, 2001). The basic principle of restoration of any water body is protection from anthropogenic disturbances. This leads to regeneration of the ecological integrity which refers to the condition of an ecosystem, particularly the structure, function and interactions with other components (Sharma et al., 2009). For the resurgence of any water body, two broad categories are classified, namely indirect or preventive methods and direct or restorative methods (Singh, 1982). Indirect methods are those methods which recognize the contaminant and check their entry to the water body, while the restorative methods influence either the waste treatment before reaching to the lake or direct intrusion in the water body. Application of traditional ecological knowledge has imparted a great effect on the conservation of water resources in Rajasthan since long period. *Johad, Bandha, Khadeens, Baories,* etc., are common structures used from ancient time for conserving and restoring rainwater for future use in the state. After independence, many policies were implemented for conservation of wetlands in India, which reveals that their conservation is indirectly influenced by legislative measures (Parikh and Parikh, 1999). It is noticed that proper implementation of such acts and policies for effective management and conservation of lakes and reservoirs is the need of the time.

Along with the Aravali ranges of Rajasthan, there are many small and large water bodies. During the rainy season, these water bodies get water through run-off from different channels from hilly slopes of Aravalis. A good number of water bodies are situated in the cities and towns like Alwar, Bharatpur, Jaipur, Ajmer, Beawar, Sirohi, Bhilwara, Udaipur, Kota Udaipur, Jaipur, etc. These water bodies are polluted or eutrophicated because of various anthropogenic activities. For restoration of such water bodies and impoundments, integrated efforts are required which include forest management in the vicinity, public participation, governance and also thorough legal actions (Fig. 14.1). Awareness about water conservation and its importance should be developed among masses.

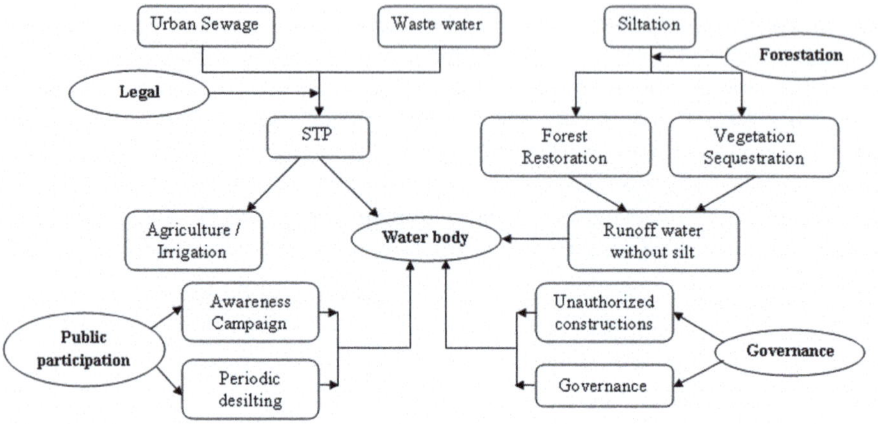


**Fig. 14.1** Integrated action for restoration of lakes and water bodies in urban areas (AbouRayan & Djebedjian, 2016)

## *Climate Change and Water Resources*

Recently, climate change has elicited attention of large number researchers; many studies have shown that climatic change is greatly impacting freshwater resources availability. Climate change is expected to have adverse impact on water resources at the local, regional and global scales. Climate change can strongly alter the magnitude of extreme events, for example, floods and droughts, and further causing great damage to society and even endanger human lives. The impact on water resources is aggravated because of increasing population pressure in the state. Destruction of forest has resulted in increased soil erosion, sediment haulage, siltation, drying-up of water bodies and lowering of groundwater table from 5–10 to 50–100 m. South Rajasthan is a case in point, which used to receive sufficient rainfall, and is experiencing erosion of water resources, loss of biodiversity, declining groundwater table, all leading to increasing pressure on existing environmental resources further aggravating the problem (Rathore & Verma, 2013).

The climatic trend has become harsher and influenced the fluctuation in temperatures, sparse and unpredictable rainfall, and uncertainty in arrival of monsoon is being persistently estimated for the state of Rajasthan. In a collaborative study of 'Swedish Meteorological and Hydrological Institute (SMHI), Sweden' and Central University of Rajasthan, the hydroclimatology of the Luni region of Rajasthan was assessed for the impacts of climate change in contrast with the present and projected fluxes. It was expected that the long-term average evaporation and precipitation change will range from (−) 20 to (+) 20%, whereas the predicted change for run-off is in the range of (−) 40 to (+) 40%. The other hydroclimatic components may also be slightly affected by the climate change, while the magnitude and shape of the flow dynamics of Luni River are verily affected (Pechlivanidis, 2015). Also, it has been estimated that the significant changes in climatic conditions will affect the major rivers of Indian subcontinent. The study indicates a (−) 20 to (+) 50% change in evapo-transpiration and long-term average precipitation (Pechlivanidis, 2016).

In view of predicted impacts of climate change, the government is focusing for sustainable development through climate change agenda. In the Policy Framework on Environment, strategies to mitigate impacts of climate change are identified under 'seven task forces viz. (a)water resources, (b) agriculture and animal husbandry, (c) forestry and biodiversity, (d) human health, (e) enhanced energy efficiency and solar energy, (f) Urban governance and sustainable habitats and (g) strategic knowledge for climate change'. The climate change impacts on water bodies may be severe as the state is already under water crisis. The various socio-economic and environmental aspects are affected by climate change, which require modifications for responding to such changes. These modifications may be in policies and laws, in behavioural and livelihood adjustment and infrastructural and institutional responses. The agro-pastoralism, water, forests and livestock centric solutions must be advocated to combat with the climate change (Kaushik & Sharma, 2015).

## Recommendations

In the wake of imminent menace, 'National Lake Conservation Plan (NLCP)' and the 'National Wetlands Conservation Programme (NWCP)' have been framed and rolled out by the 'Ministry of Environment and Forests and climate change'. About 60 lakes in 14 states have been given active backing for rejuvenation under the NLCP programme with help of relevant departments and agencies of respective state governments (MOEF 2010). From Rajasthan, six lakes including Man Sagar Lake (Jaipur), Ana Sagar Lake and Pushkar Lake (Ajmer), Fatehsagar Lake and Pichhola Lake (Udaipur) and Nakki Lake (Mount Abu) are covered under the NLCP. It is interesting to note that most of the selected lakes under this project are situated in urban or suburban regions, which is a direct indication that they are degrading due to human interferences. These lakes are conserved on the basis of integrated water resource management. 'Integrated water resources management is a process which promotes the coordinated development and management of water, land and related resources in order to maximize the resultant economic and social welfare in an equitable manner without compromising the sustainability of vital ecosystems' (Rahaman & Varis, 2005). It is a general perception that most of the water bodies located in the urban area of developing and underdeveloped countries are severely facing the threats for extinction.

Considering the threats and challenges in protection of water bodies of Rajasthan, the following recommendations are proposed for the effective management:

1. Holistic approach for effective management of water resources should be considered seriously in implementing the management plan.
2. Green belt of suitable plant species should be developed around the lakes and reservoirs to check the siltation and nutrient inflows from surroundings areas.

3. There is a need to emphasize on developing a systematic strategy for engaging different stakeholders for better coordination with the government agencies for restoration and protection programmes of water reservoirs.
4. Anthropogenic activities like motor boating, sewage dumping, bathing, washing, etc., within the water bodies as well as mining, encroachment, industrial activities, etc., in their catchment areas should be controlled.
5. Extensive public awareness programmes for conservation of water resources must be conducted, and capacity building must be used as an important tool for effective management of water bodies.

## References


AbouRayan, M. M., & Djebedjian, B. (2016). Urban water management challenges in developing countries: The Middle East and North Africa (MENA). In Younos, T., & Parece, T. (Eds.), Sustainable water management in urban environments. The Handbook of Environmental Chemistry (Vol. 47). Springer, Cham.

Andersen, F. O., & Jensen, H. S. (1992). Regeneration of inorganic phosphorus and nitrogen from decomposition of seston in a freshwater sediment. *Hydrobiologia, 228*, 71–81.

Arunpandiyan, M., et al. (2015). Impacts of urbanization on water resources and vegetation on the delta region of Tamil Nadu using remote sensing and GIS research and reviews. *Journal of Ecology and Environmental Sciences, 3*(4), 43–47.

Balai, V. K., et al. (2015). Phytoplanktonic diversity in Lake Jaisamand, Rajasthan (India). *Journal of Applied and Natural Science, 7*(2), 592–597.

Barupal, G. K., & Gehlot, R. K. (2015). Status of phytoplanktonic biomass, chlorophyll and energy content of Kolayatlake, Bikaner (Rajasthan) India. *Journal of Algal Biomass Utilization, 6*(4), 68–71.

Bhatnagar, C., & Jani, K. (2007). Vanishing habitats of aquatic birds in the city of lakes, Udaipur: A case study. *The Indian Forester, 133*(10), 1395–1402.

Cairns, J. (1988). *Rehabilitating damaged ecosystems*. CRC Press.

Cairns, J. J., & Pratt, J. R. (1995). The relationship between ecosystem health and delivery of ecosystem services. In D. J. Rapport, C. L. Gaudet, & P. Calow (Eds.), *Evaluating and monitoring the health of large-scale ecosystems* (pp. 273–294). Springer-Verlag.

Central Ground Water Board (2017). Western region, Jaipur. http://cgwb.gov.in. Accessed May 20, 2020.

Charan, P. D., Patni, A., Sharma, K. C. (2008). Pesticide contamination in Lake Anasagar of Ajmer, Rajasthan. In Sengupta, M., Dalwani, R. (Eds.), *Proceedings of TAAL, 2007, The 12th World Lake Conference*, Jaipur, pp. 360–365.

Charan, P. D., & Sharma, K. C. (2014). Assessment and monitoring of spatio-temporal changes in lake Anasagar of Ajmer affected by agricultural activities. In J. B. Khan (Ed.), *Environment management challenges and conservation* (pp. 23–34). Lap Lambert Academic Publishing.

Charan, P. D., Sharma, R., & Sharma, K. C. (2010). Gas chromatographic analysis of organochlorine pesticides in lake Anasagar of Ajmer, Rajasthan (India). *Journal of Environment Science Engineering, 52*(1), 37–40.

Chouhan, C. S., Sharma, K. C. (2009). Pollution status of lake Budha Pushkar of Central Aravali near Ajmer, Rajasthan. In *International conference on water, environment, energy and society (WEES-2009)*, New Delhi. 12–16 January 2009.

Costanza, R., et al. (1997). The value of the world's ecosystem services and natural capital. *Nature, 387*, 253–260.

Costanza, R., et al. (2011). Valuing ecological systems and services. *F1000 Biology Reports, 3*, 14.
Das, B. K., & Singh, M. (1996). Water chemistry and control of weathering of Pichola Lake, Udaipur District, Rajasthan, India. *Geo, 27*, 184–190. https://doi.org/10.1007/BF00770431
Dutta, A., & Sharma, K. C. (2001). Nutrient release potential of some macrophytes of a shallow lake of Ajmer, Rajasthan. *Bulletin of National Institute of Ecology, 11*, 43–48.
Fischer, H. B., List, E. J., Koh, R. C. Y., Imberger, J., & Brooks, N. H. (1979). *Mixing in inland and coastal waters*. Academic.
Gachter, R., & Meyer, J. S. (1993). The role of microorganisms in mobilization and fixation of phosphorus in sediments. *Hydrobiologia, 253*, 103–121.
Gao, H., et al. (2012). Global monitoring of large reservoir storage from satellite remote sensing. *Water Resources Research, 48*, 1–12.
Gardner, W. S., Lavrentyev, P. J., et al. (2001). Nitrogen dynamics in sandy freshwater sediments (Saginaw bay, Lake Huron). *Journal of Great Lakes Research, 27*, 84–97.
Gasith, A., Hoyer, M. V. (1998). Structuring role of macrophytes in lakes: Changing influence along lake size and depth gradients. In Jeppesen, E. et al. (Eds.), *The structuring role of submerged macrophytes in lakes*. Springer Science Media, New York.
Government of Rajasthan (2003). *Water resource vision-2045*. Department of Planning, Government of Rajasthan, Jaipur, India.
Gupta, J., & Sharma, K. C. (1998). Ecological studies of a semi-arid zone lake of Rajasthan. *Indian Journal of Environmental Sciences, 2*(1), 17–22.
Heck, K. L., & Crowder, L. B. (1991). Habitat structure and predator-prey interactions in vegetated aquatic systems. In S. S. Bell, E. D. McCoy, & H. R. Mushinsky (Eds.), *Habitat structure, the physical arrangement of objects in space* (pp. 281–299). Chapman and Hall.
Horne, A. J., Goldman, C. R. (1994). Limnology (p. 576). McGraw-Hill Inc.
Jain, M. K. (2000). Biology and fisheries of Indian major carps from Siliserh reservoir Alwar, Rajasthan, India. PhD thesis, Maharana Pratap University of Agriculture and Technology, Udaipur, Rajasthan (India).
Jeppesen, E., et al. (1998). *The structuring role of submerged macrophytes in lakes*. Springer-Verlag.
Kaushik, G., & Sharma, K. (2015). Climate change and rural livelihoods-adaptation and vulnerability in Rajasthan. *Global NEST Journal, 17*(1), 41–49.
Khinchi, A., et al. (2015). A study of flora and fauna of Siliserh lake with special reference to biochemical characteristic. *International Journal of Innovative Science, Engineering and Technology, 2*(4), 1280–1286.
Kimber, A., Barco, J. W. (1994). A literature review of effect of waves on aquatic plants. *NatlBiolSurv, Environment Management Technical Centre, Onalaska, 25*
Kumar, A. (2008). *Environmental science appreciation and perception*. Daya Publishing House.
Kumar, A., et al. (2016). Limnology, diatom diversity and conservation of Kishore Sagarlake, Kota, Rajasthan, India. *The Research Journal, 2*(2), 150–160.
Kumar, A., & Yadav, N. S. (2014). Study of physiochemical characteristics of Jodhpur lakes. *International Journal of Earth Sciences and Engineering, 7*(1), 376–380.
Lal, D. P., & Sharma, B. K. (2017). Water quality status of lakePichhola, Udaipur, Rajasthan. *International Journal of Fisheries and Aquatic Studies, 5*(3), 181–187.
Lodhi, K. S., et al. (2014). Water quality assessment of the river Birma: A tributaries of Yamuna riverHamirpur district (U.P.) India. *International Journal of Current Research, 6* (11), 9511–9516.
Mangal, H., Pathania, S. (2015). Aquatic resources: A case study of Udaipur 'City of Lakes', Rajasthan. In Rawat, M. et al. (Eds.), *Aquatic ecosystem: Biodiversity, ecology and conservation*. Springer, India. https://doi.org/10.1007/978-81-322-2178-4_2.
Mathur, P., et al. (2010). Assessment of physico-chemical properties of Anasagarlake of Ajmer (India). *Journal of Environmental Research and Development, 4*(3), 780–787.
McDonald, R. I., et al. (2011). Global urban growth and the geography of water availability, quality, and delivery. *Ambio, 40*, 437–446.
Melina, E. K., et al. (2005). Assessment of river water quality in northwestern Greece. *Water Resources Management, 19*, 77–94.

Miller, A. C. et al. (1989). The habitat value of aquatic macropytes for macro invertebrates. Technical Report A-89-3, U.S. Army Corps of Engineers, Washington, DC.
Mishra, P. (2012). Study on some physico-chemical characteristics of main water resources of Jodhpur city. *International Journal of Basic and Applied Chemical Sciences, 2*(4), 63–67.
Mokaya, S.K., et al. (2004). Influence of anthropogenic activities on water quality of a tropical stream ecosystem. *African Journal of Ecology, 42*, 281–288.
Moore, P. A., & Reddy, K. R. (1994). Role of Eh and pH on phosphorus geochemistry in sediments of lake Okeechobee, Florida. *Journal of Environmental Quality, 23*, 955–964.
Moore, P. A., et al. (1998). Phosphorus flux between sediment and overlying water in lake Okeechobee, Florida: Spatial and temporal variation. *Journal of Environmental Quality, 27*, 1428–1439.
Morin, J. O. (1986). *Initial colonization of periphyton on natural and artificial apices of Myriophyllum heterophyllum Michx. Freshwater Biology, 16*, 685–694.
Mwanza, D. D. (2003). Water for sustainable development in Africa. *Environment, Development and Sustainability, 5*, 95–115.
Naruka, S., & Sharma, M. S. (2017). Water quality assessment of Rajsamandlake, Rajasthan, India. *International Research Journal of Environmental Sciences, 6*(6), 22–28.
National Wetland Atlas (2011) Space Application Centre, Ahmedabad. Accessed 25 Sept 2016
Ojha, S., & Sharma, K. C. (1994). Algal diversity in Gandhi Sagarlake of Bhilwara, Rajasthan. *Geobios News Reports, 13*, 143–146.
Paliwal, P. P. (1984). An ecological study of FatehSagar lake (Udaipur) with special reference to macrophytic vegetation. PhD thesis, Mohan LalSukhadia University, Udaipur (Rajasthan), India.
Pandey, D. N. et al. (2012). Evidence-based holistic restoration of Lake Anasagar, Ajmer, Rajasthan, India. A report prepared for the submission to the High Court, Rajasthan in The Matters of DB Civil Writ Petition no. 11153/2012. Rajasthan State Pollution Control Board, Jaipur, Rajasthan, India.
Parikh, J., & Parikh, K. (1999). *Sustainable wetland. Environmental Governance 2*. Indira Gandhi Institute of Development Research, Mumbai.
Pechlivanidis, I. G., et al. (2016). Multi-basin modelling of future hydrological fluxes in the Indian subcontinent. *Water Journal, 8*(5), 1–21.
Pechlivanidis, I. G., et al. (2015). Assessment of the climate change impacts on the water resources of the Luni region, India. *Global NEST Journal, 17*(1), 29–40.
Petry, P., et al. (2003). Relationships between fish assemblages, macrophytes and environmental gradients in the Amazon river floodplain. *Journal of Fish Biology, 63*, 547–579.
Pettersson, K. (1998). Mechanisms for internal loading of phosphorus in lakes. *Hydrobiologia, 373*(374), 21–25.
Rahaman, M. M., & Varis, O. (2005). Integrated water resources management: Evolution, prospects and future challenges. *Sustainability: Science, Practice, and Policy, 1*(1), 15–21
Raina, M. (2008). Conservation and management of Mansagar lake of Jaipur: A model study. In Sengupta, M., Dalwani, R. (Eds.), *Proceedings of Taal 2007, The 12th World Lake Conference*, pp. 1944–1950
Ramachandra, T. V. (2001). Restoration and management strategies of wetlands in developing countries. *Electronic Green Journal, 15*, 1–13.
Ramachandra, T. V., & Solanki, M. (2007). Ecological assessment of lentic waterbodies of Bangalore. ENVIS Technical Report-25. Centre for Ecological Sciences, Indian Institute of Science, Bangalore.
Rao, P. S., & Durve, V. S. (1987). The structure of the phytoplankton community and the dynamics of its biomass in the lake Jaisamand, Rajasthan. *Clean: Soil, Air, Water, 15*(1), 79–91. https://doi.org/10.1002/aheh.19870150113
Rathore, D. K., et al. (2014). Studies on the periodicity of Chlorophyceae of JaitSagarlake, Bundi (Rajasthan). *South Asian Journal of Experimental Biology, 4*(5), 232–239.
Rathore, N. S., & Verma, N. (2013). Impact of climate change in the southern Rajasthan, India. *International Journal of Water Resources and Arid Environments, 2*(1), 45–50.

Rawal, I., et al. (2014). Study of water quality of lakePichhola of Udaipur (India). *Journal of Herbal Medicine and Toxicology, 8*(1), 72–75.
Reddy, C. S., Hari Krishna, P., Swetha, V., Shivakala, T., & Charan, P. D. (2010). Assessment and monitoring of spatio-temporal changes in Keoladeo Ghana National Park, Rajasthan, India using geoinformatics. *Journal of Wetlands Ecology, 4*, 33–42.
Rohilla, S. K. et al. (2017). *Policy paper on water efficiency and conservation in urban India*. Centre for Science and Environment.
RUIDP (2011). India: Rajasthan urban sector development investment program, Bundi heritage sites. Initial Environmental Examination, Project Number-40031. Rajasthan Urban Infrastructure Development Project, Government of Rajasthan, India.
Salodia, P. K. (1995). Hydrobiological studies of Jait Sagar Lake, (Bundi). PhD thesis, MDS University, Ajmer (Rajasthan), India.
Sarang, N., & Sharma, L. L. (2009). Macro benthic fauna as bio-indicator of water quality in Kishore Sagar Lake, Kota (Rajasthan) India. Paper presented at the 13th International Lake Environment Conference, Wuhan, 1–5 November 2009.
Saxena, M. M. (2007). Trophic state indices in water quality monitoring. Proc DAE-BRNS National Symposium on Limnology (NSL-07), 19–21 February 2007, Udaipur (Rajasthan), India.
Schram, H. L., Jr., et al. (1987). Epiphytic macro invertebrates on dominant macrophytes in two central Florida lakes. *Journal Freshwater Ecology, 4*, 151–161.
Sculthorpe, C. D. (1967). *The biology of aquatic vascular plants* (p. 610). Martin's Press.
Sharma, D., Babel, M. S., & Kumar, V. (2010). An approach for efficient rainwater management for agriculture in arid region of Rajasthan, India. *International Journal of Water Resources and Environmental Management, Serial Publication, 1*, 39–52.
Sharma, K. P., et al. (1978). Limnological studies of polluted freshwaters: Physico-chemical characteristics. *International Journal of Ecology and Environmental Sciences, 4*, 89–105.
Sharma, L. L., & Durve, V. S. (1991). Water clarity of 26 waters of Rajasthan in relation to phytoplankton. In *Proceedings of the second Asian fisheries forum*. Asian Fisheries Society, Manila, Philippines, pp. 915–918
Sharma, L. L., & Gupta, M. C. (1994). Some aspects of limnology of Amarchand reservoir, District Rajsamand, Rajasthan: Physical parameters. *Pollution Research, 13*(2), 169–177.
Sharma, N., et al. (2016). Changing water sources in Keoladeo National park (KNP): An alarm for the world heritage site. *Biolife, 4*(1), 172–178.
Sharma, R., & Sharma, K. C. (1993). A note on trophic state of Anasagar Lake, Ajmer. *Geobios News Report, 10*, 158–159.
Sharma, R., et al. (2009). Study of microbiology of Udaipur lakes in relation to different limnological parameter. *Pollution Research, 28*(4), 565–569.
Sharma, S., et al. (2011). Evaluation of water quality of Narmada river with reference to physicochemical parameters at Hoshangabad city, MP, India. *Research Journal of Chemical Sciences, 1*(3), 40–48.
Sharma, S., et al. (2011). Water quality status of Pushkarlake as a primary data for sustainable development. *South Asian Journal of Tourism and Heritage, 4*(2), 184–192.
Sidhu, R. K. (1991). Environmental monitoring of Kolayat Lake, Bikaner. Dissertation, Dungar College, Bikaner (India).
Singh, K. P. (1982). Lake restoration methods and feasibility of water quality management in Lake of the Woods. Illinois Environmental Protection Agency, Champaign, Illinois.
Singh, K. P., Malik, A., & Sinha, S. (2005). Water quality assessment and apportionment of pollution sources of Gomtiriver (India) using multivariate statistical techniques: A case study. *Analytica Chimica Acta, 538*, 355–374.
Sinha, A., & Rajora, N. (2014). Gaze and the picturesque landscape of Amber, India. *Studies in the History of Gardens and Designed Landscapes, 34*(4), 309–322.
Srivastava, N., et al. (2003). Study of physico-chemical characteristics of water bodies around Jaipur. *Journal of Environmental Biology, 24*, 177–180.

Sukhija, L. (2010). Hydrobiologocial studies of Foy Sagarlake, Ajmer. *Nature Environment and Pollution Technology, 9*(3), 601–604.
TERI (2010). *Draft Rajasthan state action plan on climate change.* The Energy and Resources Institute.
Theel, H. J., Dibble, E. D., & Madsen, J. D. (2008). Differential influence of a monotypic and diverse native aquatic plant bed on a macro invertebrate assemblage: An experimental implication of exotic plant induced habitat. *Hydrobiologia, 600*, 77–87.
Thomases, D. (2017). Making Pushkar paradise: Hindu ritualization and the environment. *Hindu Studies, 21*, 187–210. https://doi.org/10.1007/s11407-017-9212-2
Thomaz, S. M., & Cunha, E. R. (2010). The role of macrophytes in habitat structuring in aquatic ecosystems: Methods of measurement, causes and consequences on animal assemblages' composition and biodiversity. *Acta Limnologica Brasiliensia, 22*(2), 218–236.
Tucci, C. E. M. (2017). Urbanization and water resources. In De Mattos Bicudo, C., GaliziaTundisi, J., Cortesão Barnsley Scheuenstuhl, M. (Eds.), *Waters ofBrazil.* Springer International Publishing, Switzerland.
Verma, P. C. et al. (2008). Aquatic biodiversity of Rana Pratap Sagar Lake. In Sengupta, M., & Dalwani, R. (Eds.), *Proceedings of TAAL 2007, The 12th World Lake Conference*, Jaipur, pp. 1967–1970.
Vijayvergia, R. P. (2008). Eutrophication: A case study of highly eutrophicated Lake Udaisagar, Udaipur (Raj.), India with regards to its nutrient enrichment and emerging consequences. In Sengupta, M., & Dalwani, R. (Eds.), *Proceedings ofTAAL 2007: The 12th World Lake Conference*, Jaipur, pp. 1557–1560.
Vijayvergia, R. P., Tiagi, Y. D. (2007). Bio-indicators of environment: a case study of certain plant species in eutrophicated Lake Udaisagar and nearby polluted sites, Udaipur (Raj.). *Indian Journal of Environmental Sciences, 11*(2), 111–114.
Walker, J. L. et al. (2007). Nutrients in lakes and reservoirs: A literature review for use in nutrient criteria development. Virginia Water Resource Research Center VWRRC Special report-SR34. Virginia Polytechnic Institute and State University Blacksburg, Virginia. https://vtechworks.lib.vt.edu/bitstream/handle/10919/49481/VWRRC_sr200734.pdf?sequence. Assessed May 25, 2020.
Weston, H. N., et al. (2005). Release rates and potential fates of nitrogen and phosphorus from sediments in a eutrophic reservoir. *Freshwater Biology, 50*, 301–322.
Wetzel, R. G. (1990). Land-water interfaces: Metabolic and limnological regulators. *Internationale Vereinigung Für Theoretische Und Angewandte Limnologie: Verhandlungen, 24*, 6–24.
Wetzel, R. G., & Sondergaard, M. (1998). Role of submerged macrophytes for the microbial community and dynamics of dissolved organic carbon in aquatic ecosystems. In Jeppesen, E., Sondergaard, M., Sondergaard, M., & Christoffersen, K. (Eds.), *The structuring role of submerged macrophytes in Lakes. Ecological Studies (analysis and synthesis)* (Vol. 131). Springer.
Xu, F. L., et al. (2001). A method for lake ecosystem health assessment: An ecological modeling method (EMM) and its application. *Hydrobiologia, 443*(1–3), 159–175.
Yohannes, H., & Elias, E. (2017). Contamination of rivers and water reservoirs in and around Addis Ababa city and actions to combat it. *Environment Pollution Climate Change, 1*, 116.

# Part III
# Waste Management and Energy Issues

# Chapter 15
# Socio-Economic Implications of Waste Management

**A. D. Sawant, Sujata Warrier, and Geetanjali Pawar**

**Abstract** Very few research outputs have been addressed in the area of socio-economic implications of waste management. Although there are a number of studies conducted on technological aspects of waste management, very few have touched upon the relevance of teamwork in quality waste management. Therefore, this paper is an attempt to explore the socio-economic factors intertwined in the waste management process. It has been observed that population increase in most of the developing countries of the world has certainly posed various threats to mankind and the ecosystem as a whole. The processes of globalization, industrialization and urbanization have added to increase in the generation of large quantities of municipal solid waste (MSW). Management of MSW has become a matter of great concern for the government bodies, the concerned regulatory agencies and the pollution control departments. The same phenomenon has posed challenges for solid waste management. There are socio-economic impacts of solid waste management. There is a close relationship between solid waste management and sociocultural factors like sociocultural relations in local society, men–women relationship, education and beliefs, etc. Day-to-day activities in different work organizations create solid waste. Waste from various sources is to be collected, recycled and reused in the most efficient way. If the waste resources are used, treated and disposed properly, sustainable and balanced development is possible. This in turn will add to more production, proper distribution of goods and also efficient distribution and appropriate marketability. Quality solid waste management will help the nation to maintain a good standard of living. It clearly reveals that quality waste management will teach moral values to

WTERT is Waste to Energy Research and Technology Council, India, Established in 2011 in Association of NEERI (National Environmental Engineering Council) and Global Energy Council at Earth Engineering Center (EEC) of Columbia University, USA.

A. D. Sawant (✉)
University of Rajasthan, Mumbai University, Mumbai, India

S. Warrier
Department of Sociology, N. G. Acharya & D. K. Marathe College, University of Mumbai, Mumbai, India

G. Pawar
WTERT - India, Shree Samarth Building, J. A. Raul Marg, Prabhadevi, Mumbai, India



A. Kateja and R. Jain (eds.), *Urban Growth and Environmental Issues in India*,
https://doi.org/10.1007/978-981-16-4273-9_15

the people. It will bring about discipline and cleanliness within a nation. Sanitation and hygiene will be placed on a priority. Diseases will reduce, and people will be healthy. Quality waste management will encourage the use of new technologies in managing waste in the most efficient way. Efficient use of solid waste will help in the creation of a network society wherein different professionals and stakeholders from diverse fields will join together to form a sustainable nation.

**Keywords** Municipal solid waste · Waste generation · Waste to energy (WtE)

## Introduction

Increase in population levels, booming economy and rapid urbanization are the major issues in developing countries. Generation of municipal solid waste increases with the rapid urbanization and accelerated economic development driven by advanced technological societies (Anupam et al. 2010). The lack of an environmental control system and poor evaluation of the real impact are the reasons that lead to affecting the environmental aspects of solid waste management in developing countries. Economic development, urbanization and improved living standards in the cities of developing countries have led to an increase in the generation of large quantity municipal solid waste. The rapid urbanization and consequent growth of population in urban areas resulted in municipal solid waste becoming a grave concern for Pollution Control Departments, Government bodies and other Regulatory Agencies. By 2051, the population of India is estimated to be 1823 million, and about 300 million tons per annum of MSW will be generated that will require around 1450 $km^2$ of land to dispose it (Joshi & Ahmed, 2016).

India is the world's second highest populated country consisting of 1.21 billion people (Census of India 2011). The annual rate of growth of urban population in India is over 4%, and the proportion of population living in urban areas has increased from 17.35% in 1951 to 37% now. High rate of population growth, declining job opportunities in the rural areas and shift from stagnant and low paying agriculture sector to high paying urban occupations are the factors responsible for this growth. India has achieved multifaceted socio-economic progress during the last 73 years of its independence. However, in spite of heavy expenditure by civic bodies, management of municipal solid waste (MSW) continues to remain one of the most neglected areas of urban development in India. Piles of garbage and wastes of all kinds littered everywhere have become a common sight in urban life (Vij et al. 2012). Managing solid waste is a major concern that has reached alarming proportions for most urban local bodies in India.

Uncontrolled urbanization and improved living standards thereby lead to increased rate of per capita waste generation. Currently, in India 127,488 ton/day of municipal solid waste are being generated due to various household, commercial and institutional activities (CPCB Report, 2016). Substantial amount of this waste is extremely dangerous to the living organisms including human beings (Misra and Pandey 2005).

Solid waste can affect the groundwater quality by percolations leachate and can cause air pollution by emission of various gases which cause health problems to human beings and damage the natural environment (Alam and Ahmade 2013). In spite of these difficulties, the urban local bodies have to overcome them and have to adopt waste management practices as per the US EPA Standards. To minimize the public health hazard, the main focus is laid on reduction at source, recycling of material, technologies like biomethanation or pyrolys is to produce energy and land filling. India lacks awareness about solid waste management and its hierarchy. For the last few years, the scenario of solid waste management has been under continuous change (Nandan et al. 2017). In addition to well-known available technologies, researchers have also done work on the production of energy through cofermentation of organic waste (Gavit et al. 2017).

Globalization as a phenomenon encouraged opening up of new industrial units in almost all the areas of the countries. The industries emit huge amounts of unwanted harmful gases and chemicals from the chimneys. Globalization leads to acceleration of climate change, draw down of global stocks of cheap energy, substantial increase in air, water and soil pollution, decreases in biodiversity, including a massive loss of crop and livestock varieties, deploring ocean fisheries and significant unwanted harmful gases and chemicals from the chimneys. Similarly, urban life creates a lot of waste materials. These waste materials are generally destroyed. Either they are burnt or they are thrown in water or they are used as manure. Recycling of certain waste materials has become necessary in today's day and age. If it is not done, these create disturbances in the air and water and lead to air and water pollution. WHO states that 24 million people die each year, and the cause is directly attributed to air pollution. From this, 1.5 million of these deaths are attributed to indoor air pollution.

Thus, accumulation and management of solid waste are a matter of great concern for India's development. No doubt, the latest technologies can be utilized for quality waste management but in addition to technologies, people's mobilization also plays a crucial role in managing waste effectively. High level of awareness needs to be created among people regarding the waste disposal and proper waste management. There is an urgent need for quality waste management schemes for sustainable development. Therefore, waste management has very wide socio-economic implications, which have been hardly brought about by social thinkers. A number of studies have been conducted to understand the physical and chemical characteristics of waste management. Therefore, this paper is an attempt to explore the physical and sociological aspects involved in the issue of waste management.

## Municipal Solid Waste Generation in Metro Cities

Due to over population and an increasing level of industrialization, huge quantities of waste are being generated in different forms, such as solid, liquid, sledges and gases. Similarly, waste materials are also generated through drainage systems and chimneys. Agriculture also generates a huge quantity of waste in the form of weeds, unwanted dry grass, flowers and leaves, etc. Socio-economic factors also play a big role in the generation of MSW. The quality and quantity of MSW generated by a particular community vary according to their socio-economic status, cultural habits, urban structure, population and commercial activities, etc. (Sharholy et al., 2005; Ray et al., 2005; Jha et al., 2003; Kansal, 2002; Kansal et al., 1998; Singh & Singh, 1998; Gupta et al., 1998).

Large amount of waste is created from homes, hospitals, schools, colleges, offices and public places on a daily basis like waste paper, clothes, sanitary towels, etc. Industries are also responsible for creating a lot of waste materials in the form of chemicals, gases, plastic, papers, etc. A number of steps are involved in the proper management of solid waste arising out of human, agricultural, industrial and biomedical activities including collection from the sources, where they are produced, transportation and disposal in the right way so that they do not cause harm to the environment and human beings.

As per a recent survey carried out by Central Pollution Control Board (CPCB) of India, in the year 2016, as part of The National Action Plan for Municipal Solid Waste Management. In compliance with National Green Tribunal Order Dated 5th February, 2015 in the Matter of OA No. 199 of 2014, Almitra H. Patel and Anr. Versus Union of India and Ors. It has been shown that there is about 50,000 tons per day solid waste generated by 23 metro cities of India alone. The per capita waste generation in small, medium and large cities/towns is about 0.1 kg, 0.3 kg and 0.5 kg, respectively (CPCB, 2016). Seven metro cities, namely Ahmadabad, Bangalore, Chennai, Delhi, Hyderabad, Kolkata and Mumbai, are the major contributors of the municipal solid waste generated in India (Table 15.1) (CPCB, Report 2016 and Purohit 2000).

These wastes have to be handled efficiently. Fortunately, newer technologies have been developed by developing countries for not only treating the waste but also converting waste into energy, like gasification, incineration, compaction, pyrolysis, composting, biomethanation, refuse-derived fuels (RDF), etc. In India, the attempts are under progress for installation of these plants by the companies like Hitachi Zosen, RamkyEnviro, Jindal Steel Works, Essel Infra, Sunil hitech, Anthony Waste Handling Ltd., Malhem Pvt Ltd., etc., by importing technology from developed countries.

Waste to Energy and Research Technology (WTERT), India, was established in 2011, with an MoU signed between National Environmental Engineering Research Institute, the wing of Council of Scientific and Industrial Research of Government of India (CSIR-NEERI, India) and the Earth Engineering Centre (EEC) of Columbia University, New York (USA) and represents India in the Global WTERT Council (GWC). The main objective of the Trust is to create a technology domain for the budding solid waste management (SWM) sector against the gigantic problem of waste disposal in India. WTERT-India acts as a swivelling catalyst to funnel important decisions related to SWM through organizing conferences and to take in hand problems of municipal solid waste management starting from its characterization, calorific value evaluation and assessment of feasibility of plant and monitoring of ongoing plants.

**Table 15.1** Generation of municipal solid waste in major cities of India

| S. No. | Name of the city | Population | MSW generated (TPD) |
|---|---|---|---|
| 1 | Ahmadabad* | 6,361,084 | 2637 |
| 2 | Allahabad | 1,212,395 | 450 |
| 3 | Amritsar | 1,183,705 | 600 |
| 4 | Bangalore* | 8,520,435 | 4000 |
| 5 | Bhopal | 1,886,100 | 874 |
| 6 | Bhubaneswar | 886,397 | 600 |
| 7 | Chennai* | 8,653,521 | 4700 |
| 8 | Coimbatore | 2,136,916 | 850 |
| 9 | Delhi* | 16,349,831 | 8370 |
| 10 | Faridabad | 1,414,050 | 400 |
| 11 | Hyderabad* | 7,674,689 | 4500 |
| 12 | Indore | 2,170,295 | 968 |
| 13 | Jabalpur | 1,268,848 | 550 |
| 14 | Kanpur | 2,920,496 | 1800 |
| 15 | Kolkata* | 14,035,959 | 3800 |
| 16 | Ludhiana | 1,613,878 | 850 |
| 17 | Mumbai* | 18,394,912 | 9083 |
| 18 | Nagpur | 2,497,870 | 1100 |
| 19 | Nasik | 1,561,809 | 500 |
| 20 | Pune | 5,057,709 | 1700 |
| 21 | Surat | 4,591,246 | 1200 |
| 22 | Vadodara | 1,822,221 | 1,100 |
| 23 | Varanasi | 1,423,711 | 500 |
| 24 | Vijayawada | 1,476,931 | 550 |
| 25 | Visakhapatnam | 1,730,320 | 350 |

*Source* CPCB (2016)
*Metro cities

Municipal solid waste generation in metro cities is a matter of great concern as a lot of waste is generated in urban areas from homes, offices, schools, colleges and hospitals. The collected waste is dumped generally on an open place, which is known as dumping ground. Unfortunately, it is not ensured that the dumping ground is far away from residential areas. The solid waste from cities is dumped in the ground, and many times, the disposed waste is burnt off. It is sad to note that the officials do not notice that burning of the garbage and waste or combustion creates a lot of pollution. Burning of waste and the process of waste disposal have created health hazards for the urban residents. Residents of Deonar and Chembur in Mumbai city have shown discontent and have expressed resistance to the process of waste disposal and burning of waste garbage in the most infamous Deonar Dumping Ground.

Various researchers have revealed that Chembur is a highly polluted area in Mumbai. Majority of the residents from Chembur, the Mumbai suburb belonging to different age groups, have been suffering from bronchitis, asthma and various kinds of allergies due to air pollution prevalent in the area. E-waste is also a matter of great concern in developing countries like India. It creates health hazards for the people. E-waste includes collection of electronics waste items like outdated computers, computer parts, hardware, batteries, radios and CDs, etc. Technologies are changing rapidly, and old materials and devices become outdated. People throw away these items as garbage. Recycling of such waste is very difficult or impossible. Piling up of such waste electronic items spoils the atmosphere and affects public health. Public health has now become a matter of great concern.

## Socio-Economic Impact of Waste Management

A number of studies and researches have been conducted on waste management by several scientists and environmentalists. However, hardly any collaborative research findings have been presented with an interdisciplinary approach. It is very important to note that waste management schemes have been intertwined with sociological and economic impacts and hence must involve social scientists to develop an integrated approach to this burning issue.

Industrialization and urbanization have led to the creation of a lot of waste worldwide. It has also been revealed that the greater the economic prosperity of the country and higher the percentage of urban population, the greater will be the amount of solid waste generation. The high-income countries generate more than twice the weight of waste per capita when compared with low-income countries (World Bank, 2019). Vieira et al. (2018) in their research paper have explored the impact of socio-economic factors on municipal solid waste generation in Greater Sao Paulo, which includes 39 municipalities. They opine that inequality is an important factor in waste management. Coffey and Coad (2010) and Van de Kaundert (2001) have illustrated relationship between the problems experienced by society in relation to solid waste management. They also show that there is a close relationship between solid waste management and sociocultural factors like sociocultural relations of local society, men–women relationship, education and beliefs, etc.

Social aspects are evident where recycling is the focus in developing countries. Low and middle-low-income countries recycle via the informal sector in contrast to the formal sector in high-income countries (Gupta et al., 2015; Scheinderg et al., 2010; Wilson et al., 2013). Another study reveals that waste generation changes according to average family size (Khan et al., 2016). It has to be clearly understood that the process of waste management schemes has deep impacts in different spheres. Some of them are as follows.

## Economic Impacts

Waste from various sources is to be recycled and reused to the optimum level. Economic efficiency can be improved if the waste resources are used, treated and disposed properly. These healthy practices can certainly help in production, distribution and appropriate marketability of the new product, which can be highly valuable. This will create new job opportunities.

In the twenty-first century, due to the influence of globalization and privatization, new products were launched almost every day. There is heavy competition for its production. Waste management and the recycling of waste will help in the opening up of new units, which in turn will create more opportunities. This will also help in increasing per capita national income and gross domestic production (GDP). Therefore, quality waste management schemes can be used for capacity building and will also help in sustainable development.

## Social Impacts-Active Force Practices in Reducing Health Hazards

Quality waste management will reduce health hazards in the country. It will bring about effective results, wherein more job opportunities will be created and will certainly reduce the level of absolute poverty and relative poverty. The standard of living of people will improve; the added economic growth will help them to have nutritious food, healthy mind and healthy body. It will also improve the public health scenario of the nation.

Quality waste management will help in reducing filth and garbage from each and every corner of streets. Earlier researches on health have revealed that poor sanitation and filth lead to contamination of water, which in turn creates various kinds of harmful water borne diseases and air pollution. Thus, if the waste is dumped in a proper place and recycled in a safe area, it can reduce the pollutants in the air and the surroundings, which shall in turn reduce the occurrence of harmful diseases.

## Environmental Impacts

Quality waste management reduces the environmental pollution by reducing the amount of waste. It helps in implementing and using effective strategies to recycle and reuse resources, which will certainly eliminate adverse impacts on the environment and will help in improving the quality of air, water and will also help in the reduction of greenhouse gas (GHG) emissions.

## Teaching Moral Values

One has to always understand that as members of society, humans must respect Mother Earth. Mother Earth is to be considered as God, and each and every element from our biotic and abiotic environment has to be respected. Environmental ecology teaches a man that one has to be very careful while using natural resources. If the resources are not wasted and managed properly and reused and recycled, people will learn the value of saving rather than living extravagantly. This attitude from the citizens will help in the development of a strong and well-sustained nation. The citizens should learn to respect the natural resources and also the environment.

## Sustainable Development

Quality waste management will help in taking care of the available resources to the optimum level. The present generation will learn to save resources for the future generation, which will ultimately help in the development of a self-sustaining nation.

## Research, Innovation and Technological Development

Quality waste management will encourage the use of new technologies to manage the waste and reuse/ recycle in the most effective way. This will certainly encourage more research and experiments and detailed probing in the areas, which have not been analysed yet. Quality waste management will also encourage conferences, seminars and symposiums for better exchange of practising ideas and difficulties. Natural scientists and social scientists will get an opportunity to probe into unsolved areas of research questions. They will also help in the exchange of ideas and knowledge among reputed academicians and technocrats all over the world.

According to Manuel Castells, the twentieth and twenty-first century is a network society. In this network society, each like-minded person will expand his or her networking through constant interaction and exchange of ideas and knowledge. Similarly, it can also be observed that quality waste management strategies will help in the introduction of new technologies with the help of research instincts. These conferences, seminars and symposiums will certainly help in the signing of new MOUs and collaborations between different countries. This will help in building a sustainable planet Earth.

## Urgent Need for Quality Waste Management Schemes for Sustainable Development

There are various environmental challenges associated with waste generation and inadequate waste collection, transport, treatment and disposal that a developing country like India has to face. Even though various countries claim to have made use of efficient waste management processes, it is not always so. It should not be surprising to know that despite the best effort, there is a lot of damage to the environment. The entire planet has been injured. It affects the overall economy negatively. There is a lot of oil spill, leaching of chemicals and dumping of agricultural and other solid waste, which causes soil and water contamination. Burning of waste and charcoal results in environmental pollution and also affects animal life. Improper solid waste management has affected the weather conditions negatively. There is severe climate contamination.

Waste management infrastructure has an important role in delivering sustainable development. India's increasing population, high birth rate and phenomenon of urbanization have led to depletion of natural resources. Deforestation and soil erosion are posing threat to human life and overall ecology. The transition from waste to resources can only be achieved through quality waste management as this depends on a proper coordination of various actions like collection, recycling, reusing and proper marketability of resources. Therefore, quality waste management can bring about a Green and Clean Universe.

WTERT-India is developing a policy for MSW management focusing on waste processing, recycle, reuse and mainly for conversion of waste into energy under the principle of waste to energy (WtE). Here is a case study of development of knowledge capsule through brainstorming conference cum workshop organized by WTERT-India along with CISR-NEERI on 'Waste to Energy Technology Plants & Equipments' at Hyderabad, India, on 3 February 2017 followed by several others thereafter. This workshop was organized under the WTERT mission to identify and advance the best available waste to energy (WtE) technologies for the recovery of energy or fuels from municipal solid wastes and industrial wastes. It was inferred that to make energy plants viable, the waste should be made free from moisture to maximum extent to maintain constant combustion temperature. The important discrepancy which was discussed was the operation temperature of the combustion process at 1050 °C. It was unanimously resolved to 950 °C, to minimize loss of heat and operational cost.

The main objective of this summit was to give a platform to all waste to energy (WTE) technology and equipment vendors to present and demonstrate their capabilities in design, supply, installation and commissioning of their plants and machinery equipment in a WTE plant. The growing demand for WTE plants in the country was noticed, most of the first-generation entrepreneurs showed interest to invest in WTE as a business, and this forum helped them select appropriate technology and

**Table 15.2** Daily municipal solid waste generation of some select Indian cities and some of their characteristics

| S. No. | Name of the city | Biodegradable (%) | Recyclable (%) | Calorific value (kcal/kg) | Moisture (%) |
|---|---|---|---|---|---|
| 1 | Jabalpur | 64 | 8 | 2169 | 46 |
| 2 | Surat | 51 | 11 | 1635 | 42 |
| 3 | Bhubaneshwar | 35 | 9 | 1400 | 29 |
| 4 | Patna | 40 | 11 | 3031 | 37 |
| 5 | Kolhapur | 46 | 18 | 2797 | 35 |
| 6 | Bangalore | 74 | 13 | 1656 | 50 |
| 7 | Ranchi | 60 | 35 | 1759 | 51 |
| 8 | Anantapur (cluster) | 53 | 19 | 1687 | 50 |
| 9 | Kadappa (cluster) | 48 | 20 | 1685 | 37 |

offered facilitation for establishment of the facility. The logistic of arriving to technology selection was discussed here with excerpts from lectures given by technology providers and performing experiences as a case study (Table 15.2).

WTERT-India undertook these studies of characterization of municipal solid waste with an intention to provide feasibility report with qualitative and quantitative analysis of municipal wastes.

## Results and Discussion

The results of analysis of two–five municipal solid wastes samples each from six cities of Anantapur cluster and five towns of Kadapa district indicate that the moisture content is varying greatly from 30 to 50%, indicating the complexity of the waste. Moisture is a deterrent in combustion technology as it reduces the efficiency of power generation. Thus, operating the plants at such high levels of moisture makes it an expensive proposition. The recyclable component, on average, was found not more than 2–5% and hence does not substantially add to the operational cost recovery. The biodegradable and compostable component varies from 20 to 60%. Less than 40% compostable component makes the plant non-viable. It is important to note some city wastes have higher component of combustible material and can be useful for many technologies like pyrolysis, RDF and compost making. It is also noticed that the volatile components like chlorides, arsenic and organics are noteworthy. The chloride concentration is exceptionally higher indicating their entry through household cleaning and use of detergents. Therefore, the WTE plants will generate substantial quantity of flue gases which will have to be scrubbed, further requiring periodic cleaning of the plants. With respect to heavy metals, the concentrations vary from 1 to 25 ppm for zinc, copper, chromium, cadmium, nickel and less than 1 ppm, for arsenic. Incidentally, mercury concentrations are low, and mercury is not been explicitly handled. The concentration of iron is on higher side ranging from 25 to 45 ppm. However, it is not of toxic in nature and can be scrubbed easily. The inert material at many sites is of appreciable quantity close to 50% at someplace which necessarily requires presegregation before using MSW for any technology of WTE.

The studies have revealed that the calorific value of municipal solid waste is the most important factor as the generation of heat energy leading to the power generation is dependent on it. Higher the calorific value, more efficient shall be the WTE plant. MSW samples analysed of sites show the calorific values between 1600 and 3000 kcal/kg. This should produce about 20–25 MW electricity per hour from a sizable and standard WTE plant, and the plant can become viable after about 7 years of its operation.

The experiment carried for characterization also revealed that quality waste management is very essential in bringing about sustainable development. It is pertinent to note that the creation of a knowledge domain for the industrial sector is very much essential in the implementation of feasible and quality waste management schemes in developing countries like India.

## Amala Bharatham (Clean India Programme)

The authors are also representing here the case of social study with the exercise called Amala Bharatham Campaign, popularly known as ABC of Mata Amritanandamayi Devi (Amma), which started in the year 2010 at Quilon in Kerala. Mata Amritanandamayi Devi, who is one of the richest women saints, also known as 'Hugging Saint', implemented one of the best models of waste management and cleanliness drive. She introduced the Amala Bharatham Campaign (ABC) in India. This project was implemented from India to Switzerland. Amma says that nature and environment are manifestations of God. The project mission was called 'Clean India Project'. She, even though Chancellor of University, finds time to teach good values to students. Her dedicated teams are working hard to make Hon'ble Prime Minister Narendra Modi's dream project Swatch Bharat Mission a success. Prime Minister was also highly impressed with Amma and her team's activities.

**The aims of Amala Bharatham Campaign are summarized as follows**:

- To embrace new practical initiatives to clean India, promote health through hygiene, sort out and segregate garbage and properly dispose of waste.
- To make people more aware of the need to maintain environmental cleanliness
- Install trash cans in schools and on roads. Donate funds to the government to carry on with the cleanliness drive.
- Provide 1 million reusable handkerchiefs to school children to ward off some of contagious diseases.

Therefore, Amala Bharatham Project (http://www.amrita.edu.>amritaesd) has helped in bringing about cleanliness and educating the world that the waste is to be segregated, recycled and reused in the most systematic way so that the next generation will not have any scarcity of natural resources. Mata Amrutanandmayi Devi proclaims "One must nurture our own Earth". According to her, nature has given a lot and is still giving. However, if we do not preserve this natural gift from God, the major sufferers are going to be the next-generation children. "Therefore, to protect their life and well-being, at least now we must start taking care of our Mother Earth. If we do not do it now, our next generation will never forgive us."

## Swachh Bharat Abhiyan

Swachh Bharat Mission, which is also known as Swachh Bharat Abhiyan, is a nationwide campaign in India, which was started in the year 2014. The aim of this mission is to clean up the streets and infrastructure of India's roads, cities and towns and rural areas. It was the brainchild of Hon. Prime Minister of India Shri Narendra Modi. The target period was 5 years to accomplish 'Clean India Mission'. He encouraged the school teachers, students and college teachers to take up clean area initiatives. Students were made to organize poster exhibitions on the basic theme of clean India. The Government of India is spending a good amount to create awareness at the national level regarding clean India. Awareness has been created through slogans, posters and documentary films.

## Conclusion

Quality waste management is very essential for sustainable development. It is a crucial need of the day and has to be kept as a priority. Specific organizations have to be given the responsibilities to monitor the waste management accurately. Therefore, taking references from practising technocrat engineers and scientists, community and corporate involvement and the sustainable practices for solid waste management, the best practices can be listed as follows:

- Adopt practice of waste collection and segregation at source.
- Household and community involvement.
- Training front line workers of collection, handling, transportation and processing.
- Accurate assessment of calorific values as well as biodegradability.
- Assessment of capping fees of establishment and providing right unit cost and tipping fees and subsidy.
- Establishment of ease of interdependence of stakeholders.

The government has to encourage research works development of small-scale entrepreneurships for 'Reduce, Reuse and Recycle' of waste. Each and every part from the waste must be converted and recycled. Awareness on waste management has to be created at the grass route level of the society.

To bring about sustainable development, the government, different organizations, stakeholders, academicians, NGOs and researchers have to work together as a team and make the waste management programme in India self-sustainable. For waste-related infrastructure to be 'accepted' by communities, it is critical to engage with the communities from the outset, to determine their needs and address their concerns with such projects. If all the efforts are mobilized in the right manner, India can achieve the goal at least by 2025, and our Mother Earth can be protected from environmental disasters happening due to such contributory factors.

## References

Alam, P., Ahmade, K. (2013). Impact of solid waste on health and the environment. *International Journal of Sustainable Development and Green Economics (IJSDGE)*, *2*(1).

Amruta Project. http://www.amrita.edu.>amritaesd

Anupam, K., Yamamoto, Y., & Morioka, T. (2010). Estimation of municipal solid waste generation and landfill area in Asian developing countries. *Journal of Environmental Biology, 31*(5), 649–654.
Central Pollution Control Board (CPCB), Gov. of India, New Delhi - Report, 2016
Coffey, M., & Coad, A. (2010). Collection of municipal solid waste in developing countries. UN-Malta: HABITAT. Available at http://www.eawag.ch/fileadmin/Domain1/Abteilungen/sandec/E-Learning/Moocs/Solid_Waste/W1/Collection_MSW_2010.pdf. Accessed on January 21, 2018.
Gavit, B. K., Bhange, H. N., Ingle, P. M., & Purohit, R. C. (2017). Energy production through co-fermentation of organic waste and sewage in Nashik City, India. *International Journal of Current Microbiology and Applied Sciences, 6*(4), 2788–2798.
Gupta, N., Yadav, K. K., & Kumar, V. (2015). A review on current status of municipal solid waste management in India. *Journal of Environmental Sciences, 37*, 206–217.
Gupta, S., Krishna, M., Prasad, R. K., Gupta, S., & Kansal, A. (1998). Solid waste management in India: Options and opportunities. *Resource, Conservation and Recycling, 24*, 137–215.
Joshi, R., & Ahmed, S. (2016). Status and challenges of municipal solid waste management in India: A review. *Cogent Environmental Science, 2*, 1139434. https://doi.org/10.1080/23311843.2016.1139434.
Jha, M. K., Sondhi, O. A. K., & Pansare, M. (2003). Solid waste management—A case study. *Indian Journal of Environmental Protection, 23*(10), 1153–1160.
Kansal, A. (2002). Solid waste management strategies for India. *Indian Journal of Environmental Protection, 22*(4), 444–448.
Kansal, A., Prasad, R. K., & Gupta, S. (1998). Delhi municipal solid waste and environment—An appraisal. *Indian Journal of Environmental Protection, 18*(2), 123–128.
Khan, D., Kumar, A., & Samadder, S. R. (2016). Impact of socioeconomic status on municipal solid waste generation rate. *Waste Management, 49*, 15–25.
Misra, V., & Pandey, S. D. (2005). Hazardous waste, impact on health and environment for development of better waste management strategies in future in India. *Environment International, 31*(3), 417–431.
Nandan, A., Yadav, B. P., Baksi, B., & Bose. (2017). Recent scenario of solid waste management in India. *World Scientific News, 66*, 56–74.
Ray, M. R., Roy Choudhury, S., Mukherjee, G., Roy, S., & Lahiri, T. (2005). Respiratory and general health impairments of workers employed in a municipal solid waste disposal at open landfill site in Delhi. *International Journal of Hygiene and Environmental Health, 108*(4), 255–262.
Scheinberg, A., Wilson, D.C., Rodic, L. (Eds.) (2010). Solid waste management in the world's cities: Water and sanitation in world cities. Earthscan for UN-Habitat.
Sharholy, M., Ahmad. K., Mahmood, G., & Trivedi, R. C. (2005). Analysis of municipal solid waste management systems in Delhi—A review. In *Book of proceedings for the second international congress of chemistry and environment*, Indore, India, pp. 773–777.
Singh, S. K., & Singh, R. S. (1998). A study on municipal solid waste and its management practices in Dhanbad-Jhariacoalifield. *Indian Journal of Environmental Protection., 18*(11), 850–852.
Van de Klundert, A., & Anschütz, J. (2001). Integrated sustainable waste management—The concept. In Scheinberg, A. (Ed.), *Integrated sustainable waste management: Tools for decision-makers, set of five tools for decision-makers—Experiences from the urban waste expertise programme (1995–2001)*. WASTE. Available at http://www.rainfoundation.org/tools/downloads/tools_ISWMconcept.pdf. Accessed on January 21, 2018.
Vieira, V. H. A. D. M., & Matheus, D. R. (2018). The impact of socio-economic factors on municipal solid waste generation in São Paulo, Brazil. *Waste Management & Research, 36*(1), 79–85.
Vij, D. (2012). Urbanization and solid waste management in India: Present practices and future challenges
Wilson, D. C., Velis, C. A., & Rodic, L. (2013). Integrated sustainable waste management in developing countries. *Proceedings of the Institution of Civil Engineers: Waste and Resource Management, 166*, 52–68.
World Bank (2019). https://www.worldbank.org/en/topic/urbandevelopment/brief/solid-waste-management. Accessed on January 5, 2020.

# Chapter 16
# Solid Waste Management in Urban Areas: An Urgent Priority

**Ravinder Palakurthy and J. P. Kesari**

**Abstract** While urbanization is now considered as a global phenomenon, its adverse impacts are more pronounced in developing economies. High rate of population growth, declining opportunities in rural areas and shift from stagnant and low paying rural agriculture sector to more rewarding urban occupations largely contribute to rapid urbanization. In many cities, the rapid population growth has overwhelmed the capacity of municipal authorities to provide even basic services. Rapid escalation of population and unconstrained urbanization has given rise to critical problems like huge waste generation and unscientific methods of waste (both solid and liquid wastes) handling resulting in health and environmental hazards. Solid waste management (SWM) which has already become a huge task in India is going to be more aggravated with the ever-increasing urbanization, changing lifestyles and consumerism. The Swacch Bharat Mission—Urban (SBM-U), launched on 2 October 2014, aims at making urban India free from open defecation and achieving 100% scientific management of municipal solid waste in 4041 statutory towns in the country. The municipal bodies and other organizations responsible for waste management in urban areas have been dealing with the problem of solid and liquid waste management by adopting various technologies options available. However, the problem of collection and segregation of various types of waste generated has been a major obstacle in achieving desired results. In this paper, an attempt has been made to appraise major issues in solid waste management (SWM), treatment options available and practised in India along with some case studies of solid waste management projects in Delhi.

**Keywords** Municipal solid waste · Solid waste management · Swachh Bharat Mission · Waste generation

R. Palakurthy (✉)
Mahatma Gandhi Institute for Combating Climate Change (Govt. of NCT of Delhi), Delhi, India

J. P. Kesari
Department of Mechanical Engineering, Delhi Technological University, Delhi, India



A. Kateja and R. Jain (eds.), *Urban Growth and Environmental Issues in India*,
https://doi.org/10.1007/978-981-16-4273-9_16

## Introduction

While urbanization is now considered as a global phenomenon, its adverse impacts are more pronounced in developing economies. High rates of population growth, declining opportunities in rural areas and migration from stagnant and low-paying rural agriculture sector to more rewarding urban occupations largely contribute to rapid urbanization. In many cities, the rapid population growth has overwhelmed the capacity of municipal authorities to provide even basic services.

Rapid increase in population and unrestricted urbanization has given rise to critical problems like huge waste generation and unscientific methods of waste (both solid and liquid wastes) handling resulting in health and environmental hazards. Solid waste management (SWM) which has already become a huge task in India is expected to be more aggravated with the ever-increasing urbanization, changing lifestyles and consumerism. With this increasing population, municipal solid waste management (MSWM) in the country has emerged as a challenge not only because of the environmental and aesthetic concerns, but also because of the huge quantities of municipal solid waste (MSW) generated every day.

India's estimated generation of solid waste is of the order of 1.5 lakh Tons Per Day (TPD), with urban India contributing to 58% of the same. Waste generation in urban India is increasing by 5% every year because of increasing population (3–3.5% per year) and consumption. As per the Ministry of Environment, Forest and Climate Change (MoEF&CC), MSW generation will reach 4.5 lakh TPD by 2030 and 11.9 lakh TPD by 2050 (CSE, 2019). Indian waste generation is set to accelerate in relation to other large waste-generating nations. The increasing scale of generation would have its own complexities in terms of appropriate technology adoption, governance and institutional structures. In the case of urban SWM, there would be an added challenge of managing a large manpower spread across various locations.

## Waste Generation in India

Municipal solid waste is defined to include household waste, commercial and market area waste, slaughter house waste, institutional waste (e.g. from schools, community halls), horticultural waste (from parks and gardens), waste from road sweeping, silt from drainage and treated biomedical waste.

According to Central Pollution Control Board (CPCB, 2017), nearly 154,554.5 TPD of MSW was generated in India during 2016, with an average waste of 0.34 kg/capita/day. Of the total MSW, approximately 137,650.37 TPD (89%) were collected, while only 29,975.48 TPD (19%) were processed or treated (Table 16.1). Segregation at source, collection, transportation, treatment and scientific disposal of waste was largely insufficient leading to the degradation of environment and poor quality of life.

**Table 16.1** Details of MSW generated in India (statewise) 2016

| S. No | States | Generated (TPD)* | Collected (TPD)* | Treated (TPD)* | Waste to energy potential (MW)** |
|---|---|---|---|---|---|
| 1 | Andaman and Nicobar | 70 | 70 | 05 | – |
| 2 | Andhra Pradesh | 6440 | 6331 | 500 | 123 |
| 3 | Arunachal Pradesh | 116 | 70.5 | 0 | – |
| 4 | Assam | 7920 | 6338 | 200 | 8 |
| 5 | Bihar | 1670 | – | – | 73 |
| 6 | Chandigarh | 370 | 360 | Nil | 6 |
| 7 | Chhattisgarh | 2245.25 | 2036.97 | 828.18 | 24 |
| 8 | Daman Diu and Dadra | 85 | 85 | Nil | – |
| 9 | Delhi | 9620 | 8300 | 3240 | 131 |
| 10 | Goa | 450 | 400 | 182 | – |
| 11 | Gujarat | 10,480 | 10,480 | 2565 | 112 |
| 12 | Haryana | 4837.35 | 3103 | 188 | 24 |
| 13 | Himachal Pradesh | 276 | 207 | 125 | 2 |
| 14 | Jammu and Kashmir | 1792 | 1322 | 320 | – |
| 15 | Jharkhand | 3570 | 3570 | 65 | 10 |
| 16 | Karnataka | 8842 | 7716 | 3584 | – |
| 17 | Kerala | 1339 | 655 | 390 | 36 |
| 18 | Lakshadweep | 21 | – | – | – |
| 19 | Madhya Pradesh | 6678 | 4351 | – | 78 |
| 20 | Maharashtra | 22,570 | 22,570 | 5927 | 287 |
| 21 | Manipur | 176 | 125 | – | 2 |
| 22 | Meghalaya | 208 | 175 | 55 | 2 |
| 23 | Mizoram | 552 | 276 | Nil | 2 |
| 24 | Nagaland | 344 | 193 | – | – |
| 25 | Orissa | 2574.7 | 2283.9 | 30 | 22 |
| 26 | Puducherry | 513 | 513 | 10 | 3 |
| 27 | Punjab | 4456.2 | 4435 | 350 | 45 |
| 28 | Rajasthan | 5037 | 2491 | 490 | 62 |
| 29 | Sikkim | 49 | 49 | 0.3 | – |
| 30 | Tamil Nadu | 14,500 | 14,234 | 1607 | 151 |
| 31 | Tripura | 415 | 368 | 250 | 2 |

(continued)

**Table 16.1** (continued)

| S. No | States | Generated (TPD)* | Collected (TPD)* | Treated (TPD)* | Waste to energy potential (MW)** |
|---|---|---|---|---|---|
| 32 | Telangana | 6740 | 6369 | 3016 | – |
| 33 | Uttar Pradesh | 19,180 | 19,180 | 5197 | 176 |
| 34 | Uttrakhand | 918 | 918 | Nil | 5 |
| 35 | West Bengal | 9500 | 8075 | 851 | 148 |
| | Total | 154,554.5 | 137,650.37 | 29,975.48 | 1534 |

*Source* *CPCB (2017), **MRNE (MNRE, 2017)

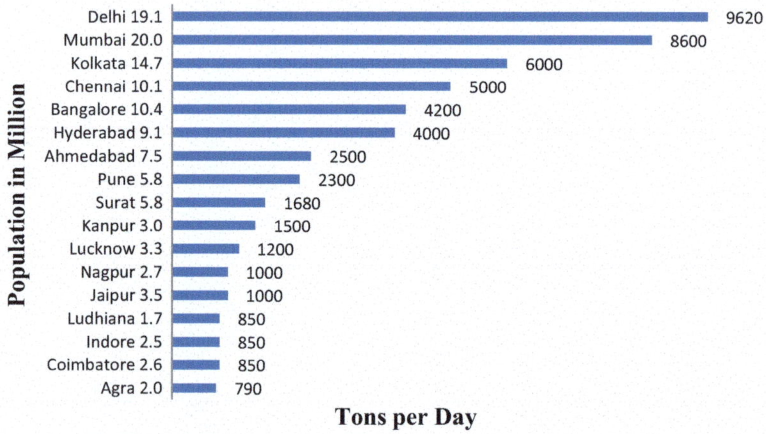


**Fig. 16.1** Top MSW-generating cities/UAs in India 2016. *Source* CPCB (2017)

Figure 16.1 presents Indian cities which were top generators of solid waste in 2016 based on the data reported by the cities themselves. The six largest metropolitan cities (Delhi, Mumbai, Kolkata, Chennai, Bengaluru and Hyderabad) generate the maximum volume of solid waste, ranging from 4000 TPD in Hyderabad to 9260 TPD in Delhi. Together they account for 21% of the total MSW generated in all Indian cities and towns, while their share of the total urban population is only 16%. Among the smaller cities with population between one million and five million, Kanpur and Lucknow are the highest waste generators, generating 1500 TPD and 1200 TPD, respectively.

Waste composition specifies the components of the waste stream as percentage of the total mass or volume. Waste composition is influenced by factors such as culture, economic development, climate and energy sources; composition impacts how often waste is collected and how it is disposed. As a country urbanizes and population become wealthier, consumption of inorganic materials (such as plastics, paper and aluminium) increases while the relative organic fraction decreases. The report of CPCB (2017) places the biodegradable component of the total MSW

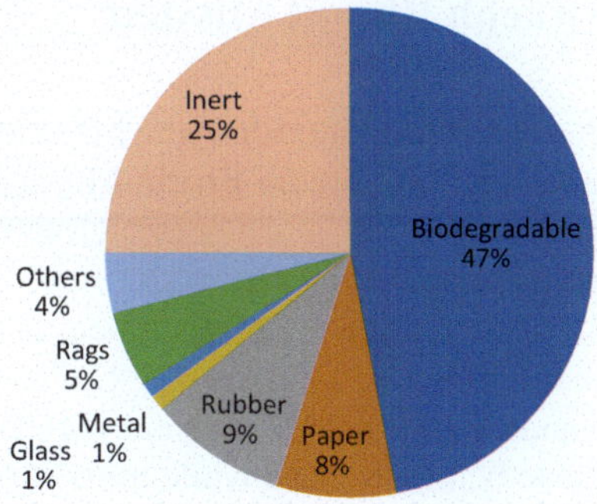


**Fig. 16.2** Composition of MSW in India. *Source* CPCB (2017)

in India at 47%, which seems to be an underestimate (Fig. 16.2). According to a report of Earth Engineering Centre (EEC), Waste-to-Energy Research and Technology Council (WTERTC), Columbia University posted on blog (http://swmindia.blogspot.in/) that population growth and rapid urbanization mean bigger and denser cities and increased MSW generation in each city. The data compiled for this report indicates that 366 cities in India were generating 31.6 million tons of waste in 2001 and are currently generating 47.3 million tons, a 50% increase in one decade. It is estimated that these 366 cities will generate 161 million tons of MSW in 2041, a fivefold increase in four decades. At this rate, the total urban MSW generated in 2041 would be 230 million TPY (tons per year; 630,000 TPD).

The major issues relating to management of MSW in the country (CPCB, 2017) relate to the following:

1. Lack of comprehensive short- and long-term plan with municipal authorities to handle MSW in accordance with the MSW Rules 2000 (which have been revised as Revised MSW Rules, 2016).
2. Majority of the municipal authorities do not have preparedness to set up waste processing and disposal facilities.
3. Waste management is being looked either for making wealth or generating revenue.
4. Cities and towns, in future, will not get wastelands for further dumping of wastes. In fact, there will be a need to go for 'total' recycling and reuse of waste and aim for negligible or 'zero waste' for landfilling, etc.

Globally, waste volumes are increasing quickly, even faster than the rate of urbanization. Current global MSW generation levels are approximately 2.01 billion TPY and are expected to increase to approximately 3.4 billion TPY by 2050 (Kaza et al., 2018). The world generates 0.74 kg of waste per capita per day, and the number is expected to increase substantially under a business as usual scenario. However, global averages are broad estimates only as rates vary considerably by region, country, city and even within cities.

## Swachh Bharat Mission

In order to give a push to the MSWM in cities, the Ministry of Urban Development launched the Swachh Bharat Mission (SBM) in 2014. The SBM seeks to promote cities as engines of economic growth through improvement in the quality of urban infrastructure, with assured service levels and efficient governance. It aims to address the challenges in the management of MSW and support cities in developing modern and appropriate systems (MUD, 2016). It emphasizes on the integrated solid waste management (ISWM) system (MUD, 2016), with an aim to reduce the amount of waste being disposed while maximizing resource recovery and efficiency. Based on this aim, an assessment of local needs and conditions should lead to the selection of an appropriate mix of processes and technologies. The preferred waste management strategies include the following.

**Reduction and reuse at source**

The most preferred option for waste management is to prevent the generation of waste at various stages including the product design, production, packaging, use and reuse. Waste prevention helps reduce handling, treatment and disposal costs. It also reduces leachate, air emissions and greenhouse gases (GHGs) which have significant impacts on the environment.

**Waste recycling**

The next preferred alternative is the recovery of recyclable material resources and strengthening a waste recycling chain through segregation, collection and reprocessing to create new products.

**Waste to composting**

The organic fraction of waste can be composted to improve soil health and agricultural production adhering to relevant norms.

**Waste disposal**

Residual waste at the end of the hierarchy, which ideally comprises inert waste, should be disposed in sanitary-lined landfills which are constructed in accordance with stipulations of the SWM Rules, 2016.

The ISWM concept is closely linked to the 3R approach (reduce, reuse and recycle), which is also aimed at optimizing MSW management from all the waste-generating sectors (households, commercial and institutional establishments, parks and gardens, construction and demolition, urban agriculture, and safety and health-care facilities) and involving all the stakeholders (waste generators, service providers, informal sector, regulators, government, and community or neighbourhoods). The adoption of the 3R concept helps to minimize the amount of waste to be handled by the municipal authority, minimizing the public health and environmental risks associated with it.

**Waste to energy**

Where material recovery from waste is not possible, energy recovery from waste through production of heat, electricity or fuel is preferred. Biomethanation, pyrolysis, gasification, waste incineration, production of refuse-derived fuel (RDF) and coprocessing of the sorted dry rejects from MSW in cement kilns are commonly adopted as waste-to-energy (WTE) technologies.

Most wastes that are generated find their way into land and water bodies without proper treatment causing severe water pollution. They also emit GHGs like methane and carbon dioxide and add to air pollution. Any organic waste from urban and rural areas and industries is a resource due to its ability to get degraded, resulting in energy generation. The problems caused by solid and liquid wastes can be significantly mitigated through the adoption of environment-friendly WTE technologies that will allow treatment and processing of wastes before their disposal. These measures would reduce the quantity of wastes, generate a substantial quantity of energy from them and greatly reduce environmental pollution. India's growing energy deficit is making the central and state governments become keen on alternative and renewable energy sources. Waste to energy is one of these, and it is garnering increasing attention from both the central and state governments. While the Indian government's own figures would suggest that the cost of WTE is somewhat higher than other renewable sources, it is still an attractive option as it serves a dual role of waste disposal and energy production.

According to the Ministry of New and Renewable Energy (MNRE, 2017), there exists a potential of about 1534 MW from urban waste (Table 16.2). The ministry is also actively promoting the generation of energy from waste, by providing subsidies and incentives for the projects. The ministry is promoting all the technology options available for setting up projects for recovery of energy from urban wastes. In developed countries, environmental concerns rather than energy recovery are the prime motivator for WTE facilities, which help in treating and disposing of wastes. Energy in the form of biogas, heat or power is seen as a bonus, which improves the viability of such projects. While incineration and biomethanation are the most common technologies, pyrolysis and gasification are also emerging as preferred options. A common feature in most developed countries is that the entire waste management system is

**Table 16.2** Waste generation: international comparison (kg/capita/day)

| Country | | Projected for | |
|---|---|---|---|
| | 2016 | 2030 | 2050 |
| India | 0.57 | 0.70 | 0.90 |
| China | 0.43 | 0.56 | 0.67 |
| USA | 2.24 | 2.40 | 2.53 |
| Brazil | 1.04 | 1.17 | 1.35 |

*Source* Kaza et al. (2018)

being handled as a profitable venture by private industry or non-government organizations with tipping fee for treatment of waste being one of the major revenue streams.

## Impact of MSW on Environment and Health

The decomposition of waste into constituent chemicals is a common source of local environmental pollution. This problem is especially acute in developing countries. Very few existing landfills in the developing countries would meet environmental standards accepted in industrialized nations, and with limited budgets there are likely to be few sites rigorously evaluated prior to use in the future. The problem is again compounded by the issues associated with rapid urbanization. A major environmental concern is the gas released by decomposing garbage. Methane is a by-product of the anaerobic respiration of bacteria, and these bacteria thrive in landfills with high amounts of moisture (Srivastava & Singhvi, 2013). Methane concentrations can reach up to 50% of the composition of landfill gas at maximum anaerobic decomposition. A second problem with these gases is their contribution to the enhanced GHG effect and climate change (Goorah et al., 2009). Waste products, like plastic and rubber, when burnt pollute the atmosphere with noxious fumes. Organic solid wastes emit obnoxious odour on their decomposition and make the environment polluted. Liquid leachate management varies throughout the landfills of the developing world. Leachate poses a threat to local surface and groundwater systems. Leachates from refuge dumps percolate into the soil and contaminate underground water. Scavengers and stray animals invade the roadside garbage and litter the waste over large area causing much aesthetic damage to the atmosphere.

Direct health risks concern mainly the workers in this field, who need to be protected, as far as possible, from contact with wastes. Inadequately disposed off or untreated waste may cause serious health problems for populations surrounding the area of disposal. There are also specific risks in handling wastes from hospitals and clinics. For the general public, the main risks to health are indirect and arise from the breeding of disease vectors, primarily flies and rats.

Uncontrolled hazardous wastes from industries mixing up with municipal wastes create potential risks to human health. Traffic accidents can result from toxic spilled wastes. There is specific danger of concentration of heavy metals in the food chain; a problem that illustrates the relationship between MSWs and liquid industrial effluents containing heavy metals discharged to a drainage/sewerage system and/or open dumping sites of MSWs and the wastes discharged thereby maintains a vicious cycle.

Some other types of problem are as follows (Alam & Ahmade, 2013):

1. Vectors like rats and insects invade refuse dumps and spread various diseases like plague, salmonellosis, trichinosis, endemic typhus, etc.
2. During handling and transfer of hospital and clinic wastes, disease transmission may take place.

3. Water and food contamination through flies causes various diseases in humans as dysentery, diarrhoea and amoebic dysentery.
4. Water supply, if gets contaminated with pathogens present in solid wastes, may result in cholera, jaundice, hepatitis, gastroenteric diseases, etc.
5. Choking of drains and gully pits by solid wastes results in water logging which facilitates breeding of mosquitoes and results in the spread of diseases like malaria and plague (Chadar & Chadar, 2017).

## Municipal Solid Waste Management in Delhi

The state of Delhi is considered to be the largest producer of solid waste in India. Of the three municipal corporations, namely Municipal Corporation of Delhi, North Delhi Municipal Corporation (NDMC) and Delhi Cantonment Board, later two together generate around 9500 metric tons of waste that is dumped at the three landfill sites of Ghazipur, Okhla and Narela-Bawana. As per the official figures of Delhi Pollution Control Committer (DPCC, 2014), the existing WTE plants at Ghazipur, Okhla and Narela-Bawana process 1500 metric tons, 1950 metric tons and 3000 metric tons of solid waste and produces 12 MW, 16 MW and 24 MW of electricity, respectively (Table 16.3).

As Delhi continues to grow, the requirement for more landfill/dumping sites is ever increasing. Over 14 landfill sites have already been used up by municipal bodies, the three disposal sites currently in use have already far exceeded their capacity. Still, they continue to receive waste.

### Some Case Studies in Delhi

An attempt is made in this paper to review two WTE projects in Delhi for the purpose of understanding their importance in addressing the problem of MSWM in big cities like Delhi.

The details of the projects whose cases have been presented in this paper are as given under the following:

1. Case I: Timarpur-Okhla Municipal Solid Waste Management project
2. Case II: Biomethanation plant of Delhi Technological University (DTU) campus

**Table 16.3** Details of MSW-processing/disposal facilities in Delhi

| | | |
|---|---|---|
| Sanitary landfill sites | 04 | – |
| Compost plants | 03 | 2200 MTD |
| WTE plants | Operational 03 | 6450 MTD |

*Source* DPCC (2014)

**Case I: Timarpur, Okhla Municipal Solid Waste Management Project**

Taking note of the grave situation of waste management in Delhi, the state government thought it is appropriate to find a solution by developing a WTE project through public–private partnership (PPP) route.

Timarpur-Okhla municipal solid waste management project is the first commercial WTE facility in India that aims to convert one-third of the Delhi garbage into the much-needed electricity. This WTE plant at Okhla is known as Delhi Timarpur-Okhla Waste Management Co Pvt Ltd. (TOWMCL) and is an initiative of M/s Jindal ITF Ecopolis. The JITF Urban Infrastructure Ltd. won the bid to develop the project on a Built Own Operate and Transfer (BOOT) basis, in a PPP with the Delhi government. The TOWMCL is a special-purpose vehicle which is responsible for the implementation of project activity.

The project is the first and the largest integrated waste management project ever being set up in the country, aiming for a sustainable solution (zero waste concept) taking MSW through an environment-friendly process to generate clean and renewable energy from MSW. The project is CDM registered with United Nations Framework Convention on Climate Change (UNFCCC) for earning carbon credits. The estimated amount of annual average GHG emission reductions of this project is 308,262 $CO_2$. Existing project site at Okhla was an 'open dump'.

The project was earlier envisaged to be developed at two different locations, that is, Timarpur and Okhla. About 650 TPD of MSW was envisaged to be processed at the Timarpur site while 1300 TPD of MSW was envisaged to be processed at Okhla site for the preparation of refuse-derived fuel (RDF). Additionally, 100 TPD of green waste (waste collected from gardens like dry leaves, cut grass, etc.) were to be utilized at Okhla site for generation of biogas through biomethanization plant. The Timarpur site has been dropped from the project, and the entire waste (1950 TPD) is now processed at the Okhla site. The project cost is estimated at ₹200 crore (PDD, 2014).

The incineration plant was commissioned in January 2012 and is designed to process 1950 TPD MSW at Okhla site for the preparation of RDF and generating 16 MW of electricity. NDMC and MCD will be providing the waste free of cost. The proposed project is an environment improvement project which treats MSW and converts waste to energy.

Preproject scenario.

According to project participants (PDD, 2014), in the absence of the project activity, the waste would have continued to be dumped in the existing landfills without any pretreatment as per the existing practice, leading to the release of GHGs over a period of time. Despite the introduction of MSW Rules, 2000 (management and handling), there was no structured system available for management and handling of MSW across most of the states.

The project activity would achieve significant reduction in GHG emission due to the following two components:

1. Avoidance of methane emission by dumping solid waste in the landfill site.

2. Avoidance of $CO_2$ emissions by producing electrical energy supplied by power plants running on MSW and supplying the electricity to the grid. Therefore, the project contributes to sustainable development.

The owner of the project activity, that is TOWMC, believes that the project activity has the potential to enhance the economic, environmental and social life of the people in the region and would have beneficial effect on the local industries and employment in the region.

**Case II: Kitchen waste biomethanization plant at DTU campus**

Biomethanization technology offers an excellent alternative for decentralized processing of solid biodegradable waste and avoids the contamination of landfill sites.

Delhi Technological University (DTU) and Bhaba Atomic Research Centre jointly took up the project of installation of Nisargruna biogas plant at campus in the backyard of college canteen having capacity to process 500 kg of kitchen waste/day. Nisargruna technology is an environment-friendly technology, which delivers two valuable products, which are methane and manure. The waste generated in DTU's main kitchen is processed in this biogas plant. If 500 kg of kitchen waste is fed in the digester per day, 50 kg of manure and 20 kg of biogas are supposed to be generated by the plant. Biogas so generated from the plant is supplied to the university's canteen and the organic manure is used for horticultural requirements of the university.

The biogas generated from this plant has better calorific value and is used for cooking purpose. The manure generated in the process is weed free and rich in organic carbon contents. Hence, it is also a soil conditioner.

The Nisargruna technology differs from the classical anaerobic digesters in various aspects. It has a broader scope to accept a variety of raw materials mentioned earlier, while the anaerobic system developed in our country is mainly used for processing animal dung. It is a high rate biomethanization process. It uses a mixer to homogenize waste with water into free-flowing slurry. Process involves pretreatment of the homogenized biodegradable waste slurry in an aerobic digester for a limited period (about 3–4 days). The waste material like, kitchen (food) waste, vegetable waste, abattoir waste, cattle dung, poultry, manure, etc., can be used to process the same for biomethanization or generating biogas in decentralized units.

## Decentralized Application of the Nisargruna Technology

Nisargruna technology offers great opportunity for decentralized processing of solid biodegradable waste generated by various sources like residential sector, vegetable and fruit mandies/markets, hotels and restaurants, and also by other processing industrial units which generate biodegradable/organic MSW. The problem of handling MSW of this nature be minimized and contributed to environmental sustainability of urban areas.

Rapid urbanization is posing a serious problem of huge solid waste generation and its eco-friendly disposal. As discussed in the paper, now there are many tested technologies/methods with which the processing of MSW can be done as per the relevant rules.

The SBM which envisages scientific way of processing and disposing of MSW now provides a great opportunity for municipal local bodies in handling MSW, by way of providing technical guidance for handling of MSW and by supporting them with appropriate policy interventions in doing so. Accordingly, the local municipal bodies and state governments are expected to take appropriate decisions to handle the MSW in various cities of India so as to minimize the public health and environmental risks associated with it.

## Solid Waste Management Chain Capital Investment Requirements for Achieving a Swachh Bharat and Clean India

India's GDP for the year ending 31 March 2015 was ₹106.44 lakh crore. Investments estimated for sanitizing India (Capital Expenditure, CAPEX) for five years in-line with SBM and associated components as in the base case have been estimated by Centre for Policy Research, CII (Dasgupta et al., 2015) to be approximately ₹389,642 crore. This sanitation expenditure is expected to be significant, and these estimates show that it could be up to 0.7% on an annual basis for the next five years. As per some estimates, this spike in investments in sanitation could lead to the mitigation of the estimated 6.4% GDP losses in India. The estimate of losses undertaken in 2007 did not monetize environmental pollution and mitigation costs. This reflects that the spike in spending over the next five years and the returns to the economy from Swachh India investments if implemented and adopted in a time-bound manner could be 3–4 times the investments made.

The benefits to the economy and society are potentially even greater as there are a number of benefits which have not been quantified, including the potential impact on the infant mortality rates and climate resilience. This effort when planned well could provide the equity and dignity issues that it will help to address.

MSW collection ranges from 70 to 90% in major metro cities and is lower than 50% in smaller cities. The equipment required for collection, transportation treatment and disposal activities in urban areas is the most significant capital cost of the MSW system and varies by the size of the city. It is assumed that a combination of tricycles and auto tippers would be used to serve half the population each along with long haul vehicles (large capacity vehicles which are used for transportation over large distances) and compactors. Estimated equipment requirement is set out in Table 16.4. Nearly, ₹20,000 crore is required for providing capital investments for urban SWM sector over a five-year period. Collection and transportation investments comprise

**Table 16.4** Capital expenditure requirements for urban SWM till 2019

| Item | Numbers | Costs (₹ crore) |
|---|---|---|
| Tricycles | 297,578 | 595 |
| Auto tippers | 53,993 | 5129 |
| Bins | 154,742 | 493 |
| Refuse compactors or vehicles | 2045 | 920 |
| Street sweeping equipment | 1,818,352 | 182 |
| Composting | – | 7938 |
| WTE | – | 2750 |
| Landfill | – | 1838 |
| Total | 2,326,710 | 19,845 |

*Source* Dasgupta et al. (2015)

37%, while treatment and landfilling comprise 54% and 9% of the investments, respectively (Table 16.4).

Dasgupta et al. (2015) have attempted to establish CAPEX requirements for a five-year period and corresponding operation and maintenance (OPEX) for a period of ten years to assess the share in overall sanitation requirements as in Table 16.5. The estimated costs for implementing SBM, both CAPEX (till 2019) and O and M expenses for ten years are approximately ₹22,815 lakh crore. Another ₹43,200 crore is expected to be spent on soft components such as IEC, capacity building and administration.

A few highlights that are evident from Table 16.5 are as follows:

1. Expenses towards sanitation are nearly 3.3 times that of MSWM.
2. Urban investments are about 42%, while those in rural India is about 58%.
3. O and M expenses for ten years are nearly 1.6 times of capital investments till 2019.

**Table 16.5** CAPEX and OPEX for SWM for 10 years

| | | SWM (crore) |
|---|---|---|
| CAPEX | Urban | 19,846 |
| | Rural | 2969 |
| | Total | 22,815 |
| OPEX | Urban | 136,489 |
| | Rural | 47,023 |
| | Total | 183,512 |
| Total | Urban | 156,335 |
| | Rural | 49,992 |
| | Total | 206,327 |

*Source* Dasgupta et al. (2015)

SWM sectors are labour intensive and would add to job opportunities across the unskilled and skilled labour markets. It is estimated that to construct individual toilets alone, 160 crore person-days are required; this is nearly 5.5% of a five-year annual average of the Mahatma Gandhi National Rural Employment Guarantee Act (MNREGA) person-months consumed. The programme also provides for ₹7700 crore of IEC and capacity-building opportunities and ₹5.5 lakh crore (for ten years) of O and M opportunities to various service providers in sanitation and SWM sectors.

## Conclusion and Suggestions

Litter-free urban areas require a significant behavioural change from all stakeholders–residents, floating population and other users. In addition to providing infrastructure, it is essential that a large-scale behavioural change programme be undertaken to create awareness for segregation of waste by households and limit littering on the streets. There is a significant value addition in terms of managing solid waste at the stage of treatment and disposal of waste, wherein the solid waste is scientifically controlled through landfills or other equivalent systems. The various processes of solid waste treatment are composting, vermin composting, incineration, power generation, fuel pelletization and biomethanation. These processes ensure that the waste is stabilized and made inert before it is disposed. However, an efficient collection and transportation management system is necessary for maximizing the benefits of a scientific treatment and disposal system. Therefore, the importance of a 100% efficient primary collection with source segregation and transportation system cannot be emphasized enough.

The Government of India has initiated a number of programmes to create civic infrastructure which have been articulated to leverage private monies and create markets. It is hoped that scale and sequencing of sanitation and SWM projects across urban and rural India will encourage industry to invest in the capacity to deliver these projects. Active efforts will be required and enable governments at the implementation level to develop innovative approaches to planning, designing and delivering projects. SBM and other schemes rightly focus on outcomes, not prescriptive solutions—this should be backed by innovative approaches to financing, given the expectations of scale.

Proper SWM has to be undertaken to ensure that it does not affect the environment and cause health hazards to the people living there. At the household level, proper segregation of waste has to be done, and it should be ensured that all organic matter is kept aside for composting, which is undoubtedly the best method for the correct disposal of this segment of the waste. In fact, the organic part of the waste that is generated decomposes more easily, attracts insects and causes disease. Organic waste can be composted and then used as a fertilizer.

Categorization of waste at collection and also at disposal point should be made and be available in public domain. Government should take initiative to encourage universities and technical institution to take up waste management in its curriculum. Assistance of academic institutions should be solicited in characterization of waste in their vicinity. Thereby most part of India would be covered and location-specific appropriate solutions for waste management can be developed. It can also help to select suitable WTE technologies for particular regions. Manufacturing of non-recyclable polyethylene bags should be banned or research should be initiated to develop biodegradable polyethylene. Protection of groundwater contamination from leachate percolation from open dump/landfill site should be made compulsory. Appropriate technological solution should be adopted to achieve this goal.

## References

AlamP, A. K. (2013). Impact of solid waste on health and the environment. *International Journal of Sustainable Development and Green Economics, 2*(2), 165–168.

ChadarSN, C. K. (2017). Solid waste pollution: A hazard to environment. *Recent Adv Petrochem Science, 2*(3), 1–3.

CPCB (2017). *Report of central pollution control board on implementation of solid waste management rules, 2016*. Ministry of Environment and Forests.

CSE (2019). *Solid waste in India*. Center for Science and Environment. Available at https://csestore.cse.org.in. Accessed on May 7, 2019.

Dasgupta, S., Ramaswamy, R., Noronha, K., Rao, S., Seetharaman, R., George, N., Singh, A., Singh, T., Dhiman, S., & Bhol, A. (2015). *Swachh Bharat: Industry engagement—scope and examples. SCI-FI: Sanitation initiative, research report*. Centre for Policy Research.

DPCC (2014). *Annual review report of Delhi Pollution Control Committee*. Govt. of NCT of Delhi.

Goorah, S., Esmyot, M., & Boojhawon, R. (2009). The health impact of nonhazardous solid waste disposal in a community: The case of the Mare Chicose landfill in Mauritius. *Journal of Environment Health, 72*(1), 48–54.

Kaza, S., Yao, L., Bhada-Tata, P., & Van Woerden, F. (2018). *What a waste 2.0: A global snapshot of solid waste management to 2050. Urban Development Series*. World Bank.

MNRE (2017). Annual report (2016–17). Ministry of New and Renewable Energy, Government of India.

MUD. (2016). *Municipal solid waste management manual*. Central Public Health and Environmental Engineering Organization.

PDD (2014). *Project design document form for CDM project activities*. The Timarpur-Okhla Waste Management Company Pvt Ltd., Integrated Waste to Energy Project.

Srivastava, S., & Singhvi, R. (2013). Impact of solid waste on health and the environment. *International Journal of Science and Research, 4*(9), 1770–1773.

# Chapter 17
# Environment Hazards from Municipal Solid Waste of Urban Space: A Global Perspective

**Amithy Jasrotia and Prama Esther Soloman**

**Abstract** Problems associated with the management of Municipal Solid Waste (MSW) have been addressed very often, but very little is known about the impacts of open dumping of the MSW on the quality of the resources and its subsequent impact on the health of the community at large. As per the Solid Waste Management Rules, 2016, notified by Governmnet of India, solid waste excludes industrial waste, biomedical and e-waste, battery waste and radioactive waste generated in the area under the local authorities. This paper focuses on the extent of the problem associated with the dumping sites. The study draws attention to the worst affected group of human population in developed and underdeveloped/developing nations of the world. The menace of the MSW is common in all parts of the globe. But the way of handling it and the perception of the population towards its management are different in different parts of the world. Environment knows no boundary and thus has the ability to affect any life form without any regard to caste, colour, creed, race, economic status or religion. This study attempts to investigate which sections of the society are the worst sufferers of the developmental processes in the developing and the developed nations. After reviewing the literature, it has been observed that lower classes are more prone and are closer to risk as compared to higher class, though lower classes are contributing less in creating these hazards. Thus, factors like socio-economic differences, environmental racism and environmental injustice play a vital role in determining the affected population in and around the municipal waste management process in both developed and developing nations.

**Keywords** Environment hazards · Municipal solid waste (MSW) · Urbanization · Risk

---

A. Jasrotia (✉)
Department of Sociology, University of Rajasthan, Jaipur, Rajasthan, India

P. E. Soloman
Indira Gandhi Centre for Human Ecology, Environmental and Population Studies, University of Rajasthan, Jaipur, Rajasthan, India



A. Kateja and R. Jain (eds.), *Urban Growth and Environmental Issues in India*,
https://doi.org/10.1007/978-981-16-4273-9_17

## Introduction

Environmental concerns have developed enormously at global level after the close observation of negative consequences of rapid urbanization and economic development. The repercussions on health have been in the form of thronged diseases and infections. However, universally, a number of initiatives have been taken by civil society as well as by self-governed bodies to tackle the problem. The twin face of urbanization is that on the one end we have bright, lustrous, clean urban space and on the other hand, after the cleaning and collection of waste of urban space, we have increased solid dumps in the form of big heaps mostly in the outskirts of major cities. The practice of dumping solid wastes without any sanitary land filling practice is happening in many developing countries (Vasanthi et al., 2008). The composition of the municipal solid waste (MSW) depends upon several factors like location, lifestyle, food habits, cultural traditions, climate and economy of a nation (Patle et al., 2014; Srivastava et al., 2014). Due to the rise in population and urbanization, the annual waste generation has increased (Idris et al., 2004). Further, the extent of the waste that produced from urban localities also acts as the function for human development index that heavily depends up on human life expectancy, gross domestic product and education indices (Chandrappa & Das, 2012).

Lot of time we blame overpopulation as the main cause of creation of dumping zones. But if we observe closely, apart from increasing population, other factors like industrialization, technology and affluent lifestyle are contributing equally. Literature showed that all dumping sites are putting enormous challenges and adverse effects on health. Further, large numbers of medical and scientific uncertainties have arrived (Nelson et al., 1987: 211). Apparently, we have three different types of waste sites; solid waste sites also known as dumps, hazardous waste sites and uncontrolled hazardous waste sites due to industrial and factories activities (Szasz & Meuser, 1997: 102). Apart from this, it has also been realized that environmental threat is not equal. Lower classes are more prone and are closer to *'risk'* as compared to higher class. In terms of Marx, and by applying NEP model**,** Catton and Dunlap (1978) emphasized that, 'in the long run, as environmental constraints become more obvious, ecologically aware "haves" are likely to opt for increased emphasis on managed scarcity to cope with them. The results would be disastrous for the "have-nots" (Catton & Dunlap, 1978: 47)'. Thus, risk has been defined as follows:

> The probabilities of physical harm due to given technological or other processes (Beck, 1992: 4).

It has often been seen that many a times open dumps (wherever land is available) are the most common and easiest way of disposing off the solid waste without regard to safety, health threats and aesthetic dilapidation. The heterogeneous nature of the waste makes the situation worse since in majority of cases the waste is not segregated. According to a report, uncontrolled growth in urban areas has led to deficiency of infrastructural services such as water supply, sewerage and MSW management in many Indian cities, thus leading to unhygienic and contaminated surroundings and supplies (Central Public Health and Environmental Engineering Organization, Solid Waste Management manual, 2000). The Central Pollution Control Board (2015) of India in their action plan for management of municipal solid waste has clearly pointed out that the municipalities fail to retain systematic data on waste generation and composition. The same document also states that the major fraction of Indian waste is comprised of organic fraction, but waste segregation is not followed at any step of waste management, and even if the waste is segregated at source, there is final mixing of all types of waste in the dustbins.

People living in the vicinity of the dumping site are the worst sufferers since they are exposed to polluted soil, air, water and pathogens. Several studies have shown relationship between health and environmental contamination at the waste dump. In most of the cases, waste management involves direct exposure of the workers to solid waste and many ragpickers are found scavenging for recyclable materials from the dumpsite. These two groups are in direct contact with the solid waste, making them prone to various types of infectious and chronic diseases (Nwanta et al., 2010). A large number of ailments like respiratory diseases, irritation of the skin, eyes and nose, gastrointestinal problems, psychological disorders, dermal infections, respiratory tract infection and allergies have been reported in many studies conducted on health of people associated with the dumpsites and waste handling (Gouveia & Prado, 2010; Sekyere et al., 2013).

The menace of the MSW is common in all the parts of the globe. But the way of handling it and the perception of the population towards its management are different in different parts of the world. This paper is going to study the extent of problems and impact on the human beings residing near the dumping sites by putting light on two cities each from developed and developing countries.

## Troubles Associated with Dumping Sites

There are numerous troubles associated with an open dump that have adverse impact on the soil quality, surface water and groundwater resources, agricultural activities, air quality and ultimately human beings who occupy the topmost position in the food chain. The extent of adulteration of surrounding soil, surface and groundwater resources and ultimately of the ecosystem depends on the characteristics and amount leachate generated (Mor et al., 2006). Studies also showed high values of electrical conductivity (EC), total dissolved solids (TDS), biochemical oxygen demand (BOD), chemical oxygen demand (COD) and chloride in both leachate and groundwater around the dumpsites (Fatta et al., 1999; Rathod et al., 2013).

Dumping of MSW also leads to the problems of acidification, nitrification and bacterial contamination of groundwater that may be used for drinking purposes. Land has always been an easily available medium for disposal of waste of all types. Therefore, the polluted soil near active dumpsites becomes great resource for many airborne diseases and after the consumption of vegetables grown there (Anekwe & Nwobodo, 2002). The unsorted MSW is a rich source of heavy metals both in numbers and amount. Metals tend to persist in the environment and could accumulate to toxic levels (Ihedioha et al., 2017). The route of entry of such metals in waste is due to the significant amounts of plastic, paper, metal garbage and batteries (Pasquini & Alexander, 2004; Zhang et al., 2002).The effects of exposure to some toxic metals on the human health as a result of unsegregated waste dumping are mentioned in Table 17.1.

The major objectives of this paper are as follows:

- To highlight the serious impacts of the MSW dumpsites on human health.
- To draw attention to the worst affected group of human population in developed and underdeveloped/developing nations of the world.
- To find out the factors governing the response of affected population towards hygiene and sanitation.

## Methodology

The current study aims to fulfil the objectives by focusing on the case studies. The sites chosen for the case studies represent the developed and the developing areas of the world. People are forced to live in unhygienic conditions around dumpsites. Ragpicking being very common exposes a significant part of the population to unhealthy conditions in developing nations. On the other hand, a landfill located in the developed economies addresses the subject of MSW management with a different approach. The rate of MSW generation has been very high, whereas ability to effectively collect and safely dispose off the waste has been very sluggish in the developing nations.

The problems affecting the residents in the vicinity were studied, and the possible reasons for their prevailing circumstances were analysed. Impacts of the dumpsite on the health of the people were also assessed. The information was obtained by the relevant publications on the related work, personal field visits of site and questionnaire analysis in one dumpsite of the developing nations and finally examining the sociological aspects of the problem.

## Case Study I

The very first case study has been taken from the secondary data and is the case of one developed nation that represents more educated people, people with more sense of sanitation and equipped with more resources to handle the waste problem. This case study had been done by Adeola in 1994. The focal point of this study was to address the critical issue of hazardous waste and the immediate health problems associated with it. In Louisiana, the waste dump known as the Devil's swamp, a hazardous waste dumpsite is considered to be the most dangerous of all the landfills. It is situated along the Mississippi River near Scotlandville community of East Baton

**Table 17.1** Harmful impact of some heavy metals on human health

| Heavy metal | USEPA* Regulatory limit (ppm) | Deadly effects | References |
|---|---|---|---|
| Ag | 0.10 | Exposure may cause skin and other body tissues to turn grey or blue–grey, breathing problems, lung and throat irritation and stomach pain | ATSDR (1990) |
| As | 0.01 | Affects essential cellular processes such as oxidative phosphorylation and ATP synthesis | Tripathi et al. (2007) |
| Ba | 2.00 | Causes cardiac arrhythmias, respiratory failure, gastrointestinal dysfunction, muscle twitching and elevated blood pressure | Acobs et al. (2002) |
| Cd | 5.00 | Carcinogenic, mutagenic, endocrine disruptor, damages lungs and make bones fragile, affects calcium regulation in biological systems | Salem et al. (2000) |
| Cr | 0.10 | Hair loss | Salem et al. (2000) |
| Cu | 1.30 | Brain and kidney damage, elevated levels result in liver cirrhosis and chronic anaemia, stomach and intestine irritation | Salem et al. (2000) |
| Pb | 15.00 | Excess exposure in children causes impaired development, reduced intelligence, short-term memory loss, disabilities in learning and coordination problems, risk of cardiovascular disease | Salem et al. (2000) |
| Ni | 0.20 | Allergic skin diseases such as itching, cancer of the lungs, nose, sinuses, throat through continuous inhalation, immunotoxic, neurotoxic, genotoxic, affects fertility, hair loss | Salem et al. (2000) |

(continued)

**Table 17.1** (continued)

| Heavy metal | USEPA* Regulatory limit (ppm) | Deadly effects | References |
|---|---|---|---|
| Hg | 2.00 | Autoimmune diseases, depression, drowsiness, fatigue, hair loss, insomnia, loss of memory, restlessness, disturbance of vision, tremors, temper outbursts, brain damage, lung and kidney failure | Neustadt and Pieczenik (2007) |
| Se | 50.00 | Dietary exposure of around 300 μg/day affects endocrine function, impairment of natural killer cells activity, hepatotoxicity and gastrointestinal disturbances | Vinceti et al. (2001) |
| Zn | 0.50 | Dizziness, fatigue, etc. | Hess and Schmid (2002) |

*USEPA: United State Environmental Protection Agency (2009)
*Source* Modified from Dixit et al. (2015)

Rouge covering an area of about 150 acres of land (Adeola, 1994; Hardy, 2017). Very high levels of polychlorinated biphenyls were detected around the area in the 1980s, and they still have the potential to affect the population residing nearby.

## *Existing Situation*

- The landfill received about 1,500 tons of hazardous waste every day.
- The site was close to Mississippi River causing severe pollution of the river as well as to the groundwater resources due to undesirable topography.
- Air and soil were also polluted with toxic waste and due to the closeness to industries involving oil refining and chemical manufacturing process.
- Scotlandville and Alsen were also known as cancer corridors due to exposure of the population to xenobiotics (chemicals that are foreign to the biosphere), polychlorinated biphenyls, etc., that acted as carcinogens, mutagens and teratogens.
- Minority groups like African–Americans were forced to reside around such areas due to poverty. Blacks constituted 90 per cent of the employees in the landfill area.
- Race and sex were more significant factors than socio-economic factors (like education and family income) in deciding the residential proximity to waste site.

### *Environmental and Health Impacts of the Dumpsites*

- People suffer from lung cancer, miscarriages among pregnant ladies, birth defects, leukaemia, immune disorder and mental disorders.
- High lung cancer rates are reported among White males, but there is no such documentation among the African–American males and other minority communities.
- The African–American males and other minority communities are obviously the worst affected population due to their exposure to occupational hazards.
- The toxic corridor is found to extend from the East Baton Rouge to the Mississippi River.

### *Management Problems Related to Social and Political Conditions:*

- Powerlessness and political alienation were the major reasons why minority communities were forced to live in such polluted and toxic environments.
- These people had no voice in the decision-making processes.
- Environmental racism was also a cause of the plight of these communities.
- Poverty, lack of knowledge, job requirements and weak political organizations for resistance were also some factors that forced the population to reside in such areas.

## Case Study II

A study conducted by Norton et al. (2007) focused on the location of solid waste treatment plants with respect to community, race and wealth in the developed economy of North Carolina, USA. The authors used census block groups to get racial and economic features, while data about the solid waste treatment plants was obtained from permit records of the solid waste facility. To conduct the study, a total of 419 solid waste treatment plants were taken into consideration.

### *Existing Situation*

- The North Carolina Division of Waste Management (NCDWM) was the agency accountable for distributing authorizations to solid waste disposal sites in the state.
- Their study showed a strong inverse relationship between the existence of allowed solid waste facilities and population density.

- The occurrence of a solid waste treatment facility was larger in block groups or areas with larger proportions of individuals of colour and lower wealth as compared to high wealth block groups.
- It was observed that new private facilities came up more in areas with larger population of non-White people.
- Most of the solid waste management facilities were controlled by the local government, while the industrial solid waste facilities were privately owned.

## *Environmental and Health Impacts of the Dumpsites*

- Most of the MSW disposal facilities had become a source of groundwater contamination.
- The bad odour and toxic gases from the landfills caused pollution.
- Municipal solid waste disposal sites were potential cause of low birth weight in children, respiratory diseases and site-specific cancers of the stomach, liver and pancreas.
- Bad odour, heavy traffic and visual pollution from landfills discouraged coming up of amenities related to community health, education, sports, etc.
- Noise pollution caused by the heavy truck traffic and daily activities at the disposal site affected well-being and induced stress.
- The population residing nearby (that included community of colour or low-income groups) was more vulnerable to the effects of waste facility due to age factor, more disease occurrence rate, less availability of healthy food and higher exposure to hazards related to occupation.

## *Management Problems Related to Social and Political Conditions*

- Being a developed economy, not many problems related to engineering facilities around the waste facilities or problems in collection were reported.
- The study indicated that population that produced the maximum amount of waste disposed it off in areas with less resources and low political influence; this raised likelihood of serious health effects on community. This also excluded the feedback assessment between production and consumption that might have helped in reducing the amounts of waste produced.
- In North Carolina, even the rich and White communities had a scarcity of waste disposal facilities.
- Thus, suggestions to build new landfills especially in the eastern Carolina had created apprehensions that the state will import large amounts of solid waste from outside the state, making the situations worse.

- Several private companies had applied for permits to operate landfills in eastern North Carolina which was inhabited by poor, deprived communities and is historically African–American area of North Carolina, and the state legislature was even considering to give them approval until one-year moratorium on new landfills was adopted in 2006.

## Case Study III

The third case study has also been undertaken from secondary sources. This study was done by Henry et al. (2006) who had conducted their study in Kenya, a low-income developing country. Their focal point was to understand the MSW management and the challenges which many developing countries are encountering in today's time. Kenya is also a developing country, having land area of about 569,137 $km^2$. The total population is around 33.4 million, and the urban population is nearly 34% of the total population. The suburban areas are by and large densely populated with inadequacy of good infrastructure, amenities and basic social services. The population in suburban areas mainly constitute the rural–urban migrants, low-earning labourers, unemployed and poor people. They reside in very poorly constructed housing structures. Similar to a number of developing countries, the socio-economic pressures on Kenya have also constantly affected the economic growth negatively.

In Kenya, local authorities are responsible for the collection and disposal of both the solid and liquid municipal wastes in the areas under control. Currently, MSW is largely disposed in open dumps without any effective pollution control measures. Open dumps are not even regularly monitored to check for accumulation and transfer of pollutants. Like other developing nations, due to lack of funds and mismanagement of the resources, Kenya also is facing many environmental problems related to waste management that is leading to environmental pollution. Thus, Kenyan plans are underway to convert its open dumpsites into sanitary landfilling. Kenya is divided into eight provinces, and one of them is Nairobi. Nairobi is the chosen area studied for the present paper, by keeping in consideration that the situation in most of the other provinces is similar to that of Nairobi.

### *Existing Situation*

- The waste disposal facility provided by the Nairobi city was found to be very ineffective, since the MSW collection vehicles were not operating to their full capacities. More than one-third of the collection trucks were not operational in all the localities.
- The collection of waste was greatly affected by the rainy season. The amount of uncollected MSW in the high- and middle-income areas was quiet high during

the rainy seasons when roads became deteriorated and transport system was hampered.

- As per the local residents (interviewed by the author), the MSW disposal was inappropriate and was done on paths, river sides, and road sides, and collection services were poor indicating the failure of the local authorities.
- Environmental impacts were not given any consideration in the selection of dumpsites. The condition in the low-income suburban areas was worse due to indiscriminate disposal of MSW.
- None of the five local authorities under study were found practising safety measures and sanitary measures like putting daily soil cover on the waste heap or putting fence around the dumpsite, etc. Not even a single dumpsite in areas under study met the fundamental provisions for protection of groundwater resources from leachate migration as they did not have any liner system.
- The leachate from the dumpsite washed pollutants in the Nairobi Dam is specially made in 1953 to store potable water for the residents, and it was observed that some slum inhabitants irrigated their food crops with that contaminated water.
- The occurrence of noxious chemicals in MSW and the dumpsite may be attributed to the absence of strict monitoring system, dumping of non-segregated waste in the dumpsites and the existence of several unauthorized dumpsites.

### *Environmental and Health Impacts of the Dumpsites*

- An incidence of diarrhoea was triggered due to the consumption of vegetables grown in a farm which used the surface water that was polluted by MSW dumped upstream.
- An abandoned sand quarry (Mwenderi, in Eldoret town) was illegally used as MSW disposal site. This site acted as a water catchment area and many small streams from here drained water into a nearby river (Sosiani River). Thus, the pollutants from this contaminated catchment area were causing pollution of the freshwater river.
- Illegal discharge of solid and liquid municipal wastes into the rivers (e.g. Kisat River) was responsible for the pollution of the rivers as well as the lakes.
- The Nairobi Dam, which was meant to supply drinking water, had been severely polluted by MSW that gets washed away along with storm water from the nearby Kibera slums.
- The mal odours and unsightly view from the dam and the dumpsites had affected value of residential property in the neighbourhood and has reduced the value of the property.
- The pollution of the lakes had affected the habitat of the flamingos leading to a drastic decline in their population.
- Polluted rivers entering into the ocean posed a threat to the recreational beaches and sites as well as the coastal ecosystem such as coral reefs.

### *Management Problems Related to Social and Political Conditions*

- The upgradation of the slums of Nairobi had not been implemented because some leaders were fearful of losing considerable number of voters who might move out of the area due to slum upgradation efforts. There were cases when such projects declined particularly for political reasons.
- The local authorities maintained overestimated numbers of poorly trained employees that drained out major part of the budget in the form of salaries and so a meagre amount was left for other services. Subsequently, collection and efficient disposal of MSW was not duly carried by the authorities. Thus, they faced a lot of financial difficulties.
- Though the legislations related to waste disposal, collection, waste management and project implementation were adequate, still the authorities were incompetent to implement them effectively.
- Rural–urban migration of the population required attention and needed to be controlled by increasing the rural incomes.
- The rapid rise in the urban population had also resulted in the incapability of the authorities in handling the waste problems effectively.
- The municipal waste management system, being centralized in Kenya, always faced a delay in the grant of approvals and hence a delay in the discharge of services.
- A waste collection system, unaffected by the seasonal variations needed to be established so that delays caused could be avoided.
- There was no data on the deteriorating quality of water in the adjacent areas to find out the effect of pollution on the affected population. No regular monitoring of the soil and water quality was done to examine the effect of the dumpsites on the water and land resources.

## Case Study IV

The fourth case study is from Rajasthan, the largest state of the Indian subcontinent. This case study has been done by authors themselves by field visit and with the help of primary data. Jaipur is the capital of Rajasthan state. Jaipur district has 47.6% rural and 52.4% urban population (Census of India, 2011). According to the Rajasthan State Pollution Control Board, by 2025 solid waste generated by the city would be 4671 metric tons/day if there is annual increase of 1.5% per capita/day in the amount of solid waste generation (Master Development Plan-2025, 2011). The Jaipur Municipal Corporation has the responsibility of the collection and disposal of the MSW.

## *Existing Situation*

- There are two major open dumpsites in Jaipur city, one is located at Mathuradaspura and the other at Sewapura. Another dumpsite at Langariya has refuse-derived fuel facility and has been handed over to a private firm for waste management.
- Field surveys indicated that MSW was dumped without segregation on the land.
- No sanitary practices were followed like the addition of the cover material on a regular basis. The waste was covered with a layer of soil only when there was strong agitation by the nearby residents and disinfection around the dumpsite was done once in three to six months.
- The waste was being dumped without any regard to the protection of the nearby water resources. There was no lining to prevent the seepage of leachate in the soil and in the groundwater.
- The land on which the dumping was taking place belonged to the government.
- As per the local inhabitants, waste was being dumped in this area for the last 18–20 years. The people residing in the surrounding areas were mostly the lower sections of the society. They were poor, and most of the families had been residing here before the area was declared as an open dump.
- Apart from dumping of waste in the dump yards sometimes the trucks also dumped the waste outside the dumpsite and along the road sides.
- Despite of the resistance and protest by the local inhabitants, waste dumping continued in the area.
- Though there was no surface water body adjacent to the dumpsite, the groundwater was found to be affected by various heavy metals due to leachate migration. This was evident because the hand pumps had been sealed by the Public Health Engineering Department (PHED), which were the only sources of water supply to the inhabitants.
- The groundwater in 10 km radius had been reported to be polluted due to the dumpsite. This water was unfit for drinking or cooking and has bad taste, and many times has bad odour too.
- 88.2 per cent of the entire population cultivated vegetables and cereals in agricultural land around the dumpsite and used underground water for agricultural activities.
- Ragpickers collected and sorted out recyclables and plastic waste from the dump.
- A large number of stray as well as domesticated animals and birds were found preying on the waste dump, and these acted as carriers of various kinds of diseases in the surrounding areas.

## *Environmental and Health Impacts of the Dumpsites*

- People faced problem in breathing and several people suffered from respiratory diseases, eye, skin and lung infections especially aged people.

- People complained that the children remained ill and suffered from skin diseases, diarrhoea, dysentery, gastrointestinal problems and vector-borne diseases like malaria, dengue, typhoid, etc.
- There was a crucial problem of flies and mosquitoes which made the day-to-day activities of local people difficult. These files and mosquitoes also spread communicable diseases.
- Diseases like tuberculosis and cancer have also been reported. According to some residents, about ten people have died due to cancer.
- One case of death due to birth defect was reported in a family residing close to the dumpsite.
- The trees showed stunted growth, and crop plants showed reduced yield and delayed flowering and fruiting around the dumpsite.
- Some people around Langariya had reported that cows and buffaloes in the nearby locality have died due to excessive consumption of plastic bags. Reduced production of milk, untimely death among domesticated animals, etc., were also common. Animals suffered from skin troubles, ticks and mites and developed tumours several times.

### *Management Problems Related to Social and Political Conditions*

- The people residing in the area were poor and belonged to the lower-income groups. They had little or no political support, thus even after strong agitation by the residents, little action was taken to solve their problems.
- Poverty sometimes makes the residents ignore serious environmental troubles. This was apparent because some local inhabitants were in favour of the dumpsite as they received money for illegal trading of the diesel that they stole from the trucks.
- There was no plan or policy implemented at the government level to rehabilitate the affected population or to relocate the dumpsite.
- The segregation of the waste was not taken care of by the local municipal authorities. The MSW dumpsite freely received sufficient amount of toxic waste in the form of used batteries, plastics, tyres, old paint cans and many times biomedical waste in form of expired and unused medicines, syringes, plasters, etc.
- With the advent of the Clean India Movement, strong focus is on the cleanliness of the urban and VIP areas of the city but the areas where the poor or marginalized sections are residing are highly ignored.

## Discussion

The problem of the waste production will increase in the future and will affect more land area and population. With the increased use of xenobiotic compounds in various industrial processes, the amount of persistent chemicals would increase in the nature. These chemicals act as carcinogens, mutagens and teratogens and cause various incurable and fatal diseases in the human populations. Though the waste for some privileged groups of the societies in both developed and developing nations may become out of sight, due to the persistent nature and ability to biomagnify in the food chain they may also cause damage to even such group indirectly. Leaching of pollutants in the groundwater aquifers, transfer of vapours and gaseous pollutants through the atmosphere may affect individuals to various extents.

But as we can analyse through the aforementioned case studies, the worst affected groups are the poorer ones who are forced to have a direct association with the polluted environments. Due to their poor economic status, they are not in a position to choose a clean and hygienic dwelling place but are mostly compelled to reside around polluted industrial areas, dumping grounds, landfill sites, polluted water bodies, etc. Unawareness and lack of sense of cleanliness a another reason, especially in the developing nations, because of which they fail to understand the implications of the polluted environments.

There are several communities in the world that experience the unbalanced influences of environmental vulnerabilities. Even in developed economies there is great disparity in impacts faced by the people due to environmental pollution, this has led to the rise in the concept of environmental racism (Cole & Foster, 2001). Cole and Foster (2001) have used environmental racism and environmental justice for explaining aspects related to people's race and class of the political economic system or political economy in the societies that face attack. In 1960, many scholars found that environmental hazards like air pollution, noise pollution, pesticide poisoning, waste disposal, etc., are not distributed equally across the various groups. They are unequally distributed in various races and income groups. In Western countries, as most developed, race is the main factor or predictor of exposure to environmental hazards. Later, few studies closely observed the sites of dumping and brought into being that mainly non-White residents lived close to these sites. In addition, study done by National Law Journal observed that 'people of colour are not protected as vigorously by enforcement of environmental laws as Whites' (ibid, 2001: 57). There was a clear racial division found in the way of US government in handling toxic waste sites and punishment to the polluters. Further, research also showed racial disparities existing in the location of toxic waste material. Thus, location of the site is also important to know the disparity. Consequently, environmental racism exists in both developed and developing nations. But the factors differ in both. In developing nations, economic background as well as the social background (e.g. caste, region of belonging) of the people forms the criterion which deprives them of their basic right to live in clean environment. Lack of knowledge as well as no or low education is also a major factor. Even if the level of awareness is increasing, still the marginalized groups are not in a state to make them heard.

## Conclusion

Thus, on the edge of modernity and development, human beings have created a bundle of toxic materials in the form of wastage. Further, this is not only life threatening to the immediate targets but also impeding the most distant ones slowly through many indirect ways. Today, almost every countryside of towns and cities is holding the encumbrance of their wastages. In *Poverty of Philosophy* (1963), Marx observed that in 'every second the application of modernity in the form of chemicals is changing the nature of soil' 'and thus the environment (Marx, 1963: 162–63)'. Kautsky in his work, '*The Agrarian Question'*, has also talked, by including a full section, on '*the exploitation of the countryside by the town*' (Kautsky, 1899). Lately, the theory of *risk society* has also pointed that the path of modernization and development is piling up the irreversible threats and risks (Beck, 1992). By analysing the existing environmental hazardous, it has also been concluded that all the problems of environment have erupted due to the imbalance existed between the cultural and political sphere. At the one hand, we want development by adopting all the faces of modernity, but on the other hand the prevalence of mismanagement in the political apparatus created irrational policies. Both Giddens and Beck have talked about the consequences of modernity in their own way. Both have focused on risks which are global in nature and possess no boundaries. Giddens, although laid greater blame on the scientific expertise for the creation of risk society. Beck has put more blame on the vested interests of political authority by releasing risky policies for using scientific paradigms in a highly irresponsible way (Beck, 2014). Consequently, structural crisis in the form of heaps of dump is the result of global industrialism and modernity. Beck, very beautifully, opined that the emergence of risk in the modern society has not arisen due to the dearth of resources or poverty of people; it is reasonably the result of overexploitation of the means, rapid economic growth and swift technological development. But then the question arises, if risk from the dumping sites is due to extreme consumerism, then why do the people who are not a part of this consumerism have to pay the price of greater risk of dumping sites. Owing to the lacunae in government policies, higher rate of inequality prevails in the case of environmental risks. Risk has close relation with class positions. By quoting Rossi (2014), we conclude that ironically the modern world has been transformed into an irrational or risk society by encapsulating the rationality. This has been further ensued due to automized, unplanned and unperceived processes of development (Rossi, 2014: 64).

## Suggestions

In order to prevent the hazardous impact of the municipal solid waste, it is essential to implement a proper and engineered landfill facility, regularly monitored for pollution control, especially in the developing nations. These landfills should be emission controlled, having tree cover around the dumpsite, leachate collection system and storm water management systems. Also, the ground and surface water pollution must be a significant part of this management system. The laws framed and facilities provided for the management of solid waste must be applicable to all the sections of the society irrespective of their caste, colour, race, gender or economic status. The approach of waste reduction is the best possible option for effective management of the waste generated, but it is hardly put into practice because the impacts of the waste dumps are not faced by the privileged class of the society. Community involvement especially from the worst affected groups in siting, management and operation of the dumpsites can go a long way in the effective management and bridging the gap between the policymakers and the affected populations (Henry et al., 2006). Community involvement will also promote formation of fairer policies (Martuzzi et al., 2010). Sensitizing the masses towards cleanliness and impacts of exposure to waste, especially in the developing/under-developed parts of the world through community-based programmes must be done. Participation of the public will make the existing system more effective and efficient. There must not be any compromise on the health concerns of the individuals involved at any level of the waste management system. Inequalities in the health impacts may be refuted by prevention, and health promotion initiatives should be undertaken in combination (Martuzzi et al., 2010).

## References

Acobs, I. A., Taddeo, J., Kelly, K., & Valenziano, C. (2002). Poisoning as a result of barium styphnate explosion. *American Journal of Industrial Medicine, 41*(4), 285–288.

Adeola, F. O. (1994). Environmental hazards, health, and racial inequity in hazardous waste distribution. *Environment and Behavior, 26*(1), 99–126.

Anekwe, M. A. N., & Nwobodo, K. C. A. (2002). Long term effect of MSW disposal on soil properties and productivity of sites used for urban agriculture in Abakaliki. *Nigeria. Bioresource Technology, 83*(3), 241–250.

ATSDR (Agency for Toxic Substances and Disease Registry) (1990). *Toxicological profile for silver*. US Department of Health and Human Services, Public Health Service.

Beck, U. (1992). *Risk society: Towards a new modernity (Theory, culture and society series)*. Sage Publications Ltd.

Beck, U. (2014). *Pioneer in cosmopolitan sociology and risk society* (Ed.) (Vol. 18). Springer.

Catton, W., & Dunlap, R. (1978). Environmental sociology: A new paradigm. *The American Sociologist, 13*(1), 41–49.

Central Pollution Control Board (2015). *Action plan for management of municipal solid waste ministry of environment, forest and climate change* (pp. 1–36).

Census of India (2011). *Rajasthan series-09, part XII-B District census handbook Jaipur, village and town wise primary census abstract (PCA)*. Directorate of Census Operations, Rajasthan, pp. 1–569.
Central Public Health and Environmental Engineering Organization (2000). *Manual on municipal solid waste management*. Ministry of Urban Development Government of India.
Chandrappa, R., & Das, D. B. (2012). Introduction. In: *Solid waste management. Environmental science and engineering (environmental engineering)*. Springer.
Cole, L. W., & Foster, S. R. (2001). *From the ground up: Environmental racism and the rise of the environmental justice movement*. New York University Press.
Dixit, R., Wasiullah, MalaviyaD., Pandiyan, K., Singh, U. B., Sahu, A., Shukla, R., Singh, B. P., RaiJP, SharmaPK., & LadeH, PaulD. (2015). Bioremediation of heavy metals from soil and aquatic environment: An overview of principles and criteria of fundamental processes. *Sustainability, 7*(2), 2189–2212.
Fatta, D., Papadopoulos, A., & Loizidou, M. (1999). A study on the landfill leachate and its impact on the groundwater quality of the greater area. *Environmental Geochemistry and Health, 21*(2), 175–190.
Gouveia, N., & Prado, R. R. D. (2010). Health risks in areas close to urban solid waste landfill sites. *Revista de Saúde Pública, 44*(5), 859–866.
Hardy, S. (2017). Sierra club calls for action on long-polluted devil's swamp in North Baton Rouge. The Advocate, February 16, 2017.
Henry, K. R., Yongsheng, Z., & Jun, D. (2006). Municipal solid waste challenges in developing countries: Kenyan case study. *Waste Management, 26*, 92–100.
Hess, R., & Schmid, B. (2002). Zinc supplement overdose can have toxic effects. *Journal of Pediatric Hematology Oncology, 24*, 582–584.
Idris, A., & InancB, H. M. N. (2004). Overview of waste disposal and landfill/dumps in Asian countries. *Journal of Material Cycles and Waste Management, 6*, 104–110.
Ihedioha, J. N., Ukoha, P. O., & Ekere, N. R. (2017). Ecological and human health risk assessment of heavy metal contamination in soil of a municipal solid waste dump in Uyo, Nigeria. *Environment Geochemistry Health, 39*(3), 497–515.
Kautsky, K. (1899). *The agrarian question* (Vols. 2). Zwan, Winchester, Mass.
Martuzzi, M., Mitis, F., & Forastiere, F. (2010). Inequalities, inequities, environmental justice in waste management and health. *European Journal of Public Health, 20*(1), 21–26.
Marx, K. (1963). *The poverty of philosophy* (pp. 162–163). International Publishers.
Master Development Plan-2025 (2011). *Jaipur region* (Vol. 1). Jaipur Development Authority (JDA).
Mor, S., Ravindra, K., & DahiyaRP, C. A. (2006). Leachate characterization and assessment of groundwater pollution near municipal solid waste landfill site. *Environmental Monitoring and Assessment, 118*, 435–456.
Nelson, N., Baker, S., Levine, S. P., Young, L., O'Connor, J., Hill, R. D., & Sarofim, W. D. G. (1987). Cleanup of contaminated sites. In UptonAC. LaveLB (Ed.), *Toxic chemicals, health, and the environment* (pp. 205–279). John Hopkins University Press.
NeustadtJ, P. S. (2007). Toxic-metal contamination: Mercury. *Integrative Medicine, 6*, 36–37.
Norton, J. M., Wing, S., Lipscomb, H. J., Kaufman, J. S., Marshall, S. W., & Cravey, A. J. (2007). Race, wealth, and solid waste facilities in North Carolina. *Environmental Health Perspectives, 115*(9), 1344–1350.
Nwanta, J. A., Onunkwo, J., & Ezenduka, E. J. (2010). Analysis of Nsukka metropolitan abattoir solid waste and its bacterial contents in South Eastern Nigeria: Public health implication. *Archives of Environmental & Occupational Health, 65*(1), 21–26.
Pasquini, M. W., & Alexander, M. J. (2004). Chemical properties of urban waste ash produced by open burning on the Jos Plateau: Implications for agriculture. *The Science of the Total Environment, 319*, 225–240.

Patle, A. V., Williams, S. P. M. P., Gabhane, J., & DharH, N. P. B. (2014). Microbial assisted rapid composting of agriculture residues. *International Journal of Science Engineering Research, 5*(5), 1097–1099.

Rathod, M., Mishra, H., & Karmakar, S. (2013). Leachate characterization and assessment of water pollution near municipal solid waste landfill site. *International Journal of Chemical and Physical Sciences, 2*, 186–199.

Rossi, I. (2014). Reflexive modernization. In: Beck, U. (Ed.), *Pioneer in cosmopolitan sociology and risk society* (Vol. 18). Springer.

Salem, H. M., Eweida, E. A., & Farag, A. (2000). Heavy metals in drinking water and their environmental impact on human health. *ICEHM 2000* (pp. 542–556). Cairo University.

Sekyere, E. O., Rev Alfred, K. B., & Yahaya, A. K. (2013). Victims of location: Health implications on household living proximal to the Owhim solid waste dump in the Kumasi Metropolitan Area (KMA) of Ghana. *Journal of Environmental Science and Water Resources, 2*(2), 53–61.

Srivastava, R., Krishna, V., & Sonkar, I. (2014). Characterization and management of municipal solid waste: A case study of Varanasi city, India. *International Journal of Current Research Academy Review, 2*(8), 10–16.

Szasz, A., & Meuser, M. (1997). Environmental inequalities: Literature review and proposals for new directions in research and theory. *Current Sociology, 45*(3), 99–120.

Tripathi, R. D., Srivastava, S., Mishra, S., Singh, N., Tuli, R., Gupta, D. K., & Maathuis, F. J. M. (2007). Arsenic hazards: Strategies for tolerance and remediation by plants. *Trends in Biotechnology, 25*, 158–165.

USEPA (2009). *Drinking water contaminants*. United States Environmental Protection Agency (EPA).

Vasanthi, P., Kaliappan, S., & Srinivasaraghavan, R. (2008). Impact of poor solid waste management on groundwater. *Environmental Monitoring and Assessment, 143*, 227–238.

Vinceti, M., Wei, E. T., MalagoliC, B. M., & Vivoli, G. (2001). Adverse health effects of selenium in humans. *Reviews on Environmental Health, 16*, 233–251.

Zhang, F. S., Yamasaki, S., & Nanzyo, M. (2002). Waste ashes for use in agricultural production: I. Liming effect, contents of plant nutrients and chemical characteristics of some metals. *Science and Total Environment, 284*(1–3), 215–225.

# Chapter 18
# Urbanization and Pharmaceutical Waste: An Upcoming Environmental Challenge

**Kritika Sharma and Garima Kaushik**

**Abstract** Human pharmaceuticals are consumed in high quantities worldwide, and this amount will continue to increase as a result of improving healthcare systems and longer life expectations. Pharmaceuticals comprise a broad range of active compounds which have complex structures, functions, and physicochemical and biological properties. Pharmaceuticals are the essential element in the field of medical practice, and their benefits and side effects on human and animal health are extensively recognized. Today, the pharmaceutical industry is regarded as one of the prime and rapidly rising industries in the whole world. The USA is the major consumer of medicines, and Europe comes on second position. In the context of India, the pharmaceutical industry is the third largest country in terms of volume. As the population is increasing, the pharmaceutical industries are also increasing worldwide due to the high demand of medication. Along with this, the problem of pharmaceutical waste is also increasing due to the discharge of industrial wastewater directly into the environment and another source is the disposal of healthcare waste. Pharmaceutical waste has become an emerging problem in last few years. Even very small concentration from microgram to nanogram per litre causes very harmful impact on the environment. Pharmaceuticals enter into the environment via sewage wastewater, and human and animal excreta, and from there they enter into the food chain and cause toxicological impact on living beings. Thus, the proper treatment methods and waste management techniques of pharmaceuticals have become the major requirement to eliminate the harmful impacts of pharmaceuticals on the environment. The chapter will provide a generous review of global pharmaceutical market status, generation of pharmaceutical waste, the occurrence of pharmaceuticals in the environment, treatment and management of pharmaceutical waste.

**Keywords** Environment · Healthcare waste · Pharmaceuticals · Urbanization

G. Kaushik (✉)
Department of Environmental Sciences, School of Earth Sciences, Central University of Rajasthan, Bandar Sindri, Rajasthan, India

K. Sharma
Department of Life Sciences, Hemchandracharya North Gujarat University, Patan, Gujarat, India



A. Kateja and R. Jain (eds.), *Urban Growth and Environmental Issues in India*,
https://doi.org/10.1007/978-981-16-4273-9_18

## Introduction

Urbanization, a social transformation driven by economic, social and environmental factors, is one of the greatest challenges that humanity is facing. As a consequence, the environmental implication of this process is impacting large sections of society and communities. However, the magnitude of this impact may vary over time depending upon the physical, social, economic and developmental scales. Still positive casual relationships are found between urbanization, water quality, energy demand, air quality, carbon emission, percentage green cover, etc. Increasingly, with the greater public participation and government decentralization processes, the perception towards urbanization is shifting from challenges to opportunities. Still there are expanding challenges with the growing population that need urgent attention and one such is pharmaceutical pollution.

In medical sciences, progress has been made during the past century and the optimistic results on the society are well renowned. Pharmaceuticals are the essential elements in the field of medical practice and their benefits and side effects on human and animal health are extensively recognized (BIO Intelligence Service, 2013). Pharmaceutical industries are directly linked to human well-being and welfare, and so it is of vital importance for good health and productivity. Today, the pharmaceutical industry is considered to be one of the largest and fastest growing global industries. It is a major source of employment generation and foreign exchange earnings for many countries around the globe. The major factor responsible for the increase in the pharmaceutical consumption is ageing of the population, which elevates the requirement of pharmaceuticals to treat and cure the various kinds of diseases. Average age of population is the major factor in determining the demand for pharmaceuticals. The extent of urbanization can be an excellent indicator of pharmaceutical waste discharge to the environment. But increase in the per capita demand for drugs is directly proportional to the age of the population. Combinations and amount of pharmaceuticals are different for ageing people rather than the young. But due to tremendous characteristics of pharmaceuticals, there may be an adverse effect on flora, fauna and humans. Active pharmaceutical ingredients (APIs) are used to cure the diseases in human beings and animals, but they can go into the environment through manufacturing practices, disposal of expired medicines, biomedical wastes from the hospitals and by animal excreta used in agriculture land treatment as manure (Daughton & Ternes, 1999; Larsson, 2014). In the environment, APIs have been distinguished frequently at diverse range (Hughes et al., 2013; Ramirez et al., 2009). However, various researches are going on regarding the effect of these compounds on human and animal health through toxicological investigations (Sharma & Kaushik, 2017). Even the trace amount of pharmaceuticals in the environment is very harmful to the biodiversity. In addition to this awareness, a new area of study is emerging and it is known as pharmacovigilance. Pharmacovigilance is related to detection, assessment, understanding and prevention of the pharmaceutical-related problems (Alshakka et al., 2015). Detection of pharmaceuticals in different environmental matrices is done by various advanced instruments such as gas chromatography, liquid chromatography and UV spectrophotometry from nanogram per litre to microgram per litre. Basically, PCs have been identified in effluents of wastewater treatment plant (WWTP), river streams, marine water, drinking water, soil and groundwater (Rivas et al., 2012). Since past decade, awareness has been raised towards this issue. So, the removal of pharmaceuticals from the environment is becoming the essential need of present time. Pharmaceutical wastewater is treated by various mechanisms consisting of aerobic and anaerobic degradation, sorption, desorption, advanced oxidation process, etc. Regulation of management plans will be the most promising to mitigate the risk caused by pharmaceuticals, such as segregation of waste at the source of origin, advancement in disposal techniques, initiatives for green pharmaceutical development, consulting with stakeholders, preventive measures for release of pharmaceuticals into the environment, etc. Regulatory bodies and industrial authorities may carry out the pharmaceutical waste mitigation and preventive measures of release into the environment (Daughton, 2003; Fatta-Kassinos et al., 2011).

This chapter discusses pharmaceutical market and consumption status, generation of pharmaceutical waste and their entrance route in the environment. Impacts of pharmaceutical waste on biodiversity and their management strategies including the pharmaceutical wastewater are also highlighted.

## Pharmaceutical Market Status: India and World

Emerging markets play a major role in future development of the industry. Leading pharmaceutical companies have expedited the efforts to reinforce their existence within these markets by way of research and development (R and D) investment, accrediting the deals, purchasing and other partnerships. Contrary to demographic and disease trends, changes in market supply and demands, development of regulatory needs, it has become very difficult for manufacturers to formulate the strategies which are required for taking over in these areas. The developed countries are chief manufacturers of pharmaceutical drugs. According to World Health Organization (WHO), majorly five countries in the world, namely USA, Japan, Germany, UK and France, make two-third part of the global pharmaceutical producers, whereas China and India are major low-cost producers of pharmaceuticals. In terms of volume, Indian pharmaceutical industry is the third largest country, and in terms of value, it is at thirteenth rank in the world (India Brand Equity Foundation, 2017). In the context of India, pharmaceutical market accounts for about 2.4% of the global pharmaceutical market and 10% in terms of volume. India is the second most populated country in the world with a population of approximately 1.3 billion. In terms of volume, India is the largest provider of generic medicines which accounts for 20% of global market and USA is the largest consumer of Indian drugs attaining drugs worth 1.4 billion USD during the year 2007 (Mason, 2009). According to Organization of Pharmaceutical Manufacturers of India, Indian pharmaceutical manufacturing companies export their products to more than 65 countries and USA is their largest customer. Globally, the average consumption of pharmaceuticals is 15 g/capita/year. The annual consumption of pharmaceuticals is estimated to be approximately 100,000 tons at the global level (India Brand Equity Foundation, 2017). USA and Europe are the major consumers (55% and 24%, respectively) of medicines, while Japan comes at third rank with 14% consumption of world's total amount (Académie Nationale de Pharmacie, 2008). In Europe, consumption of human pharmaceutical products differs between 50 and 150 g API/capita/year. Germany and France take part in two-thirds of the whole European consumption. Although the population in Germany is higher than France, it has higher per capita consumption than Germany (BIO Intelligence Service, 2013). In USA, more than 10 million women use oral contraceptives, and Canada consumes high amount of drugs like ibuprofen, naproxen, carbamazepine, etc. (Boxall, 2004).

The Indian pharmaceutical industry comprises approximately 250 large-scale and 8000 small-scale units producing various formulations consisting of 350 different kinds of bulks of drug (Siringi & Naaz, 2017). Transformations in socio-economic conditions are the major factor contributing to the important modifications in India's epidemiological profile. Working adults account a major part of the whole population, and most people are becoming prosperous and long lived, but still, the users of Indian health services are confronted with challenges correlated with prevention and treatment of general chronic diseases like heart disease, obesity, cancer and diabetes. Although Indian market is leading in generic medicines, growing incomes, improved medical facilities and insurance could provide a helpful opportunity to manufacturers (Kuchenreuther & Sackman, 2014). India is gradually picking up the production and consumption of pharmaceutical drugs. According to Corporate Catalyst India (CCI) report, Indian pharmaceutical market can reach up to 55 billion USD by 2020 (Corporate Catalyst, 2014).

## Generation of Pharmaceutical Waste and Global Status

The production of a new drug involves the supportive efforts of a large group of skilled personnel specialized in medicinal, analytical and organic chemistry, biochemistry, microbiology, pharmacology, toxicology, chemical engineering and pathology. Consequently, with the diverse characteristics of pharmaceutical R and D, chemical and biological waste is generated at large extent (USEPA, 1991). Pharmaceutical waste is the result of the pharmaceutical manufacturing procedure. Pharmaceutical waste contains liquids, solids and gases, but at present, the main focus is on the liquid part which is generally known as pharmaceutical waste water. Effluents generated from pharmaceutical industries differ from manufacturer to manufacturer. Hence, due to such variety of pollutants, it is very difficult to depict a particular effluent. Pharmaceutical wastewater generation takes places during various stages of pharmaceutical manufacturing (as shown in Table 18.1) and is generally estimated in terms of pH, total suspended solids (TSS), chemical oxygen demand (COD), biological oxygen demand (BOD), oil and grease, sulphates and chlorides (Parmar et al., 2015). In pharmaceutical industries, wastewater is generally produced by the washing

**Table 18.1** Wastewater and residual waste generation during various processes in pharmaceutical industry

| Process | Wastewater | Residual wastes |
|---|---|---|
| Chemical synthesis | Process wastewater with organic contaminations | Solid crystals |
| Reaction | Process wastewaters with spent solvents, catalyst, reactants, high BOD, COD, TSS, pH range 1–11 | Reaction residuals and reactor bottom wastes |
| Separation | Spills, leaks, spent separation solvents | Separation residues |
| Purification | Spillage, leakages, spent separation solvents | Purification residues |
| Drying | Spillage, leakages, spent separation solvents | – |
| Natural product extraction | Equipment cleaning, spills, leaks, spent solvents, low BOD, COD, TSS and pH of 6–8 | Spent raw materials (plants, roots, etc.) |
| Fermentation | Spent broth during fermentation, wastewater containing sugar, nutrients, etc., high BOD, COD and pH range 4–8 | Waste filter cake, fermentation residues |
| Formulation | Tools cleaning, spillage, leakage, spent solvents, low BOD, COD, TSS and pH range 6–8 | Particulate matter, waste packaging, refused tablets, capsules, etc. |

*Source* Rana et al. (2017)

process of equipment. Although released wastewater is in small volume, it contains highly contaminated compounds. The quantity of wastewater pollution varies from the size of the industry and their type of practices. From the past few decades in India, pharmaceutical industries have increased rapidly. In pharmaceutical industries, manufacturing units contain raw material and a variety of medicines which generate the effluent containing harmful ingredients (Rana et al., 2017). Some lower-income countries are manufacturing pharmaceuticals for their growth in international market. The manufacturing process includes the synthesis of APIs and formulation into doses which creates the point source risk associated with pharmaceutical industrial waste discharge (Fick et al., 2009). Patancheru Industrial Estate in Hyderabad, India, has 90 manufacturing units which together generate 1.5 mega litres of wastewater daily (Larsson et al., 2007).

The majority of the human population resides in Asia and Africa (approximately 75%), but the major regulatory advancements regarding pharmaceuticals in the environment are witnessed in the North American and European countries (Hughes et al., 2013). Besides this, the vast population lives in these areas and there is a steady shifting of manufacturing sector from few years to low-income countries on a contractual basis.

## Entrance of Pharmaceuticals in the Environment

Over the past few years, pharmaceuticals have been attaining intense attention as increasing bioactive substances in the environment. They are regarded as emerging pollutants in the environment (Kümmerer, 2009). Pharmaceuticals enter into the environment frequently by various ways and are widely distributed in small concentrations which can disturb the water quality and definitely affect the drinking water supply, biodiversity and humans (Kolpin et al., 2002; Sirés & Brillas, 2012). The concentration of pharmaceuticals in the environment is detected between the ranges of ng $L^{-1}$ to $\mu g\ L^{-1}$. Though the presence of pharmaceuticals in the environment has been from decades, quantification of their degrees into the environment started only few years ago (Jjemba, 2006). The key sources of pharmaceuticals into the environment are pharmaceutical manufacturing units, healthcare units and households. Also, if wastewater treatment plants do not have the efficiency to properly eliminate the pollutants in treatment process, the released contaminated wastewater leads to landfill leachate which may cause groundwater contamination (Daughton & Ruhoy, 2009; Jørgensen & Halling-Sørensen, 2000; Nagarnaik et al., 2012; Phillips et al., 2010), etc., as shown in Fig. 18.1.

Pharmaceutical manufacturing units are the major contributors of pharmaceutical pollution in the environment as they produce a large quantity of wastewater

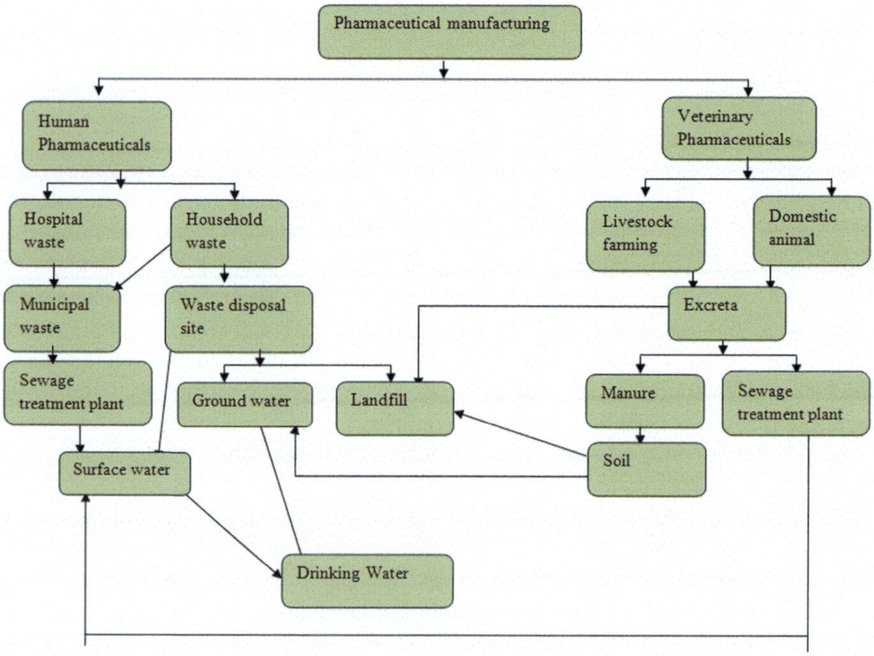


**Fig. 18.1** Various pathways of pharmaceutical compound entrance into the environment

during various stages and release it into the water bodies (Daughton & Ruhoy, 2009; Nagarnaik et al., 2012). Various studies have reported the presence of pharmaceuticals in the aquatic environment (Li, 2014; Zuccato et al., 2010). The role of households and healthcare units is disposal of expired or unused human as well as veterinary medicines, disposal of metabolized and unmetabolized medicine, washing of equipment on which pharmaceutical products were applied which can undergo the raw sewage and wastewater treatment plant (Glassmeyer et al., 2009; Ruhoy & Daughton, 2007). The effluent in wastewater is either released in water bodies directly or reclaimed for irrigation, whereas biosolids are reprocessed for soil amendment like manure or to landfill disposal (Ruhoy & Daughton, 2008). Pharmaceutical compounds can leach to groundwater via contaminated soil, and they are also found in coastal marine water (Bottoni et al., 2010; McEneff et al., 2014; Sabourin et al., 2009). Thus, the occurrence of pharmaceuticals in the aquatic ecosystem demonstrates the incomplete degradation and removal in wastewater treatment process and slow process of natural decomposition in the environment (Sharma et al., 2019, 2020).

## Impact of Pharmaceutical Waste on Humans and Biodiversity

Pharmaceuticals are complex chemicals to eliminate, and their presence in effluents and treated wastewater are potentially toxic. Even in low concentrations these micro-pollutants can cause detrimental effects to humans by consuming food and water contaminated with pharmaceuticals (Caracciolo et al., 2015). Various studies have demonstrated the chronic and acute negative impacts of environmental exposure of pharmaceuticals on humans and other living beings (Caracciolo et al., 2015; Macherius et al., 2014). Collectively, breaking down of pharmaceutical compounds and their combination with various bioactive substances may have significant effect on the environment. So, in general, pharmaceutical compounds have harmful impact on both humans and biodiversity.

### *Impact on Humans*

Although pharmaceutical compounds are present in micro to nano quantities in environmental matrices but they are biologically active compounds also and potentially active to cause great impact in human body. In a study (Sharma et al., 2020), more than 30–100 numbers of pharmaceutical compounds were found in surface waters in India. Such water becomes the part of food chain, and hence, human beings face great risk due to these micro-pollutants. Ingestion of these contaminants causes serious impacts on human health such as reproductive disorders, respiratory diseases, cancer, depression and mental retardation, cardiovascular, gastrointestinal disorders and physical abnormalities (Kim & Aga, 2007). According to WHO (2013), even the trace amount of pharmaceutical residues in drinking water may trigger substantial harmful effects on human health after their long-term exposure (WHO, 2013). Alteration in the reproductive capabilities of humans, such as decline in male fertility, birth defects and breast and testicular cancers, may be caused by the long-term pharmaceutical consumption through contaminated drinking water (Nikolaou et al., 2007).

## *Impact on Biodiversity*

As the pharmaceutical compounds enter into the waters bodies, they may disrupt any target organ, tissues or cells of the lower aquatic animals and plants (Fent et al., 2006). Drugs affect individual fitness and persistence which in turn affects the functioning of the ecosystem (Fahrig & Merriam, 1994; Smith and Blumstein, 2008). The potential effect of pharmaceuticals appears on the function of wildlife, population dynamics and ecosystem performance (Brodin et al., 2013; Dzieweczynski & Hebert, 2012; Kidd et al., 2007). In addition to direct effects, transformation in individual fitness may also cause the indirect effects. These indirect effects can take place by changes in interactions among species, alteration in population size and modifications in community composition and species richness (Brodin et al., 2013; Kidd et al., 2014).

According to the WHO (1998), emergence of antimicrobial resistance is a major issue for future health policies. The foremost problem related to the presence of antibiotic compounds in the environment is the probability for selection of pharmaceutical-resistant species. Antibiotics present in the sewage water can promote the selection pressure and endorse the transformation of a resistant gene from non-dangerous microbes to lethal pathogens (Byrne-Bailey et al., 2009; Knapp et al., 2011). Diclofenac is an extensively used the veterinary drug in South Asia that is used to treat domestic livestock (Green et al., 2004). The veterinary drug 'Diclofenac' was blamed to for driving the vulture population to the brink of its extinction as it was highly toxic leading to kidney failure. According to International Union for Conservation of Nature (IUCN), population of three *Gyps* vulture species in the Indian subcontinent has diminished since the early 1990s and now they are about to extinct (IUCN, 2004).

Pharmaceutical compounds have acute and chronic toxicities in aquatic organisms. In some studies, sertraline (antidepressant), thioridazine and diphenylamine are observed as encompassing extreme potential to acute toxicity in algae, invertebrate and fish populations. Along with this, with regard to chronic toxicity, oestrogens, selective serotonin, carbamazepine, sulphadimethoxine, clofibrate, etc., demonstrated high level of chronic toxicity towards the aquatic flora and fauna (Brausch et al., 2012). For example, exposure of endocrine disrupters (EDC) which are the by-products of manufacturing units is associated with abnormalities in thyroid functions, drop in fertility rate, declined hatching fulfilment, alteration in immune function of birds and feminization of male fish (Colborn et al., 1997). Pharmaceutical residual waste indirectly lowers the agricultural productivity, alters the agricultural patterns and leads to death of livestock (Chander et al., 2016).

## Pharmaceutical Waste Management

The risk of pharmaceuticals in the environment varies with the geography of the region, population demography, type of manufacturing sites and products, the level of pharmaceuticals use, disposal practices, environmental and climatic conditions, and removal potential of wastewater treatment plants. The risk may change due to increase in the urbanization, changes in demography and effluent instream flows, and development of advanced technologies (Brooks et al., 2006). With better knowledge to recognize the drivers for pharmaceuticals exposure in different regions, it may be possible to classify the areas which are at the higher risk. It will be more effective to control the pollution from the site of its origin and use of mitigation measures accordingly (Boxall et al., 2012). New concepts should be adopted for waste minimization for the preservation of high quality of environment. New technologies are being designed to inhibit the waste emission at origin source itself. Development of low-cost technologies for wastewater treatment and advanced methods has been found more effective for pharmaceutical industry wastewater. Cleaner production is one of the best strategies for environmental management which reduces the waste before it is generated and improvement in the resources use (Parmar et al., 2015).

The regulatory bodies of pharmaceutical manufacturing units must regularly monitor and enforce the laws for appropriate disposal of pharmaceutical waste for inhibiting environmental deterioration. In India, Central Pollution Control Board (CPCB) and State Pollution Control Boards (SPCBs) take the regulatory measures. The CPCB has provided the Minimum National Acceptable Standards (MINAS) for each industry. Standards provided for the pharmaceutical industries are shown in Table 18.2. According to that, SPCBs are required to implement the laws for pharmaceutical industries (Maria, 2013). If an industry fails to fulfil the requirements of the Water Act, 1974, the State Pollution Control Boards have the power to cut the power and water supply, close the industry or even initiate public interest litigation in front of the Supreme Court (CPCB, 1989). The Food and Drug Administration (FDA) (2013) has provided the guidelines for the appropriate disposal of drugs that are used, unused or expired. They should not be flushed into drains. Solid drugs should be assorted with litter or coffee grounds, put into the disposable containers and disposed of by community drug take-back programmes, while liquid medicines should be combined with salt, flour, charcoal and powdered spices. Packing of medicines should be multilayered and placed into an opaque container, and the lid of the container should be sealed (FDA, 2013).

**Table 18.2** Pharmaceutical (manufacturing and formulation) industry

| Parameter (effluent standards) | Standards limiting concentration in mg/l, except for pH |
|---|---|
| *Compulsory parameters* | |
| pH | 6.0–8.5 |
| Oil and grease | 10 |
| BOD (3 days at 27 °C) | 100 |
| COD | 250 |
| Total suspended solids (TSS) | 100 |
| Total dissolved solids (TDS) | – |
| Bioassay test | 90% survival after 96 h in 100% effluent |
| *Additional parameters* | |
| Mercury | 0.01 |
| Arsenic | 0.2 |
| Chromium (hexavalent) | 0.1 |
| Lead | 0.1 |
| Cyanide | 0.1 |
| Phenolics | 1.0 |
| Sulphides | 2.0 |
| Phosphate | 5.0 |

*Source* CPCB (1989)

## Good Governance and Policy Recommendations

Governance is about a process in which an organizational structure creates laws and policies, implements them, evaluates and modifies them whenever necessary and at the same time is responsive and accountable to citizens in the process of governing. Good governance is making this a sustainable model for a flourishing society characterized by quality of decision-making process (Zargar & Sheikh, 2018). In order to achieve this, a number of laws and acts exist in India (Water Act, 1974; Environment Protection Act, 1986; Hazardous Waste Rules, 2016; Bio-medical Waste Rules, 1998, 2016) but these policies are somewhere failing leading to dramatic range of pollution episodes in our ecosystems. To implement good governance in the pharmaceutical sector, a holistic approach is required involving regulatory and non-regulatory sector which can work towards public centric approach with a focus not only on economic growth but also on enhanced public participation and knowledge sharing. Interventions are required to create community awareness programme sand educating people in low-income and developing countries towards good disposal practices of pharmaceuticals such as drug take-back programme. Better ways should be adopted to curb the level of pollutants at the end of pipe treatment facility. Regulations for controlling the pollution must be stricter and firmly implemented. The concept of green pharmacy should be proposed as chiral pharmaceuticals can be a good option to minimize pharmaceutical concentrations in the environment. Innovative research should be started for the monitoring of pharmaceuticals in the environment. In this manner, ecopharmacovigilance can be a good option. Environmental risk and hazard assessment studies should be conducted. Wastewater treatment plants should be designed to eliminate the pharmaceutical compounds, and standards should be designed to check the pollution levels at public water systems.

Urbanization is an ongoing societal transformation process, although the challenges associated are far being successfully resolved. There are increasing evidences that highlight the linkages between different environmental processes such as various socio-economic and cultural aspects that are demonstrated in waste management plan of a city. Exploring and evaluating these linkages will definitely help out in framing urban policies that can curb the pollution effects. With evolving concept of smart cities, eco-cities and low-carbon cities, policy research is increasing getting recognized. So along with the process of attaining the Sustainable Development Goals (SDGs), national government is supporting urban governance in shaping urban trajectory and setting up concrete management plans for such pharmaceutical pollution loads in urban centres and creates successful models as sustainable cities.

## References

Académie Nationale de Pharmacie. (2008). Medicaments et environment. *Rapport*, 103. https://www.acadpharm.org/dos_public/1_Rapport_Med_Env_version_JMH_def_JPC.pdf. Accessed May 7, 2017.

Alshakka, M., Abdorabbo, A., Basaleem, H., Jha, N., Thamir, A., & Ravi Shankar, P. (2015). Knowledge, attitude beliefs and practices of community pharmacy dispensers in Aden, Yemen towards adverse drug reaction reporting. *World Journal of Pharmaceutical Sciences, 3*(10), 2111–2118.

BIO Intelligence Service. (2013). *Study on the environmental risks of medicinal products*. Final report prepared for Executive Agency for Health and Consumers. Paris. https://ec.europa.eu/health/sites/health/files/files/environment/study_environment.pdf. Accessed February 5, 2017.

Bottoni, P., Caroli, S., & Caracciolo, A. B. (2010). Pharmaceuticals as priority water contaminants. *Toxicological and Environmental Chemistry, 92*, 549–565.

Boxall, A. B. A., Rudd, M. A., Brooks, B. W., Caldwell, D. J., & Choi, K. (2012). Pharmaceuticals and personal care products in the environment: What are the big questions? *Environmental Health Perspectives, 120*(9), 1221–1229.

Boxall, A. B. A. (2004). The environmental side effects of medication: How are human and veterinary medicines in soils and water bodies affecting human and environmental health? *EMBO Reports, 5*, 1110–1116.

Brausch, J. M., Connors, Brooks, Rand, G. M. (2012). Human pharmaceuticals in the aquatic environment: A review of recent toxicological studies and considerations for toxicity testing. *Reviews of Environmental Contamination and Toxicology, 218*, 1–99.

Brodin, T., Fick, J., Jonsson, M., & Klaminder, J. (2013). Dilute concentrations of a psychiatric drug alter behavior of fish from natural populations. *Science, 339*, 814–815.

Brooks, B. W., Riley, T. M., & Taylor, R. D. (2006). Water quality of effluent-dominated ecosystems: Ecotoxicological, hydrological, and management considerations. *Hydrobiologia, 556*(1), 365–379.

Byrne-Bailey, K. G., Gaze, W. H., Kay, P., Boxall, A. B., Hawkey, P. M., & Wellington, E. M. (2009). Prevalence of sulfonamide resistance genes in bacterial isolates from manured agricultural soils and pig slurry in the United Kingdom. *Antimicrobial Agents and Chemotherapy, 53*(2), 696–702.

Caracciolo, A. B., Topp, E., & Grenni, P. (2015). Pharmaceuticals in the environment: Biodegradation and effects on natural microbial communities, a review. *Journal of Pharmaceutical and Biomedical Analysis, 106*, 25–36.

Chander, V., Sharma, B., Negi, V., Aswal, R. S., Singh, P., Singh, R., & Dobhal, R. (2016). Pharmaceutical compounds in drinking water. *Journal of Xenobiotics, 6*, 5774.

Colborn, T., Dumanoski, D., & Myers, J.P. (1997). *Our stolen future: Are we threatening our fertility, intelligence and survival? A scientific detective story (with new epilogue by the authors)*. New York: Penguin Group.

Corporate Catalyst India Pvt. Ltd. (2014). *A brief report on pharmaceutical industry in India*. www.cci.in/pdfs/surveys-reports/pharmaceutical-industry-in-india.pdf. Accessed January 5, 2017.

CPCB (Central Pollution Control Board). (1989). *Minimal national standards: Pharmaceutical manufacturing and formulation industry*. New Delhi: Central Pollution Control Board.

Daughton, C. G., & Ruhoy, I. S. (2009). Environmental footprint of pharmaceuticals: The significance of factors beyond direct excretion to sewers. *Environmental Toxicology and Chemistry, 28*, 2495–2521.

Daughton, C. G. (2003). Cradle-to-cradle stewardship of drugs for minimizing their environmental disposition while promoting human health. I. Rationale for and avenues toward a green pharmacy. *Environmental Health Perspectives, 111*, 757–774.

Daughton, C. G., & Ternes, T. A. (1999). Pharmaceuticals and personal care products in the environment: Agents of subtle change? *Environmental Health Perspectives, 107*, 907–938.

Dzieweczynski, T. L., & Hebert, O. L. (2012). Fluoxetine alters behavioral consistency of aggression and courtship in male Siamese fighting fish, Betta splendens. *Physiology and Behavior, 107*, 92–97.

Fahrig, L., & Merriam, G. (1994). Conservation of fragmented populations. *Conservation Biology, 8*, 50–59.

Fatta-Kassinos, D., Meric, S., & Nikolaou, A. (2011). Pharmaceutical residues in environmental waters and wastewater: Current state of knowledge and future research. *Analytical and Bioanalytical Chemistry, 399*, 251–275.

Fent, K., Weston, A. A., & Caminada, D. (2006). Ecotoxicology of human pharmaceuticals. *Aquatic Toxicology, 76*(2), 122–159.

Fick, J., Soderstrom, H., Lindberg, R. H., Phan, C., Tysklind, M., & Larsson, D. G. J. (2009). Contamination of surface, ground, and drinking water from pharmaceutical production. *Environmental Toxicology and Chemistry, 28*, 2522–2527.

Food and Drug Administration (FDA). (2013). Consumer health information: How to dispose of unused medicines. United States of America. https://www.fda.gov/files/aboutpercent20fda/published/How-to-Dispose-of-Unused-Medicines-(PDF).pdf Accessed February 8, 2017.

Glassmeyer, S. T., Hinchey, E. K., Boehme, S. E., Daughton, C. G., Ruhoy, I. S., & Conerly, O. (2009). Disposal practices for unwanted residential medications in the USA. *Environment International, 35*, 566–572.

Green, R. E., Newton, I., Shultz, S., Cunningham, A. A., Gilbert, M., Pain, D. J., & Prakash, V. (2004). Diclofenac poisoning as a cause of vulture population declines across the Indian subcontinent. *Journal of Applied Ecology, 41*, 793–800.

Hughes, S. R., Kay, P., & Brown, L. E. (2013). Global synthesis and critical evaluation of pharmaceutical data sets collected from river systems. *Environmental Science and Technology, 47*, 661–677.

India Brand Equity Foundation. (2017). *Indian Pharmaceutical industry*. https://www.ibef.org/industry/pharmaceutical-india.aspxIndia. Accessed on May 9, 2017.

International Union for Conservation of Nature (IUCN). (2004). *IUCN Red list of threatened species*. Gland, Switzerland: IUCN. http://www.iucnredlist.org. Accessed on May 9, 2017

Jjemba, P. K. (2006). Excretion and ecotoxicity of pharmaceutical and personal care products in the environment. *Ecotoxicology and Environmental Safety, 63*(1), 113–130.

Jørgensen, S. E., & Halling-Sørensen, B. (2000). Drugs in the environment. *Chemoshpere, 40*(7), 691–699.

Kidd, K. A., Paterson, M. J., Rennie, M. D., Podemski, C. L., Findlay, D. L., Blanchfield, P. J., & Liber, K. (2014). Direct and indirect responses of a freshwater food web to a potent synthetic oestrogen. *Philosophical Transactions of the Royal Society B, 369*, 20130578. https://doi.org/10.1098/rstb.2013.0578.

Kidd, K. A., Blanchfield, P. J., Mills, K. H., Palace, V. P., Evans, R. E., Lazorchak, J. M., & Flick, R. W. (2007). Collapse of a fish population after exposure to a synthetic estrogen. *Proceedings of the National Academy of Sciences, 104*, 8897–8901.

Kim, S., & Aga, D. S. (2007). Potential ecological and human health impacts of antibiotics and antibiotic-resistant bacteria from wastewater treatment plants. *Journal of Toxicology and Environmental Health Part B, Critical Reviews, 10*(8), 559–573.

Knapp, C. W., McCluskey, S. M., Singh, B. K., Campbell, C. D., Hudson, G., & Graham, D. W. (2011). Antibiotic resistance gene abundances correlate with metal and geochemical conditions in archived Scottish soils. *PLOSONE, 6*(11), e27300.

Kolpin, D. W., Furlong, E. T., Meyer, M. T., Thurman, E. M., Zaugg, S. D., Barber, L. B., & Buxton, H. T. (2002). Pharmaceuticals, hormones and other organic wastewater contaminants in U.S. streams, 1999–2000: A national reconnaissance. *Environmental Science Technology, 36*(6), 1202–1211.

Kuchenreuther, M., & Sackman, J. E. (2014). *India: Still a growing market, biopharm international*. http://www.pharmexec.com/india-still-growing-market. Accessed on April 17, 2017.

Kümmerer, K. (2009). Antibiotics in the aquatic environment: A review, part I. *Chemosphere, 75*(4), 417–434.

Larsson, D. G. J. (2014). Pollution from drug manufacturing: Review and perspectives. *Philosophical Transactions of the Royal Society B, 369*, 20130571.

Larsson, D.G., de Pedro, C., & Paxeus, N. (2007). Effluent from drug manufactures contains extremely high levels of pharmaceuticals. *Journal of Hazardous Materials, 148*(3), 751–755.

Li, W. C. (2014). Occurrence, sources, and fate of pharmaceuticals in aquatic environment and soil. *Environmental Pollution, 187*, 193–201.

Macherius, A., Lapen, D. R., Reemtsma, T., Römbke, J., Topp, E., & Coors, A. (2014). Triclocarban, triclosan and its transformation product methyl triclosan in native earthworm species four years after a commercial-scale biosolids application. *Science of the Total Environment, 472*, 235–238.

Maria, A. (2013). *The costs of water pollution in India* (pp. 30–31). Paris: CERNA.

Mason, M. (2009). World's highest drug pollution levels found in Indian stream. http://usatoday30.usatoday.com/tech/science/environment/2009-01-26-drug-india-stream_n.htm. Accessed on August 20, 2017.

McEneff, G., Barron, L., Kelleher, B., Paull, B., & Quinn, B. (2014). A year-long study of the spatial occurrence and relative distribution of pharmaceutical residues in sewage effluent, receiving marine waters and marine bivalves. *Science of the Total Environment, 476*, 317–326.

Nagarnaik, P. M., Batt, A. L., & Boulanger, B. (2012). Healthcare facility effluents as point sources of select pharmaceuticals to municipal wastewater. *Water Environment Research, 84*, 339–345.

Nikolaou, A., Meric, S., & Fatta, D. (2007). Occurrence patterns of pharmaceuticals in water and wastewater environments. *Analytical and Bioanalytical Chemistry, 387*, 1225–1234.

Parmar, R. B., Patel, R. S., & Mistry, K. P. (2015). Review on wastewater effluents from pharmaceutical companies. *Indian Journal of Applied Science, 5*(2), 221–223.

Phillips, P. J., Smith, S. G., Kolpin, D. W., Zaugg, S. D., Buxton, H. T., Furlong, E. T., Esposito, K., & Stinson, B. (2010). Pharmaceutical formulation facilities as sources of opioids and other pharmaceuticals to wastewater treatment plant effluents. *Environmental Science and Technology, 44*(13), 4910–4916.

Ramirez, A. J., Brain, R. A., Usenko, S., Mottaleb, M. A., O'Donnell, J. G., Stahl, L., Wathen, J. B., Snyder, B. D., Pitt, J. L., Perez-Hurtado, P., Dobbins, L. L., Brooks, B. W., & Chambliss, C. K. (2009). Occurrence of pharmaceuticals and personal care products in fish: Results of a national pilot study in the United States. *Environmental Toxicology and Chemistry, 28*(12), 2587–2597.

Rana, R.S., Singh, P., Kandari, V., Singh, R., Dobhal, R., & Gupta, S. (2017). A review on characterization and bioremediation of pharmaceutical industries' wastewater: An Indian perspective. *Applied Water Science, 7*, 1–12.

Rivas, F. J., Gimeno, O., & Borallho, T. (2012). Aqueous pharmaceutical compounds removal by potassium monopersulfate: Uncatalyzed and catalyzed semicontinuous experiments. *Chemical Engineering Journal,192*, 326–333.

Ruhoy, I. S., & Daughton, C. G. (2008). Beyond the medicine cabinet: An analysis of where and why medications accumulate. *Environment International, 34*, 1157–1169.

Ruhoy, I. S., & Daughton, C. G. (2007). Types and quantities of leftover drugs entering the environment via disposal to sewage revealed by coroner records. *Science of the Total Environment, 388*, 137–148.

Sabourin, L., Beck, A., Duenk, P. W., Kleywegt, S., Lapen, D. R., Li, H., Metcalfe, C. D., Payne, M., & Topp, E. (2009). Runoff of pharmaceuticals and personal care products following application of dewatered municipal biosolids to an agricultural field. *The Science of the Total Environment, 407*, 4596–4604.

Sharma, K., & Kaushik, G. (2017). NSAIDS in the environment: From emerging problem to green solution. *Annals of Pharmacology and Pharmaceutics, 2*(14), 1077.

Sharma, K., Kaushik, G., Thotakura, N., Raza, K., Sharma, N., & Nimesh, S. (2019). Fate of ibuprofen under optimized batch biodegradation experiments using Micrococcus yunnanensis isolated from pharmaceutical sludge. *International Journal of Environmental Science and Technology*, 1–14.

Sharma, K., Kaushik, G., Thotakura, N., Raza, K., Sharma, N., & Nimesh, S. (2020). Enhancement effects of process optimization technique while elucidating the degradation pathways of drugs present in pharmaceutical industry wastewater using Micrococcus yunnanensis. *Chemosphere, 238*, 124689.

Sirés, I., & Brillas, E. (2012). Remediation of water pollution caused by pharmaceutical residues based on electrochemical separation and degradation technologies: A review. *Environment International, 40*, 212–229.

Siringi, R. K., & Naaz, F. (2017). A case study on marketing strategies of the pharma industry with reference to smilax Labs, Hyderabad. *IOSR Journal of Business and Management (IOSR-JBM)*, 39–45.

Smith, B. R., & Blumstein, D. T. (2008). Fitness consequences of personality: A meta-analysis. *Behavioral Ecology, 19*, 448–455.

USEPA. (1991). *U.S environmental protection agency. Guilds to pollution prevention: The pharmaceutical industry.* EPA/625/7-91/017

WHO. (2013). *World Health Organization water sanitation health information sheet: Pharmaceuticals in drinking water*. WHO. https://www.who.int/water_sanitation_health/diseasesrisks/risks/info_sheet_pharmaceuticals/en/. Accessed on April 15, 2017.

World Health Organization. (1998). World Health Assembly (fifty-first). Emerging and other communicable diseases: Antimicrobial resistance. WHA51.17, 1998, agenda item 21.3.

Zargar, T. A., & Sheikh, M. A. (2018). Good governance in India: Challenges and prospects. *IOSR Journal of Humanities and Social Science, 23*(2), 60–64.

Zuccato, E., Castiglioni, S., Bagnati, R., Melis, M., & Fanelli, R. (2010). Source, occurrence and fate of antibiotics in the Italian aquatic environment. *Journal of Hazardous Materials, 179*, 1042–1048.

# Chapter 19
# Resource Sustainability by Electronic Waste Recycling

**P. Parthasarathy**

**Abstract** Electronic waste (e-waste) is one of the most rapidly growing solid waste problems of the world. E-waste contains hundreds of valuable resources, which can be recycled, however some of which create serious environmental and health impacts due to unscientific processing. Recycling results in economic impacts in terms of resource recovery of metals, plastics, environmental impacts by energy and water conservation and also social impacts in providing employment opportunities especially in developing countries like India. The paper discusses current practices being followed in India for recycling of e-waste. Despite the fact that informal sector recycling practices involve unscientific methods of handling and recycling of e-waste resulting in health and environmental consequences, over 95% of e-waste in India is being handled by unorganized sector as it offers much cheaper option. Formal recycling of e-waste needs to be carried out in order to create eco-friendly environment. It is a critical time to lay out well thought of management plans to substantially reduce e-waste generation and scientific disposal of WEEE. A new vision for the production and consumption of electronic and electrical goods is needed. The present study shows that e-waste recycling can be economically viable, environmentally sustainable and socially important.

**Keywords** EEE · E-waste · Informal sector · Resource recovery · Urban mining · WEEE

## Introduction

Earth resources are depleting very fast due to overexploitation by humanity. One of the important manifestations of changing consumption pattern and lifestyle of global population is increasing use of electrical and electronic equipment (EEE). With growing urbanization and enhancing living standard, use of EEE is persistently increasing. The Global E-waste Monitor 2020 (Forti et al., 2020) estimates that the total weight of global EEE consumption (excluding photovoltaic panels) on

P. Parthasarathy (✉)
E-Parisaraa Private Limited, Bengaluru, Karnataka, India



A. Kateja and R. Jain (eds.), *Urban Growth and Environmental Issues in India*,
https://doi.org/10.1007/978-981-16-4273-9_19

average increases annually by 25 million metric tons (Mt). Today, life without technology is unimaginable and EEE has become indispensable for modern way of life. In addition to the day-to-day use of EEE in households and businesses, their usage is also increasing in other sectors like health, transport, security, education and entertainment to name a few. Increasing usage of EEE is contributing towards swelling e-waste generation. The e-waste generated in 2019 weighed close to 53.6 million metric tons (mt), i.e. equivalent to 7.3 leg per capita (Forti et al., 2020). E-waste is the world's fastest growing solid waste stream. It may represent only 2% of the solid waste stream, yet it can represent 70% of the hazardous waste that ends up in landfills. It is estimated that under a baseline scenario, the amount of e-waste will more than double by 2050, to reach approximately 111 million Mt per year (Parajuly et al., 2019). Another projection (WEF, 2019) pegs the estimate at 120 million Mt annually. Thus, the approximate rate of increase in the global quantity of e-waste is a shocking 2 Mt per year.

Table 19.1 shows that the maximum amount of e-waste came from Asia in 2019, while the highest kg-per-capita waste was attributed to Europe. Europe and the USA contribute to almost one half of the total e-waste generated annually. The e-waste collection and recycling rate is low in all continents except Europe.

With surging IT industry and electronic industry, consumption of EEE has dramatically gone up and so is the case with WEEE. India generated 3.2 million tons of e-waste in the year 2017, ranking third after China (10.1 million tons) and the USA (6.9 million tons) (ASSOCHAM-EY Joint Report, 2018). The study also identified computer equipment and mobile phones as the principal waste generators in India. The report also notes that in spite of the stress on Swachh Bharat Abhiyan and Smart Cities by the Indian government, India generates more e-waste than China, USA, Japan and Germany. Among the Indian states, Maharashtra generates the most (19.8%) e-waste but recycles only about 47,810 tons per annum (TPA) whereas as Tamil Nadu (13%) recycles about 52,427 TPA, Uttar Pradesh (10.1%) recycles about 86,130 TPA, West Bengal (9.8%), Delhi (9.5%), Karnataka (8.9%) and Gujarat (8.8%). In 2016–17, India treated only 0.036 MT of its e-waste. About 95% of India's e-waste is recycled in the informal sector and in a crude manner (WEF, 2019).

**Table 19.1** E-waste generation in 2019

| Continent | E-waste generated | | E-waste document to be collected and recycled | |
|---|---|---|---|---|
| | Total (Mt) | Per capita (kg) | Amount (Mt) | Per cent of total |
| Asia | 24.9 | 5.6 | 2.9 | 11.7 |
| Africa | 2.9 | 2.5 | 0.03 | 0.9 |
| America | 13.1 | 13.3 | 1.2 | 9.4 |
| Europe | 12 | 16.2 | 5.1 | 42.5 |
| Oceania | 0.7 | 16.1 | 0.06 | 8.8 |

*Source* Forti et al. (2020)

## What Is E-Waste?

E-waste is defined as EEE that its owner discards and does not plan on reusing (Step Initiative, 2014). Products that use batteries or power supply are considered e-waste including consumer electronic items like television, personal computers, laptops, tablets and phones, as well as industrial appliances like electronic machines, and communication and instrumentation systems. E-waste is also referred to as waste electrical and electronic equipment (WEEE), e-scrap or electronic waste. EEE contains many precious metals including gold, silver, copper, palladium, etc., that is why that same e-waste represents a huge opportunity in material terms. One example can be the introduction of mobile phones and computers which are deeply coupled with the use of different base metals, precious metals and rare metals. This has indirectly put lot of pressure on the metal stock in society (UNEP, 2010). The resources have to be conserved for sustainable development, and recycling of obsolete product waste is one of the best options available. This involves mining of metals from the product waste such as electronic waste (e-waste) which is called 'Urban mining'. Electronic waste contains several valuable resources (Meskers et al., 2009) in the form of various metals, some of which are critical for future supply (Buchert et al., 2012) such as gold, silver, various types of recyclable plastics and different types of glasses. There is hundred times more gold in a ton of mobile phones than in a ton of gold ore (WEF, 2019). The report cites that one recycler in China already produces more cobalt (by recycling) than what the country mines in one year. The largest contributors to e-waste in terms of the raw material weight are iron, aluminium and copper. The recovery of precious metals assumes more importance than its material worth as a considerable amount can be reused in manufacturing, leading to a reduction in the extraction of virgin material.

E-waste also contains substances that are hazardous to human health, such as mercury, cadmium, lead and chemicals such as chromium brominated flame retardants (BFRs), chlorofluorocarbons (CFCs), hydrochlorofluorocarbons (HCFC) (Alex, 2011). It requires special handling, storage and disposal. A total of 50 t of mercury and 71 kt of BFR plastics are found in globally undocumented flows of e-waste annually (Forti et al., 2020). E-waste is not biodegradable, accumulates in the environment and can pollute water resources and food supply chains. Of the total amount, 40 million tons of e-waste is discarded in land fill (Balde et al., 2017), burned or illegally traded and treated in a substandard way every year. Studies (Greenpeace, 2005, 2009; Toxic Link and Greenpeace, 2002) have shown that most of the e-waste recycling activities carried out by unorganized informal sector involve physical breaking using tools like hammer, chisel, screw driver, etc. All these activities are undertaken with bare hands and involve toxic chemical processes which not only create occupational hazards for the workers but also damage the environment permanently (CSE, 2015). Basic recycling techniques to burn EEE expose workers and their families to many toxic substances. Women and children make up to 30% of the workforce in informal, crude e-waste processing and are, therefore, particularly vulnerable. Workers in the informal sector engaged in e-waste collection, recycling and processing are exposed to toxic elements, compounds that are carcinogenic. Long-term exposure to the chemicals and pollutants generated in large quantities during the unsafe recycling of e-waste causes health problems including damage to the nervous systems, blood systems, kidneys and brain development, disorders in the respiratory system and skin, bronchitis, lung cancer, heart, liver and spleen damage.

WEF (2019) reports consumers discard 44 million tons worth of electronics each year, of which only 20% is recycled sustainably. Rest 80% is not documented; it either goes to landfills or travels down to unregulated markets. It can pollute water, soil, air and living things. Majority of e-waste disposal and recycling jobs are unsafe and not protected by formal regulation (Brett, 2009).

Production and consumption of EEE and disposal of WEEE have implications for environment too. By 2040, the e-carbon emissions from the production and consumption of EEE will reach 14% of total emissions, which is one half of the total global transport sector. Furthermore, harvesting the recycled e-waste resources leads to significantly reduced carbon dioxide emissions in comparison with mining in the earth's crust (WEF, 2019). The report affirms that recycled metals are also 2–10 times more energy-efficient than metals smelted from mined ore concentrates new vision for the production and consumption of EEE and scientific disposal of WEEE is needed.

## Who Pays for E-Waste?

True e-waste recycling has become economically unviable due to the reasons of low prices for recycled metals and plastics because of the price fluctuations in crude oil. For example, per Kg price of scrap mild steel has come down to ₹22.00 from ₹40.00, scrap aluminium to ₹70.00 from ₹105.00, scrap copper to ₹350.00 from ₹600.00, etc.

On the other hand, increasingly paying higher prices have to be paid for buying e-waste from MNC corporates, producers, bulk consumers, software companies and telecom companies. Ever since state pollution boards are empowered with the authorization procedures transferred from CPCB during 2008, a large number of recyclers have cropped up in the country without any knowledge, facility or standards. For example, there are almost 300 authorized dismantlers and recyclers, most of them without appropriate knowledge/infrastructure in Bangalore, Karnataka. This has increased competition, and even major companies are giving importance to the price quote only and not any standard or certifications except in the State Pollution Control Boards Authorization. Government of India (GOI 2011) has brought out a report by (The Library, Reference, Research, Documentation and Information Service) LARRDIS submitted a report to the Rajya Sabha E-waste in India towards a query by one of its members.

## Resource Exploitation and Sustainability

According to US Geological Survey's annual report (2008) and UN statistics on global population, world population is approaching 7 billion and is growing at about 80 million/year; these studies have enabled to estimate the effect of increase in living standards for key minerals to run out against time period.

The exponential increase in population and rising standards of living are putting demand pressure on the materials that only earth can provide. For example, there is a great demand of platinum for use in catalytic converters of automobiles and futuristic fuel cells. It is estimated that if all the 500 million automobiles in use today are retrofitted with fuel cells, all the world's available resources of platinum would be exhausted in 15 years. Similarly, rare metals such as indium which is being consumed in unprecedented quantities for making LCDs of laptops, computers and TVs, and tantalum-based capacitors used in cell phones and computer circuit boards are becoming a threat. Even the demand for common elements such as zinc, copper and nickel is outstripping the extractable quantities.

Mountains of tailings from worked-out mines that contain small amount of valuable minerals are becoming economical to extract due to the demand and rising cost of recovery of gold from gold ore tailing. When the resources run short, civil wars and conflict between nations are certainty.

The spider chart in Fig. 19.1 is simplified and presented for some of the e-waste metals (Table 19.2).

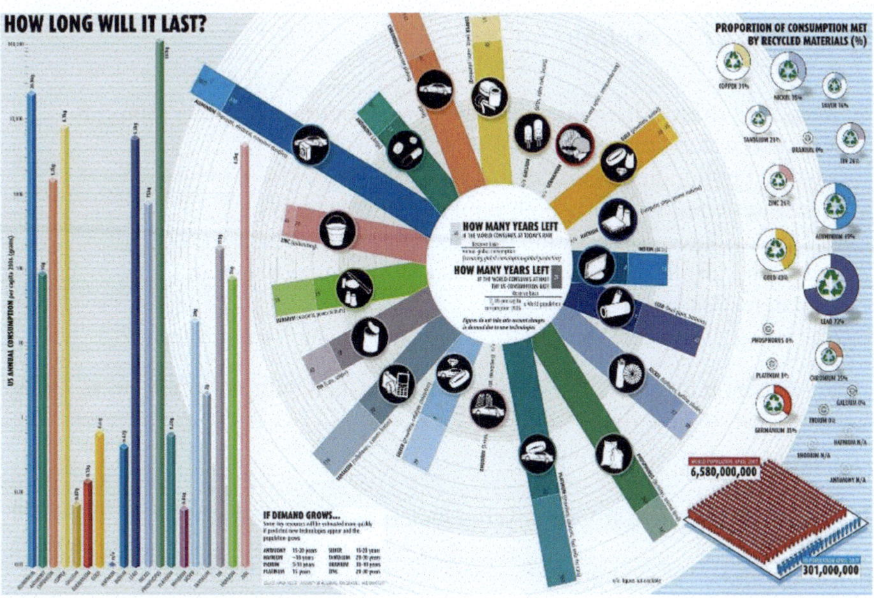


**Fig. 19.1** Resources exploitation and sustainability. *Source* US Geological Survey's annual report (2008)

**Table 19.2** Resources exploitation and sustainability

| S. No. | Material | Number of years it may last if the world consumes at today's rate | Application | Per cent of recycling |
|---|---|---|---|---|
| 1 | Indium | 13 | LCDs | 0 |
| 2 | Silver | 29 | Jewellery, catalytic converter | 16 |
| 3 | Antimony | 30 | Drugs | Data not available |
| 4 | Tin | 40 | Cans, solder | 26 |
| 5 | Lead | 42 | Batteries, paints | 72 |
| 6 | Gold | 45 | Jewellery, dental (tooth implants) | 43 |
| 7 | Zinc | 46 | Galvanizing | 26 |
| 8 | Uranium | 59 | Weapons, power stations | 0 |
| 9 | Copper | 61 | Wire, coins, plumbing | 31 |
| 10 | Nickel | 90 | Batteries, turbine blades | 35 |
| 11 | Tantalum | 116 | Cell phones, camera lenses | 20 |
| 12 | Chromium | 143 | Chrome plating, paint | 25 |

*Source* Parthasarathy (2010)

## *Importance of Recycling*

1. Recycling drastically reduces energy consumption and carbon footprint.
2. Reduced water consumption.
3. Reduced pollution of air, water and soil.
4. Generates employment in both rural and urban sectors.
5. Affects resource conservation by preventing dumping into landfills.
6. Recycling also directly helps in 'Make in India and Made in India' to achieve zero defect and zero effect.
7. Recycling also meets the requirements of 'Swachh Bharat Mission'.

Future economic, environmental and social sustainability depends on recycling of so-called waste materials.

**Economic sustainability**

1. Revenue generation from recovered materials.

**Environmental sustainability**

1. Natural resource conservation.
2. Reduction in pollution.

**Social sustainability**

1. Employment generation and skill development.
2. Awareness in informal sector.

Total stock of e-waste with domestic consumers may exceed 9.7 million metric ton. It is valued at about ₹200 billion (₹20,000 crore) (E-Parisaraa Report, 2015). According to sample survey by E-Parisaraa on generation of household e-waste in Bangalore city involving students, about 21 kg e-waste is generated per capita/year by an average middle-class family in India. The Composition of E-Waste materials received at a recycling facility during the course of three years is presented in Table 19.3. And the Typical E-waste sector wise is presented in Table 19.4.

## Restriction of Hazardous Substances (RoHS): An Introduction

E-waste contains a variety of toxic metals and organic compounds. To address this toxicity, the Restriction of Hazardous Substances Directive (RoHS Directive 2002/95/EC) was introduced in the European Union alongside the waste electric and electronic equipment (WEEE) directive.

This directive requires control of six toxic substances: 0.1% (1000 ppm) for lead, mercury, hexavalent chromium, polybrominated biphenyls (PBB), polybrominated diphenyl ethers (PBDE) and 0.01% (100 ppm) for cadmium.

### Hazard Identification and Risk Assessment (ISO 45001:2018)

Hazard identification, risk assessment aspects and impacts of the e-waste recycling process, which are important for design and development, have been studied and presented in Table 19.5.

Activities of every person coming into the premises, including that of visitors and contractors, are considered while identifying the hazards. Infrastructure, equipment and materials provided by the organization or others, processes, installations, operating procedures, changes or proposed changes, work organizations and human capabilities are considered while identifying the hazards. Any adverse impact on personal health, minor or major, resulting from the each hazard is identified and documented. Periodic health monitoring of the Employees shall be given priority.

## Informal Sector Recycling

Huge informal sector involved in recycling poses the major challenge of e-waste management in India. Over 95% of e-waste generated is managed by the informal sector (ASSOCHAM-EY Joint Report, 2018; MATT-GTZ-IMRB, 2007). Informal sector offers cheaper options; hence, it is a major threat to the formal recycling

**Table 19.3** E-waste composition

| S. No. | E-waste | Composition in % |
|---|---|---|
| 1 | Desktop computers | 15.73 |
| 2 | Laptop with adaptor | 1.05 |
| 3 | Mobile phones | 1.37 |
| 4 | Printers | 0.29 |
| 5 | Copying machines | 0.71 |
| 6 | Television | 30.13 |
| 7 | Telephones (landline and cordless) | 2.60 |
| 8 | Refrigerator | 23.88 |
| 9 | AC | 14.07 |
| 10 | Washing machine | 7.55 |
| 11 | Lamp waste FTL + CFL | 2.154 |
| | *Total* | 99.534 |

*Source* Parthasarathy (2010)

**Table 19.4** Typical E- waste Sector wise

| Heavy metals and other metals | | |
|---|---|---|
| Arsenic | Small quantities in the form of gallium arsenide with light-emitting diodes | Acutely poisonous and on a long-term perspective injurious to health |
| Barium | Getters in CRT | May develop explosive gases (hydrogen) if wetted |
| Beryllium | Power supply boxes which contain silicon controlled rectifiers, beamline components | Harmful if inhaled |
| Cadmium | Rechargeable NiCd batteries, fluorescent layer (CRT screens), printer inks and toners, photocopying machines (photo-drums) | Acutely poisonous and injurious to health on a long-term perspective |
| Chromium VI | Data tapes, floppy discs | Acutely poisonous and injurious to health on a long-term perspective causing allergic reactions |
| Gallium arsenide | Light-emitting diode (LED) | Injurious to health |
| Lead | CRT screens, batteries, printed wiring boards | Causes damage to the nervous system, circulatory system, kidneys |

process. In fact, compliance of regulatory procedures and recycling practices by formal sector would bring benefits in the form of human health and value of considerable amount of raw materials re-entering the manufacturing chain. However, poor infrastructure, legislation and framework are accountable for waste of diminishing

**Table 19.5** Hazard identification and risk assessment various metals in E-waste

| Substance | Occurrence in e-waste | Environmental and health relevance |
|---|---|---|
| Polychlorinated biphenyls (PCB) | Condensers, transformers | Adversely impacts health including cancer, damage to immune, reproductive, nervous and endocrine systems. Persistent and bioaccumulatable |
| Tetrabromobisphenol-A (TBBA)<br>Polybrominated biphenyls (PBB)<br>Polybrominated diphenyl ethers (PBDE) | TBBA is currently the most popular substance for controlling fire from plastics such as thermoplastic components, cable insulation, printed wiring boards and covers | Responsible for long-term period injuries to health<br>Highly poisonous upon inflammation |
| Chlorofluorocarbon (CFC) | Cooling unit, insulation foam | Burning is linked with poisonous emissions |
| Polyvinyl chloride (PVC) | Cable insulation | High temperature processing of cables may release chlorine, which is converted to dioxins and furans |

*Source* ISO (2018)

natural resources, irreparable damage of environment and health of the people working in industry (Eguchi et al., 2009, 2012).

Informal sector does not conform to any legal business compliance requirements in India such as Shop and Establishment Act, Factory Act and Environmental Protection Act and does not pay any form of taxes (China et al., 2011; Khullar, 2009). There are major players in e-waste trading whose trading chains supply e-waste to various small traders each having their own specialization. Normally, two to three groups are involved in trading chain. These groups collect of e-waste through government auctions and from corporate sectors as well as from individual scrap collectors and sometimes in illegal imports (Parthasarathy, 2010). To restrict the informal sector engaged in e-waste recycling, the Government of India has brought out new E-waste Management Rules, 2016 (E-Waste Management Rules, 2016), by which those engaged in e-waste recycling shall conform to the Environmental Protection Act.

Figure 19.2 shows the involvement of informal sector engaged in backyard practices. It is estimated that Moradabad area alone accounts for about 150,000 people engaged in this business. The informal sector in India following unscientific backyard practices in some of the major cities such as Bangalore and Delhi has been studied and well documented in the literature. The huge demand for copper by wire drawing/insulated wire as well as PVC wire and cable industry, largely clustered in Delhi region, has promoted Moradabad region to carry on the unscientific processing of e-waste by de-soldering, burning, acid leaching, etc., and also wire and cable burning especially on rooftops followed by melting in small ill-equipped furnaces

**Fig. 19.2** E-waste recycling in informal sector

(TOI, 2015). These burning processes are mostly clandestinely done during night hours. Across the country, it is also well known that such wire cable burning operations are carried out in graveyards, brick kilns and remote forest areas causing air, water and soil pollution besides health damage.

## E-Waste (Management) Rules, 2016

- Manufacturer, dealer, refurbisher and producer responsibility organization (PRO) have been introduced as additional stakeholders, in the rules.
- The applicability of the rules has been extended to components, consumables, spare parts of electrical and electronic equipment (EEE).
- Compact fluorescent lamp (CFL) and other mercury-containing lamps are brought under the purview of rules.
- Collection mechanism-based approach has been adopted to include collection centre, collection point, take-back system, etc., for collection of e-waste by producers under extended producer responsibility (EPR).
- Option has been given for setting up of PRO, e-waste exchange, retailer, deposit refund scheme as additional channel for implementation of EPR by producers to ensure efficient channelization of e-waste.

- Provision for pan-India EPR authorization by CPCB has been introduced replacing the state-wise EPR authorization.
- Collection and channelization of e-waste in EPR: Authorization should be aligned with the description of targets in Schedule III of the rules. There should be four phases of e-waste collection, measured either by number or by weight based on the amount of waste generated described in the EPR plan. The first three phases are of two years each. In the first phase, i.e. year one and two, it should be 30%. In the second phase, i.e. year three and four, it should be 40%. It should be 50% in the third phase, year five and six. In the final fourth phase, it should be 70% for year 7 onwards.
- Another economic incentive has been created in the form of deposit refund scheme. The consumer pays a deposit when purchasing the EEE, which the seller refunds, with interest, when the consumer returns the EEE after the end of its life.
- The rules also provide for an independent e-waste exchange market. This exchange assists authorized entities (identified by the rules) in sale and purchase of e-waste resulting from end-of-life EEE e-waste.
- The rules also specify that it is the responsibility of the manufacturer to collect the e-waste generated as part of the EEE manufacturing and recycle or dispose it, seeking appropriate authorization from State Pollution Control Board (SPCB).
- Upon delegation of the collection responsibility by the producer, the dealer needs to provide the customer with a box for collecting e-waste and send it to the producer.
- The dealer (which would be an offline or online retailer) is also responsible for refunding the consumer based on the take-back system or deposit refund scheme defined by the producer.
- Regarding the e-waste produced during refurbishing, the refurbisher is responsible for collecting it and channelling it to an authorized dismantler or recycler via its collection centre.
- The rules also specify the part of the state governments towards ensuring that the workers involved in the dismantling and recycling operations have opportunities for developing new skills and have safe and healthy working conditions.
- To that end, the state government can delegate the task and authorize a state agency like the state department of industry. This agency needs to earmark and allocate industrial space for the e-waste to be dismantled and recycled. This space can be allocated in either existing or new industry estates, parks or clusters.
- The state government can also delegate and authorize a government agency like the state department of labour with the welfare of the workers. This agency would be responsible for ensuring that the workers are recognized and registered, helping the workers form groups to simplify the creation of dismantling spaces, providing opportunities for developing skills and ensuring worker safety and health.
- It would be the responsibility of the state government to integrate the various aspects and ensure that they are implemented effectively. The progress and status are to be reported annually to the Ministry of Environment, Forest and Climate Change.

- The manifest system specifies the rules for e-waste transportation. It requires that the sender prepares a document containing the details and that the transporter has three copies of this document.
- The rules also specify the monetary liability for the environmental or third-party damages resulting from the mismanagement of e-waste including violating the provisioned rules.
- For orphaned e-waste, urban local bodies (municipal committee/council/corporation) have been authorized for collection and channelling of such e-waste to authorized dismantling or recycling facilities.

## Formal Recycling in India

Formal recycling in India started in the year 2004, and consent to establish was given to some of the recyclers in the 2005, although there were no e-waste rules notified at that time. The first draft of E-waste (Management and Handling) Rules was circulated during the 2008 and notified in 2011 after several rounds of dialogues, workshops and meetings with multistakeholders. As of 2016, CPCB had notified 178 units as Registered E-Waste Recyclers in India with an installed capacity of 438,086 metric tons per annum (Handbook of Electronic Waste Management, 2016). The new e-waste management rules referred earlier focus on EPR target-based approach for collection under EPR and defined responsibilities for various stakeholders. The performance of various producers, manufacturers and brand owners are classified based on their performance in Time to Reboot III by Toxic link report (2019).

Considering the backyard practice scenario in e-waste recycling, there is a need to develop a state of the art of e-waste recycling technology suitable for the developing countries in general and particularly for typical Indian conditions. This necessitates the design and developments of methods and procedures that are economically viable and environmentally sound which maximize the resource recovery, minimize landfills, generate employment opportunities, improve the living standards and reduce poverty.

One example of state-of-the-art recycling with best practices is E-Parisaraa Pvt. Ltd. (www.ewasteindia.com), a pioneer registered recycler which started in 2004 and is presently processing about 2000 metric tons per annum, serving a large customer base, audited and approval by many branded producers. It offers several innovative solutions for e-waste recycling technology (E-Parisaraa, 2014).

Typical e-waste recycled is presented in Table 19.6. The recycler is located in Bangalore, popularly known as the information and communication technology (ICT) hub of India, and therefore, most of the e-waste is from the IT industry. It is interesting to know that the household e-waste is less than 1% and this is typically due to lack of collection or take-back from consumers.

Closed-loop consideration is the topmost important aspect for maximum resource recovery of e-waste in the design and development of state-of-the-art recycling technology. The valuable resources of e-waste recycling are metals and non-metals. The

**Table 19.6** Hazard identification and risk assessment for various organics in E-Wastes

| | | |
|---|---|---|
| Lithium | Li batteries | May develop explosive gases (hydrogen) if wetted |
| Mercury | Is found in the fluorescent lamps that provide backlighting in LCDs, in some alkaline batteries and mercury wetted switches | Acutely poisonous and injurious to health on a long-term perspective |
| Nickel | Rechargeable NiCd batteries or NiMH batteries, electron gun in CRT | May cause allergic reactions |
| Rare earth elements (yttrium, europium) | Fluorescent layer (CRT—screen) | Skin and eye irritant |
| Selenium | Old photocopying machines (photo-drums) | Adverse health effects upon high exposure |
| Zinc sulphide | Is used on the interior of a CRT screen, mixed with rare earth metals | Poisonous |
| Toxic organic substances | Condensers, liquid crystal display | |
| Toner dust | Toner cartridges for laser printers/copiers | Dust spread via explosion has adverse health consequences |
| Radioactive substances americium | Medical equipment, fire detectors, active sensing element in smoke detectors | Can potentially lead to cancer |

*Source* ISO (2018)

metal takes the maximum economic value in the recycling mainly because of the precious metal content.

Manual dismantling and segregation are efficient ways of recycling of the resources. The staff and employees are being trained continuously on environmental health and safety aspects and hazards in the working environment, and the need for using personal protective equipment. This has resulted in social change among the employees and improved their standards of living (Fig. 19.3).

Metals from the major constituents of recycled materials largely consist of mild steel, cabinets, chassis of computers and servers which are baled for volume reduction and sent locally for melting into Tor steel generally used in construction industry. Metals such as aluminium and copper are in good demand in various industries as raw materials. Plastics are recycled locally into granules and reused for manufacture of other components (Fig. 19.6). Glass is also recycled effectively due to the demand for coloured picture tube industry. Data shows that the landfill in a scientific hazardous waste facility is less than one per cent. The lead–acid batteries are mainly dismantled from server racks, uninterrupted power supply (UPS), etc., and are recycled by authorized downstream recyclers. Over the years, shredded printed circuit boards are exported to a world-class recycling facility in Belgium in compliance with

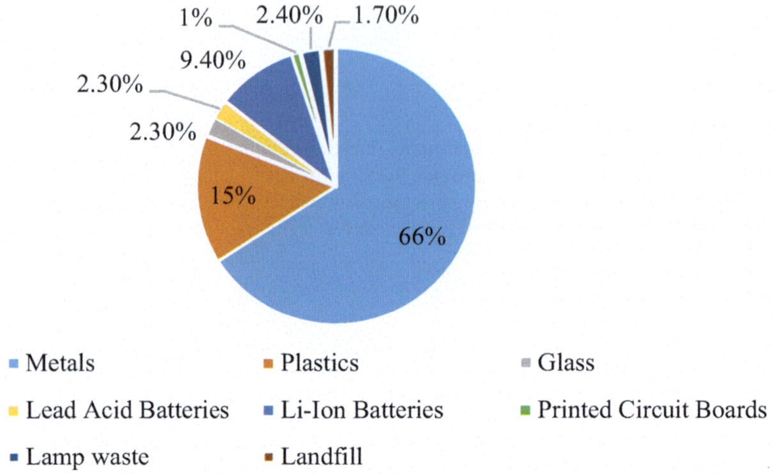


**Fig. 19.3** Major constituents of recycled materials. *Source* Parthasarathy (2010)

Basel Convention norms. The recovered values of copper, palladium, silver and gold are paid as per London Metal Exchange rates. Li-ion batteries are also recycled at the same facility. The recyclers also operate the lamp waste recycled in a pilot plant scale for its customers. Periodic monitoring of air quality and health monitoring are in place due to stringent audits by producers.

**Plastics Recycling**

The latest development in this field is the use of recycled glass for making bangles and other articles instead of throwing them in landfill (Fig. 19.7). Also recently, the analysing of phosphor is made and found that it contained rare metal yttrium and rare earth metals such as terbium and cerium. In this connection, Centre for Materials for Electronics Technology (C-MET) under the Ministry of Electronics and Information Technology (MeitY) has also proposed a project with the Department of Science and Technology (DST) for the recovery of these metals along with the help of E-Parisaraa (Figs. 19.4, 19.5, 19.6 and 19.7).

## Hazardous Materials: Identification, Isolation, Storage and Disposal Methods

Risks associated with each hazard are categorized under normal, abnormal or emergency condition. The available control methods for risks are categorized as proactive and administrative controls. Proactive control methods are verified by checking the availability of engineering controls and regular health check-ups. Administrative

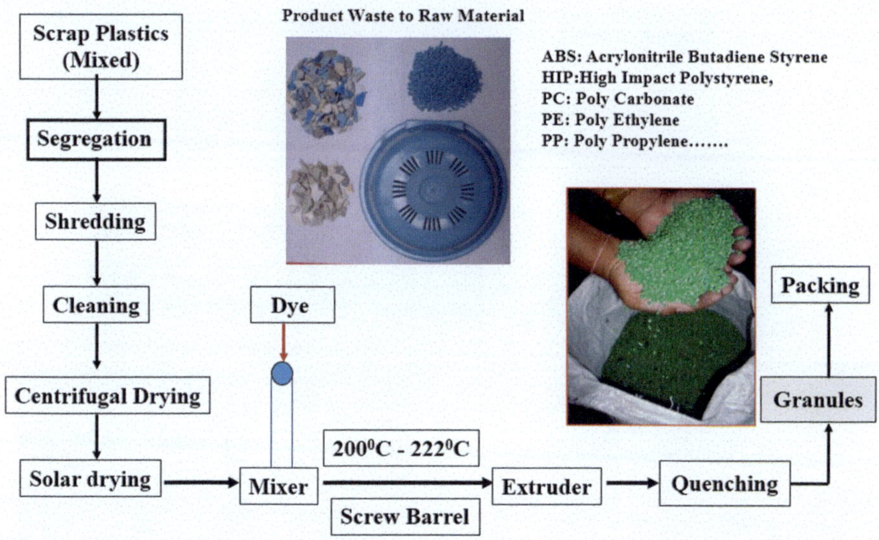


**Fig. 19.4** Product waste Plastics to raw material Granules

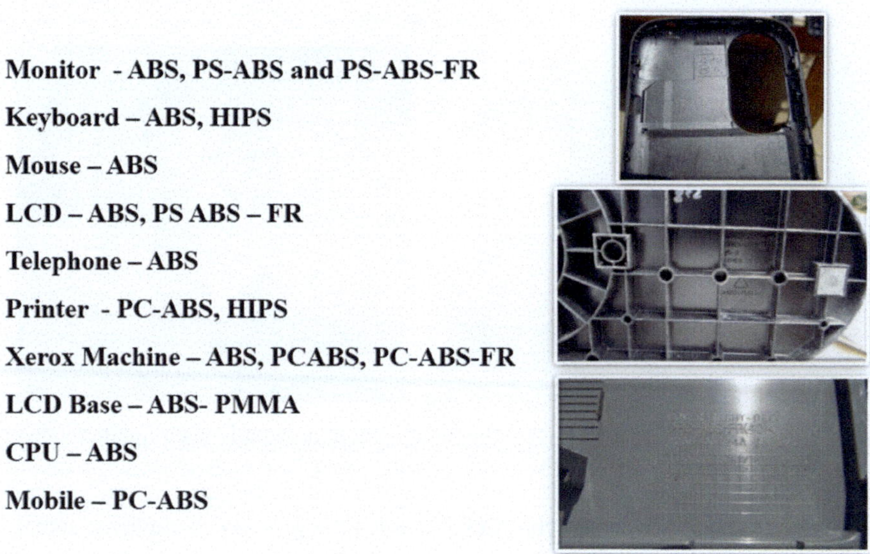


**Fig. 19.5** Types of plastics found in various e-waste products

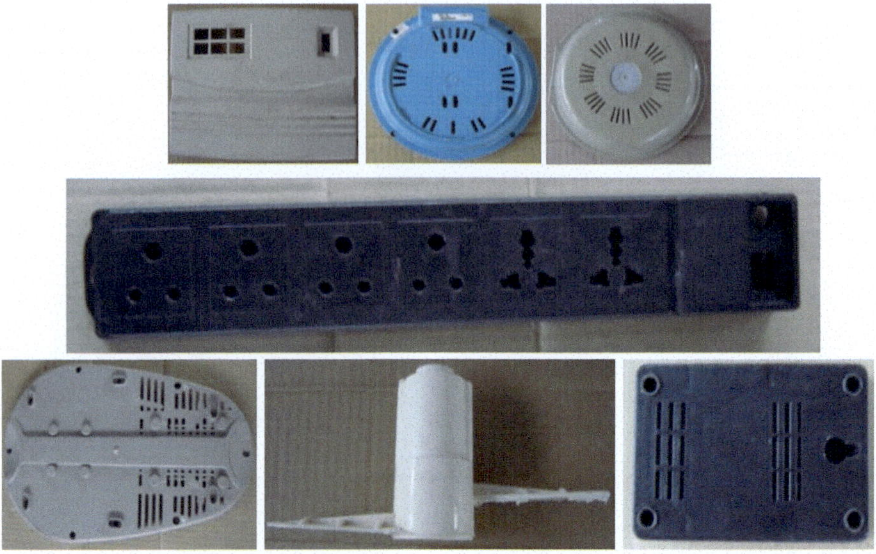

Fig. 19.6 Components made out of recycled e-waste plastics

Fig. 19.7 Bangles made out of recycled glass from lamp waste using traditional artisans

control methods are verified by checking the availability of legal compliance, emergency control, Material Safety Data Sheet (MSDS) compliance, competent person, adequacy of operational control procedure (OCP) and appropriate personal protection equipment (PPE).

**Evaluation of significance**

The scale, severity and probability of occurrence of various risks are evaluated and rated on a scale from 1 to 5. The rating range for scale of risks (SCL), severity of risk (SEV) and the probability of occurrence (POC) is as mentioned in Table 19.7.

**Table 19.7** Typical electronic waste

| Equipment category | Per cent composition (by weight) |
|---|---|
| Computer equipment | 75 |
| Telecommunication equipment | 13 |
| Entertainment equipment | 3 |
| Electrical equipment | 4 |
| Medical equipment | 4 |
| Other equipment including household e-waste | 1 |

*Source* Parthasarathy (2010)

The product of SCL, SEV and POC will give the risk potential number, based on which the risks are further evaluated and categorized as trivial, tolerable, moderate, substantial and intolerable risks (Table 19.8).

Based on the risk potential number, risks are evaluated. If the risk potential number is 28 or above or all emergencies and legal non-compliances, then that risk shall be considered as unacceptable and shall be immediately addressed. In addition, all other risks are considered as acceptable risks. The risks classified with moderate, substantial and intolerable significance shall be attended on priority with appropriate risk control measures.

**Table 19.8** Classification of risks based on risk potential

| Range of risk potential number (SCL × SEV × POC) | Classification of risk | Details of classification |
|---|---|---|
| 1–4 | Trivial | No action or monitoring |
| 5–8 | Tolerable | To be monitored for existing control methods |
| 9–27 | Moderate | To be assigned and monitored |
| 28–64 | Substantial | Urgent action to be taken and activity shall not be carried out till the allotted control measures are implemented |
| 65–125 | Intolerable | Work to be stopped immediately, and alternate measures shall be explored to reduce the hazard/risks to tolerable or moderate significance |

**Table 19.9** Control methods for significant risks

| | | |
|---|---|---|
| A | Elimination | Modify the design to eliminate the hazard |
| B | Substitution | Substitute with less hazardous material/process |
| C | Engineering controls | Insulation of systems, providing ventilation, guarding, interlocks, sound enclosures, etc. |
| D | Signage/warning | Insulation of alarms, safety procedures, equipment, administrative controls, inspections, access controls, etc. |
| E | Personal protective | Like safety glasses, face shield, mask, gloves, equipment, shoes, etc. |

## Risk Control Methods

The control methods for significant risks shall be considered as per the following hierarchy (Table 19.9).

An action plan shall be for implementing the risk control methods. The action plan shall include the description of actions to be taken to reduce the severity of risks, responsibility assigned for various related works and time frame to finish the action.

## Conclusion

E-waste comes from many resources including households, business and governments. As more and more Indians are spending on EEE, e-waste generation is exponentially increasing. Computer appliances account for almost three-fourth of e-waste followed by telecommunication (12%) and electrical (8%) equipment. Global E-Waste Monitor 2020 (Forti et al., 2020) shows that consumers discarded 53.6 million tons worth of electronics in 2019 globally, up 20% in 5 years. Global annual e-waste is worth at least $62.5 billion which is three times more than the annual output of the world's silver mines and more than the GDP of many countries (Balde et al., 2017).

At present, the rate at which e-waste is generated in India exceeds the country's e-waste recycling and dismantling capacity by a factor of 4.56. This has led to the e-waste being disposed and dumped illegally and improperly, exacerbating the damage to the environment and health (ASSOCHAM-EY Joint Report, 2018).

E-waste is a high technology waste, and it needs deep scientific knowledge base to manage it effectively. It requires unique innovative way to manage the e-waste rather than trading it with informal sector. Recycling of e-waste is very typical. Quite often, technology is not available, and many a times despite the available technology, there is no economic incentive for recycling materials from e-waste. Most of developing countries do not have proper legislative procedure and proper infrastructure to handle e-waste. It is critical time to lay out well thought of management plans to substantially reduce e-waste generation and disposal. A new vision for the production and consumption of electronic and electrical goods is needed. It is necessary for countries to formalize the environmentally sound management of e-waste and to take advantage of business opportunities it offers. India adopted e-waste legislation in 2016; however like many other countries, national e-waste management laws are not enforced seriously.

The present study shows that e-waste recycling can be economically viable, environmentally sustainable and socially important. The rural employment could be the best alternative for low technology adoption as they are underprivileged and are deprived of employment opportunities in an organized sector. This would result in social change among the people by improving standard of living and woman empowerment.

## References

Alex. (2011). *Hazardous substances*. https://ewasteguide.info/hazardous-substances/. Accessed February 13, 2019.

ASSOCHAM-EY Joint Report. (2018). *Electronic waste management in India*. Available at https://www.assocham.org/newsdetail.php?id=6850. Accessed January 29, 2021.

Balde, C. P., Forti, V., Gray, V., Kuehr, R., & Stegmann, P. (2017). *The global E-waste monitor 2017*. UNU, ITU, ISWA.

Brett, H. R. (2009). E-waste: An assessment of global production and environment impacts. *Science of the Total Environment, 408*(2).

Buchert, M., Manhart, A., Bleher, D., & Pingel, D. (2012). *Recycling critical raw materials from waste electronic equipment*. Oko Institute.

China, X., Streicher-Porte, M., Wang, M. Y. L., & Reuter, M. A. (2011). Informal electronic waste recycling: A sector review with special focus on China. *Waste Management, 31*, 731–742.

CSE. (2015). *Recommendations to address the issues of informal sector involved in e-waste handling: Moradabad, Uttar Pradesh*. Report by Centre for Science and Environment. http://www.cseindia.org/userfiles/moradabad-e-waste.pdf. Accessed April 10, 2019.

Eguchi, A., Nomiyama, K., Devanathan, G., Subramanian, A., Bulbule, K. A., Parthasarathy, P., Takahashi, S., & Tanabe, S. (2012). Different profiles of anthropogenic and naturally produced organohalogen compounds in serum from residents living near a coastal area and e-waste recycling workers in India. *Environment International, 47*, 8–16.

Eguchi, A., Nomiyama, K., Subramanian, A., Parthasarathy, P., Bulbule, K. A., Takahashi, S., & Tanabe, S. (2009). Organohalogen and metabolite contaminants in human serum samples from Indian e-waste recycling workers. In T. Isobe, K. Nomiyama, A. Subramanian, & S. Tanabe (Eds.), *Interdisciplinary studies on environmental chemistry: Environmental specimen bank* (pp. 167–74).

E-Parisaraa. (2014, July). Bold pursuits-recycling today. http://www.recyclingtoday.com/article/rtge0714-eparisaraa-electronics-recycling-profile/. Accessed April 18, 2019.

E-Waste Management Rules. (2016). http://www.moef.gov.in/sites/default/files/EWM%20Rules%202016%20english%2023.03.2016.pdf

Forti, V., Balde, C. P., Kuers, R., & Bel, G. (2020). *The global e-waste monitor 2020; Quantities flows and the circulars economy potential*. United Nations University (UNU)/United Nations Institute for Training and Research (UNITAR)—Cohosted SCYCLE Programme, International Telecommunication Union (ITU) & International Solid Waste Association (ISWA).

GOI. (2011). *E-waste in India*. Report by research unit (LARRDIS). Rajya Sabha Secretariat, Government of India. https://rajyasabha.nic.in/rsnew/publication_electronic/E-Waste_in_india.pdf. Accessed February 17, 2019.

Greenpeace. (2005). *Recycling of electronic waste in China and India: Work place and environmental contamination*. Green Peace International. https://www.greenpeace.org/international/publication/7051/recycling-of-electronic-waste-in-india-and-china-summary/. Accessed April 10, 2019.

Greenpeace. (2009). *Where does e-waste end up?* Green Peace International Report. https://books.google.co.in/books?id=wte_DwAAQBAJ&pg=PA279&lpg=PA279&dq=2009+Where+does+e-waste+end+up?+Green+Peace+International+Report&source=bl&ots=HIh9tDHR50&sig=ACfU3U1iR1IrJ6PgwNyrebofk-1e6vHJFQ&hl=en&sa=X&ved=2ahUKEwj2muXDqY3qAhW57HMBHdPdBCcQ6AEwEXoECAgQAQ#v=onepage&q&f=false. Accessed April 10, 2019.

*Handbook of electronic waste management*. (2016). http://cpcb.nic.in/List_of_E-waste_Recycler_as_on_29.12.2016.pdf. Accessed April 15, 2019.

ISO. (2018). *ISO 45001-2018: Classification of risks based on risk potential*. https://www.iso.org/obp/ui/#iso:std:iso:45001:ed-1:v1:en. Accessed April 18, 2019.

Khullar, M, (2009). *Scrap equals survival* (pp. 38–47). Recycling International.

*MAIT-GTZ-IMRB*. (2007). Quantitative understanding of generation, disposal and recycling of electronic waste in India. In *E-waste assessment in India: MAIT-GTZ-IMRB study*. http://www.nswai.com/docs/e-Waste%20Assessment%20in%20India%20-%20Specific%20Focus%20on%20Delhi.pdf. Accessed April 14, 2019.

Meskers, C. E. M., Hagelüken, C., & Van Damme, G. (2009). Green recycling of EEE: Special and precious metal recovery from EEE (Umicore). In *TMS Annual Meeting*.

Parajuly, K., Kuehr, R., Awasthi, A. K., Fitzpatrick, C., Lepawsky, J., Smith, E., Widmer, R., & Zeng, X. (2019). *Future E-waste scenarios*. UNV. VIE SCYCLE (Bonn) & UNEP/ETC (Osaka).

Parthasarathy, P. (2010). *Ecologically efficient resource recovery of electronic waste for Indian conditions* (PhD thesis). Lambert Academic Publishing. ISBN-13-978-3-659-47128-5.

Step Initiative. (2014). *One global definition of e-waste*. United Nations University. Available at https://collections.unu.edu/eserv/unu:6120/step_one_global_definition_amended.pdf. Accessed April 27, 2019.

TOI. (2015, January 30). Media report. *Times of India Newspaper*.

Toxic Link and Greenpeace. (2002). *Exporting harm: The high-tech trashing of Asia by the Basel Action Network (BAN) and Silicon Valley Toxics Coalition (SVTC) with contributions by Toxic Link, India, Scope (Pakistan) and Greenpeace, China*. http://svtc.org/wp-content/uploads/technotrash.pdf. Accessed April 4, 2019.

Toxics Link Report. (2019). Time to reboot III. http://toxicslink.org/docs/Time%20to%20Reboot%203.pdf. Accessed February 25, 2019.

UNEP. (2010). *Metal stocks in society: Scientific synthesis*. http://wedocs.unep.org/bitstream/handle/20.500.11822/8438/Metal_Stocks_EN.pdf?sequence=1&isAllowed=y

WEF. (2019). *A new circular vision for electronics: Time for a global reboot*. World Economic forum.

# Chapter 20
# Renewable Energy in India: Potential and Prospects

**Chandrakanta Mall and Prem Prakash Solanki**

**Abstract** All economic activities in a modern economy depend on energy to flourish and sustain. A well-established energy system fuels all sectors from agriculture to high technology-intensive IT sector, from manufacturing to medicine, thereby making development impossible without fuelling the engine of growth. During the last two decades, India improved access to electricity covering 75 crore additional population. In the current century, India is also planning to make a move from non-renewable sources to renewable sources, viz. wind, solar, hydro-energy, biomass and geothermal. There is no denying that India has made significant progress in shifting energy use mix from conventional to non-conventional sources. However, in coming decades, with development and further population growth energy demand is expected to increase multifold. Presently, renewable energy sources make up only 18% of the total mix bag. Increasing this substantially is the need of the hour as it is inexhaustible and also clean. So to meet the challenge, the Government of India has come up with major policy and programmes. The focus of all those programmes is energy sustainability, efficiency, security and affordability.

**Keywords** Energy access · Renewable energy resources · Solar energy · Wind energy · Sustainability

## Introduction

Energy is essential for all major economic activities including production, transportation, marketing, etc. No country can grow without developing its energy sector. The world now uses energy at a rate of approximately $4.1 \times 10^{20}$ J/yr. It is equivalent to a continuous power consumption of 13 trillion watts or 13 terawatt (TW). An increase in the earth's population to nine billion people, accompanied by rapid technology development and economic growth worldwide, is projected to witness sharp increase in demand for energy. One in every three persons in the world today lacks access to

C. Mall · P. P. Solanki (✉)
Department of Chemistry, Banaras Hindu University, Varanasi, Uttar Pradesh, India
e-mail: ppsolanki@bhu.ac.in

 
A. Kateja and R. Jain (eds.), *Urban Growth and Environmental Issues in India*,
https://doi.org/10.1007/978-981-16-4273-9_20

energy and two, having access, and depends on conventional sources like wood, coal, charcoal, dung, kerosene, etc. India, with 17% of world population, is moving on a higher growth trajectory with faster industrialization and urbanization. It may well be expected that the country will be an important player in global energy demand, around one quarter of the total. During the last two decades, India improved access to electricity covering 75 crore additional population taking the figure of population having access to electricity to 85% (IEA, 2020). As of 2016, over 800 million people globally did not have access to electricity, and India accounted for 25% of this section with more than 207 million people in the country going without electricity. If we were to look at the role of urban areas in energy consumption, we find that while they occupy just 3% of the earth's land, they account for 60–80% of energy consumption and at least 70% of carbon emissions. Keeping all these facts in mind, it is necessary to generate more and more renewable energy which has low environmental impact and is naturally replenished (United Nations, 2015).

Globally, the renewable energy share of total final energy consumption gradually increased from 16.6% in 2010 to 17.5% in 2016. Even though the absolute level of renewable energy consumption has grown by more than 18% since 2010, only since 2012 has the growth of renewables outpaced the growth of total energy consumption.

## Energy Production: Global and Indian Trends

In 2016, out of the total world energy production, 80% was derived from fossil fuels, 10% from biofuels, 5% from nuclear and 5% from renewable sources (hydro, wind, solar, geothermal, etc.). The consumption details are as follows—only 18% energy was in the form of electricity, while the other 82% was used for heat, transportation, etc. (IRENA, 2018).

The consumption of energy from non-renewable and renewable resources at global level (Fig. 20.1) shows that among all fuel type energy resources, oil (34.2%) is used maximum followed by coal (27.6%), natural (23.4%), renewable (10.4%) and nuclear energy (4.4%). In renewable energy resources, the highest and lowest energy production is from hydroelectric (65.4%) and solar (7.1%) power resource, respectively.

In the current century, steady rise has been witnessed in the share of renewable energy sources (RES). In 2018, the global figure stood at 26%. Developed countries including USA, China and European Union, have been advocating for climate policies. It is heartening to note that these countries are also supporting developing nations in expansion of RES.

Renewables now cover 36% of the power mix in Europe, 26% in China and around 18% in the USA, India and Japan (Global Energy Statistical Yearbook, 2019).

In the world energy production scenario in 2017, China was chief producer followed by Russia and Iran. India is placed at fifth position in energy production from non-renewable and renewable energy resources at global level (Global Energy Statistical Yearbook, 2018). China is the world's largest energy consumer too. India

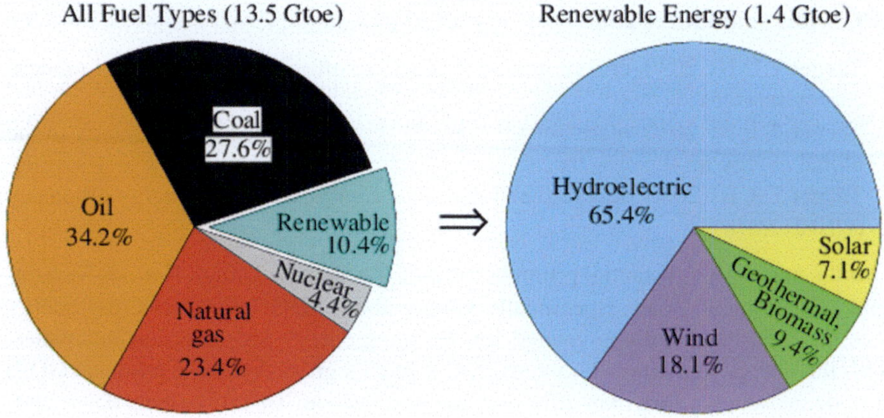


**Fig. 20.1** Global energy consumption. *Source* IEA (2018)

has third place in the countries which consumed more energy after China and the USA (Global Energy Statistical Yearbook, 2018).

Resources of energy can be primarily divided into renewable and non-renewable energy resources on the basis of their derivation. Non-renewable contains fossil fuels (oil, coal, natural gas and uranium) which form over a very long time period, and they will be completely exhausted near 2052 (EIA, 2020). Therefore, energy has to be derived from renewable energy resources which are limitless and eco-friendly. Renewable energy can be divided into six types, that is, ocean, solar, biomass, geothermal, wind and hydroelectric energy.

## Energy and Sustainable Development

Energy is a prerequisite for human and economic development. Lack of access to clean and cheap energy significantly hampers development opportunities of the people. On the other hand, it is the major source of air pollution and greenhouse gas (GHG) emissions. The then Secretary General of the United Nations, Ban Ki-moon, had said, 'Energy is the golden thread that connects economic growth, social equity, and environmental sustainability. With access to energy, people can study, go to university, get a job, start a business—and reach their full potential' (United Nations, 2015).

Energy features prominently in United Nations Sustainable Development Goals (SDGs) (IEA, IRENA, UNSD, World Bank, WHO, 2020) . At least, three SDGs are most closely related to energy:

SDG 7 Universal energy access.
SDG 3.9 Reducing impact of air pollution.
SDG 13 Tackling climate change.

**SDG 7 has again three goals:**

- **Target 7.1**: By 2030, ensure universal access to affordable, reliable and modern energy services.
- **Target 7.2**: By 2030, increase substantially the share of renewable energy in the global energy mix.
- **Target 7.3**: By 2030, double the global rate of improvement in energy efficiency (UNEP, 2018).

India has made substantial progress towards meeting the United Nations Sustainable Development Goals. Specifically talking about SDG 7 (on delivering energy access), India has been appreciated for declining energy and emission intensities. Both the energy and emission intensities of India's gross domestic product (GDP) have decreased by more than 20% over the past decade. This represents commendable progress even as total energy-related carbon dioxide ($CO_2$) emissions continue to rise. Today, India's per capita emissions of $CO_2$ (1.6 tons) are well below the global average of 4.4 tons, while its share of global total $CO_2$ emissions is some 6.4% (IEA, 2020).

## Non-renewable Energy and Environmental Consequences

Energy is the most important input for economic development of any country. Increase in per capita energy consumption is one of the indicators of economic development. Developed countries use more energy per unit of economic output and far more energy per capita than less developed countries. With adoption of more efficient technologies of energy production, energy use per unit of output gradually declines. Development brings expansion in economic activities, drives labour and financial markets, supports modernization of all sectors including agriculture, facilitates infrastructure like transportation and communication and provides more opportunities for social services such as education and health. All these offshoots of development involve and heavily depend on energy use, so much so that development story is very much parallel to energy growth story. With economic development changes the source of energy generation. More developed countries can afford alternative sources which are cleaner, though expensive in comparison with conventional sources, thereby paving way to sustainable development. Sustainable energy generates opportunity—it transforms lives, economies and the planet.

All energy sources have some effect on the environment, human health and well-being. Conventional energy sources based on fossil fuels are particularly harmful for the environment as the generation of energy using fossil fuels emits poisonous gases like carbon dioxide, carbon monoxide, sulphur dioxide, etc. Coal and oil are main culprits for air pollution and smog which are seriously detrimental for human health. Emission of greenhouse gases accelerates global warming and contributes to climate change, one of the most pressing issues humanity faces today. On the top of all, depletion of fossil fuels poses threat in the form of unsustainable situations.

Renewable energy sources, on the other hand, are infinite and environment-friendly. They are unlimited in supply causing little or no pollution. However, one decisive factor is cost of technology required for adoption of RES. It is the prohibitive cost, in terms of money and time, the most important barrier that hinders the adoption of renewable energy sources. Other factors include market-related, geographical and sociocultural factors.

Despite 100% increase in energy use in India since 2000, per capita consumption is merely one-third of the world average. A big portion of the country's energy demand is met by fossil fuels; coal occupies the top rank with over 70% share in power generation. Coal, in fact, is the most plentiful domestic fossil fuel resource, although dependence on coal imports has grown in recent years. India was the world's third largest importer of crude oil in 2014, but is also a major exporter of oil products, thanks to a large refining sector (CSE, 2019).

The installed capacity of energy production till 31 July 2019 indicates that the maximum energy production is from coal followed by renewable (small hydropower project, biomass gasifier, biomass power, urban and industrial waste power, solar and wind), hydro, gas, nuclear, lignite and oil (Ministry of Power).

During 2007–17, India's total primary energy supply (TPES) increased by 55%. While energy is essential for all developmental activities, it gives rise to environmental issues like air pollution and greenhouse gas emission. For example, coal constitutes 44% of TPES and despite the fact that it is important cause of air pollution, its share has gone up by 10% during last decade. As a proactive member of Paris Agreement, the Government of India aims to make a significant move from conventional sources to non-conventional sources. In coming decades, with development and further population growth energy demand is expected to increase multifold. Presently, renewable energy sources make up only 18% of the total mix bag. Increasing this substantially is the need of the hour as it is inexhaustible and also clean. So to meet the challenge, the Government of India has come up with major policy and programmes. The focus of all those programmes is energy sustainability, efficiency, security and affordability.

During 2007–2017, India's total final consumption (TFC) increased by 50%. Major growth was experienced in the industrial sector, which accounted for 42% of TFC in 2017, including non-energy consumption. The rapid growth in TFC, as well as in power generation to supply rising electricity demand, has led to a rapid increase in TPES. In 2017, coal continued to dominate TPES, followed by oil and natural gas (GOI, 2020a).

India is the 3rd largest $CO_2$ emission country in the world on the basis of Enerdata, 2017, as given in Fig. 20.2.

Data shown in Fig. 20.3 estimates that global energy-related $CO_2$ emissions rose by 6040 million metric tons (21%) between 2005 and 2017 at an annual rate of 1.6%. This rate of growth is projected to slow to 1% in 2018 and then to remain essentially flat in 2019. Growth in global energy-related $CO_2$ emissions from 2005 to 2017 was led by China, India and other countries in Asia, which collectively increased by 6260 million metric tons, while emissions in the rest of the world collectively decreased by 220 million metric tons. Energy-related $CO_2$ emissions in most regions

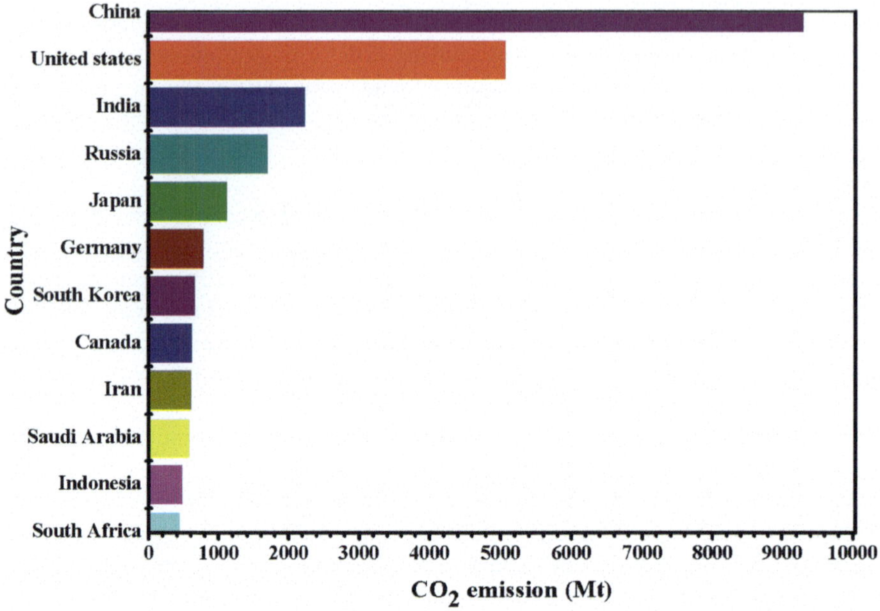


**Fig. 20.2** $CO_2$ emission at global level (2017). *Source* Global Energy Statistical Yearbook (2018)

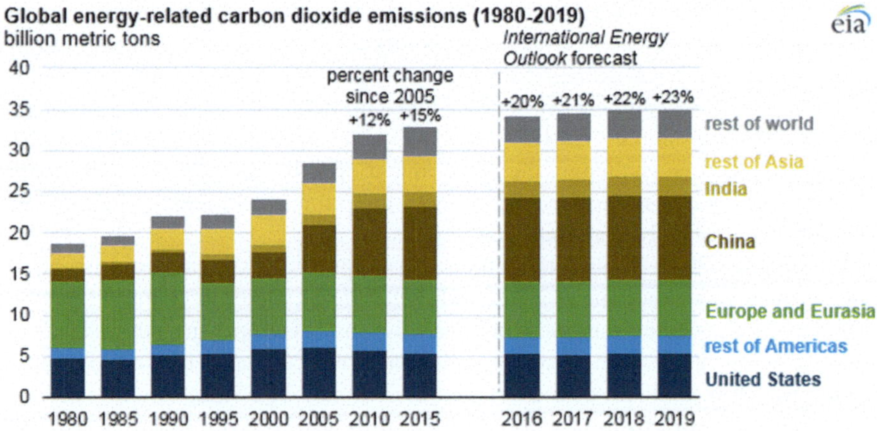


**Fig. 20.3** Global energy-related carbon dioxide emissions (1980–2019). *Source* EIA (2020)

are projected to stay relatively flat through 2019 with slight growth in India, the USA and China.

It is well known that $CO_2$ emission causes climate change, environmental pollution, acid rain, ozone layer depletion, etc. Therefore, due to the thrust of time, it is important to treat renewable energy as an alternate source, boost its production for our future needs, overcome the environment destruction and make energy available for a long span of time. However, it should be noted that the concept of renewable energy is not new but it is just an emerging trend against the fossil fuel and other unsafe energy sources. Hence, it is necessary to address properly renewable energy resources and develop cost-effective and efficient various harvesting technologies with more honestly.

## Renewable Energy Resources

The global renewable generation capacity increased by 167 GW and reached 2179 GW worldwide by the end of 2017. This represents a yearly growth of around 8.3%, the average for seven straight years in a row. Maximum energy at a global level has been obtained from mixed hydro and ocean power followed by solar, wind, bioenergy/biomass and geothermal (IRENA, 2018).

Figure 20.4 shows the leading countries which were installed with renewable energy capacity in 2017. This statistic gives a ranking of the leading countries in installed renewable energy capacity worldwide in 2017. China and India were ranked first and fifth with renewable energy capacity of around 618.8 and 106.28 GW, respectively, in 2017.

India is one of the countries with the largest production of energy from renewable sources. All renewable sources (except large hydro) accounted for up to 20% electricity of the total installed power capacity. Total electricity generation in the country,

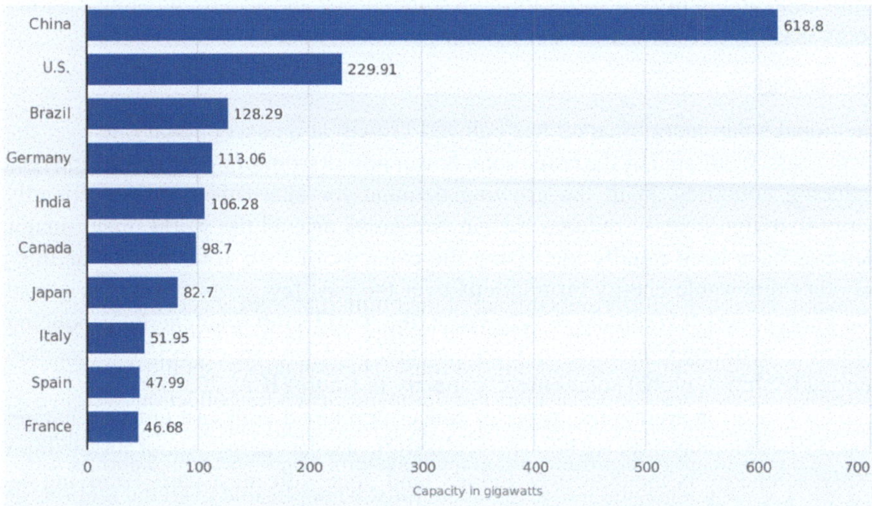


**Fig. 20.4** Leading countries in installed renewable energy capacity worldwide in 2017 (in GW). *Source* IRENA (2018)

from utilities and non-utilities taken together during 2018–19, was 1,546,779 GWh. Out of the total electricity generated through utilities, 1,072,314 GWh was generated from thermal, 134,894 GWh was from hydro, 37,813 GWh was generated from nuclear and 126,759 GWh was from other renewable sources. Total output from non-utilities was 175,000 GWh (GOI, 2020a).

The large hydro-installed capacity was 45.29 GW as on 31 March 2018 and contributes to 13% of the total power capacity. The large hydropower comes under the older Ministry of Power instead of the Ministry of New and Renewable Energy (MNRE) that is why it is not included in all renewable sources. If one includes hydropower with other renewable sources, then the total installed power capacity from renewable sources of India becomes 33%. India has installed renewable energy capacity across 328 million hectares and is sufficiently bestowed with renewable resources of energy at global level (GOI, 2020b).

India has an estimated renewable energy potential of about 900 GW from commercially exploitable sources, viz. 102 GW wind (at 80 m pole height), 20 GW small hydro, 25 GW bioenergy and solar power 750 GW, assuming 3% wasteland is made available. The ministry had taken up a new initiative in 2014 for implementation of wind resource assessment in uncovered new areas with an aim to assess the realistic potential at 100 m level in 500 new stations across the country under the National Clean Energy Fund (NCEF) (MNRE, 2020).

National Institute of Wind Energy/Centre for Wind Energy Technology (NIWE/C-WET) has used advanced modelling techniques and revised the estimate the wind power potential 302 GW at 100 m. Preliminary estimates of offshore wind energy potential indicate potential in Tamil Nadu and Gujarat. Under off-grid applications, there exists significant potential for meeting hot water requirement for residential, commercial and industrial sector through solar energy and also for meeting cooking energy needs in the rural areas through biogas (GOI, 2020b).

***Solar energy***

The global solar energy market has enjoyed growth at an exceptional rate over the recent years, facilitated by the rising solar power output from world's top solar energy-producing countries. With the growing demand for alternative and eco-friendly energy that significantly reduces carbon emissions around the world, many major countries have been rapidly increasing the capacity of their solar power facilities and other renewable energy installations over the past few years. While the global solar energy market continues to surge/rush, the world's top solar energy-producing countries, including China, Japan, Germany and the USA, are expected to maintain their leadership in global solar energy capacity in future (IEA, 2020).

India's location around the tropic of cancer combined with vast territory having about 250–300 sunny days is a boon for realization of solar energy. About 5000 trillion kWh/year energy is incident over India's land area with most parts receiving 4–7 kWh/m$^2$/day. Hence, both technology routes for conversion of solar radiation into heat and electricity, namely solar thermal and solar PVs, can effectively be harnessed providing huge scalability for solar in India. Geographically difficult areas such as hilly regions, forests and islands are generally not suitable for other sources of energy.

Rajasthan, a desert state of India, presents huge possibility for solar thermal hybrid power plants bringing India into leader position in the utilization of solar energy. 140 MW solar thermal hybrid power plants with 35 MW solar through component will be constructed in Rajasthan raising India into the second position in the world in utilization of solar thermal energy. Grid interactive solar PV power projects aggregating to 2490 kW have so far been installed, and other projects of 800 kW capacities are under installation (GOI, 2020a).

India is the seventh largest country in the world for the production of solar energy as per data given by IEA in 2016. The top 10 largest solar energy-producing countries in the world in decreasing order given by the IEA are China, Japan, Germany, USA, Italy, UK, India, France, Australia and Spain (Fig. 20.5).

The production of solar energy in India differs by states. Rajasthan has first place in solar power production followed by Jammu and Kashmir, Maharashtra, Madhya Pradesh, Andhra Pradesh, Gujarat and other states.

***Wind energy***

Wind energy is an infinite source having massive possibilities. It is a clean source of energy as it does not cause environmental pollution. Wind energy, in combination with solar energy, can be used for perennial source of energy making it sustainable, affordable, efficient and secure. Large turbines need to be erected for wind energy, the land around which can be used for farming. Long coastal lines have been found practically useful for this particular source of energy as extract of energy from waves is more efficient in comparison with direct collection of power from wind. Among the renewable sources of energy, it is considered to be the most economical and efficient

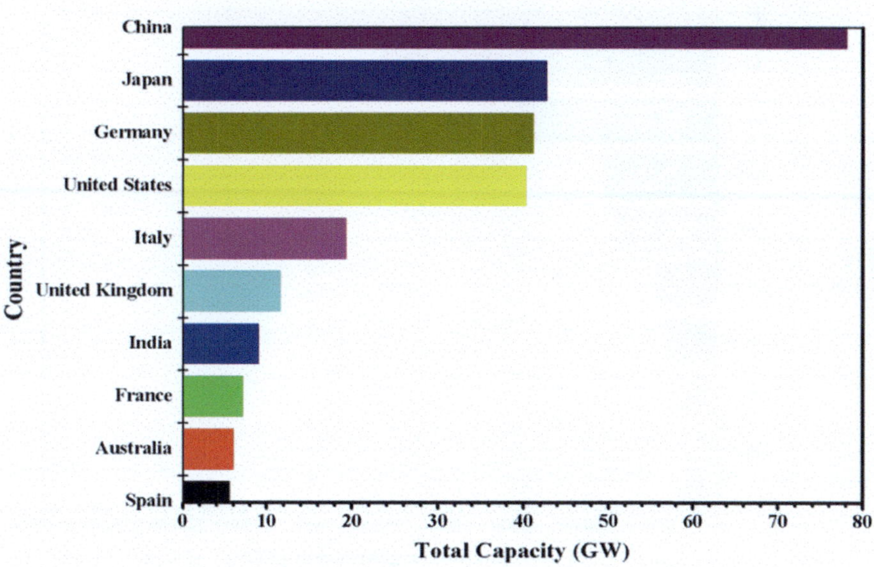


**Fig. 20.5** Top 10 largest solar energy-producing countries in the world. *Source* IEA (2018)

source. In addition to being non-polluting, it is affordable with low establishment period and minimum investment required.

India is ranked fourth in the global wind power installed capacity index with cumulative installed wind power generation capacity of 25,088 MW in 2015 (GOI, 2020a). The index was topped by China, followed by USA and Germany with cumulative installed wind power generation capacity of 145,362, 74,771 and 44,947 MW, respectively, as shown in Fig. 20.6.

The 11th plan proposal foresees a total installed wind energy capacity of 40,000 MW by harnessing around 88% of the wind energy potential in India by 2022. Recently, the country has the fourth highest wind installed capacity in the world with total installed capacity of 34.98 GW as in October 2018 against a target of 60 GW by 2022. Further, around 9.4 GW capacities are under implementation or have been tendered out.

The wind power development in India was initiated in the early 1990s followed by the introduction of 100% accelerated depreciation benefit for wind projects in 1994. The present wind power installed capacity is over 32.7 GW, and wind energy constitutes around 55% of the total renewable capacity. In India, wind energy accounts for around 70% of the renewable energy generation capacity.

As of 2016, data suggests that Gujarat has maximum wind power capacity followed by Andhra Pradesh, Tamil Nadu, Karnataka, Maharashtra (CSE, 2019).

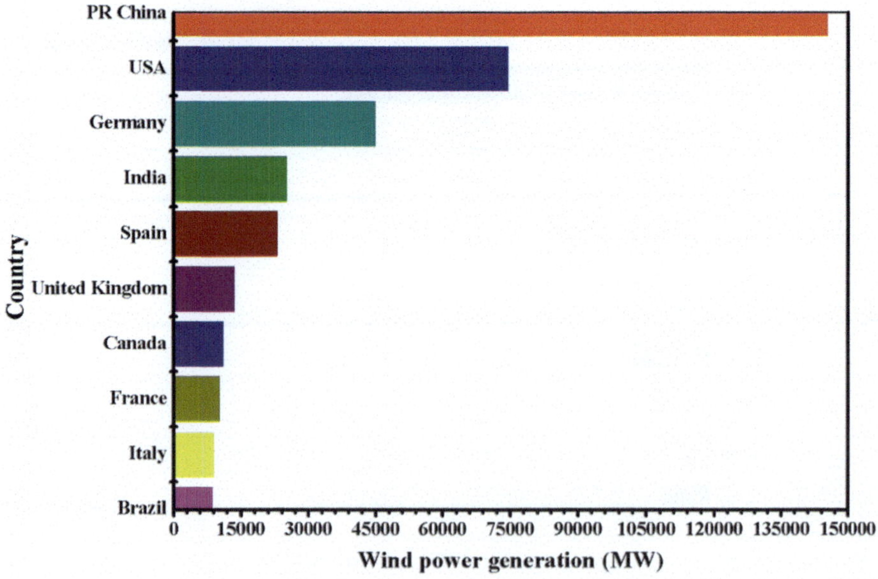


**Fig. 20.6** Top ten countries of world producing electricity from wind power. *Source* IRENA (2018)

However, the speed of wind fluctuates each day and this can be a big disadvantage for wind turbines. The location for wind turbines must have sufficient wind supply. Because of this, wind turbines are installed on top of high elevations like hills, mountains, off shores, etc. Wind turbines are also capital intensive, and this factor inhibits the adoption of wind energy. Other drawbacks include the requirement of deep digging into the earth which also has negative effects on the underground habitats. The most significant disadvantage is noise pollution which is created by activities. The installation of wind turbines also requires a large area which affects agriculture in a negative way.

***Hydroelectric/hydropower***

Hydropower is a clean source of energy as it does not lead to air pollution like coal. Another benefit of hydropower is that through wires and cables, it can easily be transmitted to long distances. But a major disadvantage is that it cannot be stored for future, so consumers have to be fixed before the production. It has been a source of energy for ages. Man started utilizing water flow and water falling from height to produce energy through construction of dams long ago. Continuous flow of water is utilized to make turbines move which enables realization of kinetic energy.

The hydroelectric potential of India ranks fifth in terms of exploitable hydro-potential on global scenario. India is gifted with economically exploitable hydropower potential to the tune of 148,700 MW of installed capacity. In addition, 56 pumped storage projects have also been identified with probable installed capacity of 94,000 MW. In addition to this, hydro-potential from small, mini- and micro-schemes has been estimated as 6782 MW from 1512 sites. Thus, in totality, India is endowed with a hydro-potential of about 250,000 MW. It is the most widely used form of renewable energy. India ranked sixth in hydroelectricity generation globally after China, Canada, Brazil, USA and Russia in the year 2013. The installed capacity of hydropower was 40,730 MW as in June 2014. India is the seventh largest producer of hydroelectric power in the world as in 2017. As on 30 April 2017, India's installed utility scale hydroelectric capacity was 44,594 MW of its total utility power generation capacity.

From the data of MNRE, Karnataka has maximum capacity followed by Himachal Pradesh, Uttarakhand, and Jammu and Kashmir which generate electricity from small hydropower plants (GOI, 2020a).

Along with the opportunities of hydroelectric power, there are several challenges with these plants. These plants are usually constructed across the rivers, and this can affect the life of aquatic animals. Another major issue associated with construction of such power plants has been acquisition of land and resulting displacement of people. It requires large areas of land to be cleared leading to deforestation on a mass scale, removal of top soil, all combined, have severe impacts on natural ecosystem. Construction of large-scale dams also poses threat of geological damage. A case in point is geological damage caused by the construction of Hoover Dam in USA which caused earthquakes and led to depression of earth's surface in the area. Though, hydroelectricity is relatively clean source of energy, environmental issues surrounding erection of huge dams, power plants, submergence of expansive areas with vanishing biodiversity, resettlement and rehabilitation of displaced local people have given rise to controversies, agitations and movements.

On the basis of all aforementioned observations, it can be said that the India may become a leader of solar and wind energy resources in the world, if India becomes successful in harvesting the total falling solar energy in all sunny days and all wind flow in a year.

## Recommended Renewable Energy Sources for India

After study of Section Renewable Energy Resources, it can be concluded that no single renewable energy resource works best at all times in all conditions. Wind usually performs well at specific geographic regions with good wind speed all around the year. While hydropower plants have long gestation period, the photovoltaic (PV) has the benefit of no moving parts, no maintenance, high reliability and long life averaging about 25 years or more for solar panels (CSE, 2019).

Therefore, on the basis of geographical structure of India, one can say that India can generate maximum electricity from solar and wind power plants. Since, maintenance of solar energy is low and wind energy is high, maintenance charge of solar and wind power plants will be compensated by each other. Thus, India can become the leader of solar and wind power in the world if it is successful in harnessing the solar energy of all sunny days and wind speed of most part the year.

The comparison of different renewable energy resources (ocean/wave, solar, biomass, geothermal, wind and hydroelectric resources) on the basis of some characteristics (location, maintenance, lifespan, reporting time, power output, installation, noise, project cost, alternate power source, global rank of India and others) is given in Table 20.1. On the basis of comparison, it indicates that solar energy is the best resource for energy production because solar power plant can be installed anywhere, is low maintenance, silent, and simple with 25 years lifespan. Thus, if India tries to harness the maximum solar power to generate electricity by developing suitable techniques, India can achieve energy independence by the year 2030.

**Table 20.1** Comparison of various renewable energy resources on the basis of some characteristics

| Characteristics | Renewable energy resources | | | | | |
|---|---|---|---|---|---|---|
| | Ocean/wave energy | Solar energy | Biomass energy | Geothermal energy | Wind energy | Hydroelectric/hydropower |
| Location | Can be installed of near sea | Anywhere | Anywhere | Prime sites are often far from population site | With good wind speed for most part of the year | Near the resources of water |
| Maintenance | High | Low | High | High | High | Extremely high |
| Lifespan (years) | 15–20 | 25 | 30 | 30 | 25 | 30 |
| Reporting time | – | Monthly | – | – | Every hour | Every 15 min |
| Power output | Continuous | Intermittent (depend on sunshine) | Provide steady and dependable electricity | – | Intermittent (depend on nature of wind) | Continuous |
| Installation | – | Simple | | – | Simple | Very complex |
| Noise | Much noisier than waves naturally | Silent | Noise as loud as average street traffic | Noise level as a domestic refrigerator | Considerable | – |
| Project cost | – | High | High | – | High | Very high |
| Alternate power source | – | In some areas, use of a large battery bank and/or alternate power source is necessary | – | Geothermal heat pumps are needed as power source | Required an alternate power source | Not required |
| Global rank of India | – | 7th | 2nd | – | – | 7th |

(continued)

**Table 20.1** (continued)

| Characteristics | Renewable energy resources | | | | | |
|---|---|---|---|---|---|---|
| | Ocean/wave energy | Solar energy | Biomass energy | Geothermal energy | Wind energy | Hydroelectric/hydropower |
| Others | Cost of electric power generation per kWh is very high; moving parts replaceable | No moving part | Require the preservation of biomass because it is seasonal | Affects stability of land and causes earthquake, providing both heating and cooling in one system | | High plant load factor, quit and often can be made unremarkable, moving parts usually wear out, etc. |

*Source* Author's compilation

## Conclusion

India has not only experienced commendable progress in access to electricity but also shifted to clean and inexhaustible sources. As a result, India could achieve around 20% reduction in the energy and emission intensities of its GDP. Despite the significant progress, the fact remains that around 207 million people still lack access to electricity, and as the population continues to grow, the demand for cheap and clean energy is expected to grow. Renewable energy resources are the ultimate option to fulfil ever-growing energy demand. In India, solar and wind power are the best renewable energy resources due to 300 clear sunny days, over a dozen perennial rivers and a coastline of more than 7500 km with its territorial waters extending up to 12 nautical miles into the sea. Investing in solar, wind and thermal power, increasing energy productivity and ensuring energy for all are vital if SDG 7 is to be achieved by 2030. To increase renewable and clean energy access, expanding infrastructure and upgrading technology is a must. India stands at the cusp of a momentous shift in the energy sector. Last decade has proven to be transformational for the renewable energy industry in the country. Renewable energy has started playing an increasingly important role in the augmentation of grid power, providing energy access, reducing the consumption of fossil fuels and helping India pursue its low-carbon development path.

## References

Centre for Science and Environment. (2019). https://www.cseindia.org/state-of-renewable-energy-in-india-2019-9243. Accessed June 18, 2020.

EIA. (2020). *Annual energy outlook 2020*. US Energy Information Administration, US Department of Energy.

Global Energy Statistical Yearbook. (2018). Enerdata. https://yearbook.enerdata.net/. Accessed May 23, 2020.

Global Energy Statistical Yearbook. (2019). Enerdata. Available at https://yearbook.enerdata.net/renewables/wind-solar-share-electricity-production.html.

GoI. (2020a). *Energy statistics 2020*. National Statistical Office, Ministry of Statistics and Programme Implementation, Government of India.

GoI. (2020b). *Annual report 2019–2020*. Ministry of New and Renewable Energy, Government of India.

IEA. (2018). *World energy outlook 2018*. International Energy Agency. https://webstore.iea.org/download/direct/2375?fileName=World_Energy_Outlook_2018.pdf.

IEA. (2020). *India 2020 energy policy review*. International Energy Agency.

IEA, IRENA, UNSD, World Bank, WHO. (2020). *Tracking SDG 7: The energy progress report*. World Bank.

IRENA. (2018). International Renewable Energy Agency. https://www.irena.org/404?item=%2fnewsroom%2fpressreleases%2f2018%2fapr%2fglobal-renewable-generation-continues-its-strong-growth-new-irena-capacity-data%e2%80%a6&user=extranet%5cAnonymous&site=IrenaLive. Accessed December 29, 2018.

MNRE. (2020). *Annual report 2019–20*. Ministry of New and Renewable Energy, Government of India.

UNEP. (2018). https://wedocs.unep.org/bitstream/handle/20.500.11822/25762/SDG7_Brief.pdf?sequence=1&isAllowed=y. Accessed June 17, 2019.

United Nations. (2015). *Sustainable development goals*. https://www.un.org/sustainabledevelopment/cities/. Accessed June 3, 2020.

# Chapter 21
# Necessity of Paradigm Shift from Non-renewable Sources to Renewable Sources for Energy Demand

**Alka Kataria and T. I. Khan**

**Abstract** The demand for energy in the country has increased with increase in population. The source and methods of energy production have always been insufficient. To meet the energy need, we have been dependent on fossil fuels such as coal, oil and natural gas. While producing and using energy, we have witnessed adverse impacts on environment in general and human health specifically. There is a necessity to abide by the international commitment of stabilizing the increase of temperature by 2 °C as the $CO_2$ level has already crossed 409.8 ppm in 2019. To deal with this, the shift from traditional sources (fossil fuels) to renewable sources of energy that are considered to as clean sources of energy is desirable. The optimum use of renewable sources of energy minimizes the harmful effects on living and non-living components of environment especially human beings. Internationally as well as nationally, regulations and policies have been developing to encourage the people and industrialists to adopt the sustainable renewable techniques to meet the needs of energy especially in untouched rural areas with minimum environment deterioration. This article evaluates all possible options to make the fullest use of renewable sources of energy. The literature related to the reasons behind the need of shift from non-renewable to renewable energy sources has also been summarized. Besides this, the initiatives taken by the Government of India in terms of the availability, current status and policies in bridging the demand–supply gap of energy sources have also been discussed.

**Keywords** Energy production · Fossil fuels · Non-renewable energy sources · Renewable energy sources

A. Kataria (✉)
Department of Environmental Science, IIS (Deemed to be University), Rajasthan Jaipur, India

T. I. Khan
Indira Gandhi Centre for HEEPS, University of Rajasthan, Jaipur, Rajasthan, India

 
A. Kateja and R. Jain (eds.), *Urban Growth and Environmental Issues in India*,
https://doi.org/10.1007/978-981-16-4273-9_21

## Introduction

Energy has always played a vital role in shaping society, be it industrial revolution, electrification or information and communication technology, energy growth story through nineteenth and twentieth century and now, has become story of human progress. With the passage of time, growing population and economic growth (increased commercialization and industrialization), countries across globe have been witnessing exponentially growing demand for energy. Developing countries like India have been trying to fulfil its primary energy demand by increasing the share of fossil fuels (mainly coal) since 2000 (Fig. 21.1). Still in many rural areas, people do not have access to energy resources, and as a result, the demand of primary energy has increased from 6 to 11% (BP Energy Outlook, 2019), but due to limited resources and available technologies, the demand–supply gap has been widening. The energy shortage is estimated to have a demand–supply gap of around 8%, with demand being 30% greater than supply during peak hours (TERI, 1998), and it will be doubled by 2040, due to increase in appliance ownership and cooling needs (IEA, 2020). To bridge this gap, various alternate energy resources like solar energy, wind energy, etc., must be employed along with conventional energy resources.

## Non-renewable Energy Resources

Traditional power generation techniques include the use of fossil fuels such as coal, oil and gas. In 2019, fossil fuels account for around 84% of all primary energy consumption worldwide (Rapier, 2020). In India, 56% of non-renewable energy is used in the industrial sector (IEA, 2020). When compared with other types of energy generation methods, the per capita consumption of coal in world and India

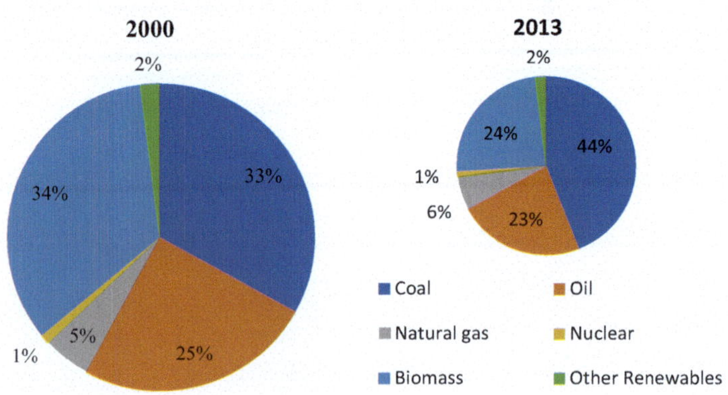


**Fig. 21.1** Primary energy demand in India by fuel. *Source* India Energy Outlook (2015)

is found to be the highest and is still increasing with the population and demand increase (Figs. 21.2 and 21.3). The World Energy Forum has forecasted that fossil-based reserves will be depleted in less than ten decades due to their extensive and unsustainable use (Kumar et al., 2010). Also, combustion of fossil fuels and their mining releases many greenhouse gases, and one such gas is $CO_2$, the level of which has already crossed 409.8 parts per million in 2019 (Lindsey, 2020). According to the International Energy Outlook (IEA, 2016), world energy-related $CO_2$ emissions will rise from 32.2 billion metric tons in 2012 to 43.2 billion metric tons in 2040—an increase of 34% over the projection period.

Apart from $CO_2$, combustion of fossil fuels also releases other greenhouse gases like sulphur dioxide, ozone, nitrogen dioxide and particulate pollutants. Most of the primary particulate matter released is generally smaller than 2.5 μm ($PM_{2.5}$), 1 μm

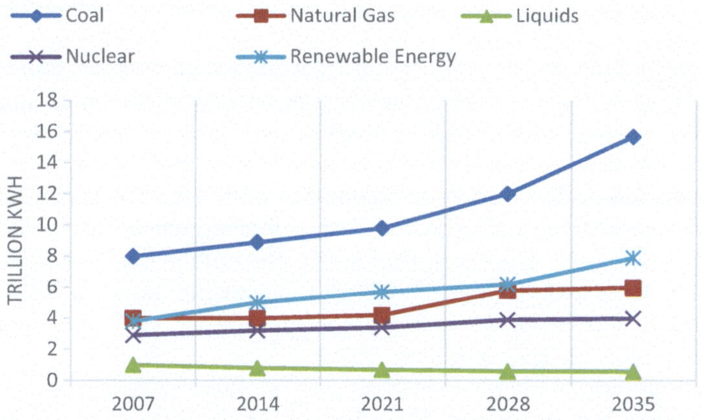


**Fig. 21.2** Global net electricity generation by fuel (trillion kWh) during the period 2007–35. *Source* Annual Energy Outlook (2010)

**Fig. 21.3** India in top five coal consumers. *Source* Kumar et al. (2014)

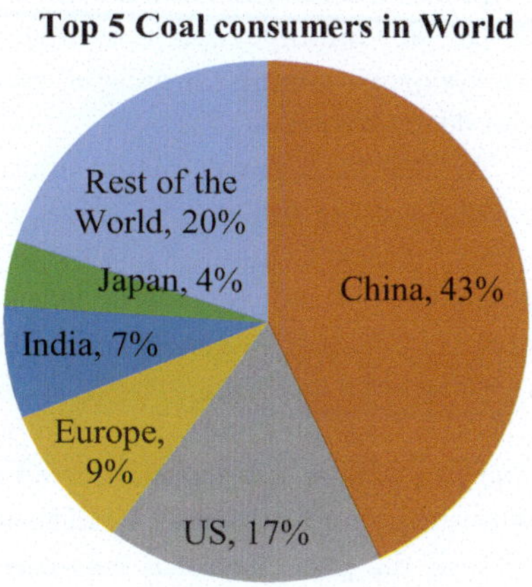

($PM_1$) and ultrafine particles in urban areas (Lighty et al., 2000). These pollutants not only cause air pollution (Prasad et al., 2006) in big cities (partly by coal-fired power plants and partly by emission from vehicles) but also cause various other problems like ash disposal (Carlson & Adriano, 1993), land degradation, water pollution (Beerbaum & Weinrebe, 2000; Poornima et al., 2005), etc. The careless disposal or dispersion of these pollutants can cause harm to living organisms such as animals, plants and microorganisms and can show negative impact on buildings, monuments, etc. Emission of greenhouse gases in the atmosphere is one of the major causes of global warming and global climate change. So it has become a need of hour to have an alternative which is sustainable like renewable resources.

## *Coal*

Coal reserves in India are roughly 10% of the world's total reserves. India stands at the third position after USA and China as far as underground mining is concerned (GMI, 2015). Despite India's rich coal reserves, the quality of the coal is considered to be low because it contains high ash content (45%), high moisture content (4–20%), low sulphur content (0.2–0.7%) and low calorific values (between 2500–5000 kcal/kg) (IEA, 2002). As a result, India is facing difficulties in balancing the demand with the domestic need. India produced nearly 340 million tons of coal equivalent (mtce) in 2013, but it also imported 140 mtce, accounting for about 12% of global coal imports (61% from Indonesia, 21% from Australia and 13% from South Africa) (India Energy Outlook-Special Report, 2015).

Coal deposits are mostly found in the country's eastern and south central regions. Jharkhand, Odisha, Chhattisgarh, West Bengal, Madhya Pradesh, Telangana and Maharashtra have 98.09% of the country's total coal reserves. As of 31 March 2019, the state of Jharkhand had the largest share (25.88%) of the country's total coal reserves (Energy Statistics, 2020).

In India, different methods are being followed for mining of coal—some of them are surface mining, captive mining (around 7% of national production comes from it), open cast mining (at present, more than 90% of coal in India is produced by it) and deep mining.

At coal-fired power plants, the combustion converts the coal into useful heat energy, but the Indian coal sector is facing a challenge such as low-grade quality of coal reserves. Another challenge is low to medium calorific values and high ash content of most of the available hard coal, which require burning of coal in large quantity for energy generation. The high ash content produced after combustion requires large land area for its disposal, which more often either results in deforestation at a large scale or causes percolation of hazardous elements into groundwater. The groundwater gets contaminated when ash disposal in ash ponds comes in contact with surface or rain water. Apart from the aforementioned issues, another significant issue is the great distance between the location of hard coal reserves and power plants. This increases the cost of transporting coal to power plants after mining. On an average, coal has to travel more than 500 km before it is converted to electricity. Not only this, the burning of coal in coal-based power generation plants emits particulate matter and greenhouse gases like $SO_2$, $NO_X$ and CO. If there is no control on the emission of pollutants in the atmosphere, a 500 MWe plant using coal with 2.5% sulphur (S), 16% ash and 30,000 kJ/kg heat content would emit 200 metric tons of sulphur dioxide ($SO_2$), 70 metric tons of nitrogen dioxide ($NO_2$), 500 metric tons of fly ash, 500 metric tons of solid waste and about 17 gigawatt-hours (GWh) of thermal discharge per day (Tripathi et al., 2016). The emission of $SO_2$, $NO_2$ and other gases at a very high concentration may contribute for global warming and can result in acid rain. Also some coal-derived nano-pollutants not only affect abiotic components but also affect the biotic components, causing acute respiratory (attack of asthma) and cardiovascular diseases (coagulability of blood) (Seaton et al., 1995). Oliveira et al. (2014) found that the nano-minerals (quartz, silica and some sheet silicates) found in coal and coal ashes cause occupational diseases such as lung cancer (Zhao et al., 2006).

### *Oil*

The crude oil production capacity of India is less than the demand as a result India heavily depends on imports of crude oil. The crude oil production of India is just over 900 thousand barrels per day (kb/d), far insufficient to meet the refinery capacity of 4.4 mb/d (India Energy Outlook, 2015). The maximum reserves of crude oil in India are found in the western part of the country, particularly in Rajasthan and offshore areas near Gujarat and Maharashtra, followed by Assam (Energy Statistics, 2020). About two-thirds of crude oil in India is produced by state-owned enterprises like Oil and Natural Gas Corporation Limited (ONGC) and Oil India Limited (OIL), with the rest coming from joint ventures with national oil and gas firms.

The leakage during the oil and natural gas drilling, processing, transportation and utilization may cause different types of pollution (mainly pollution of air and water like aquifer contamination, toxicity of water) and accidental fires that may continue burning for days or weeks. Certain activities like drilling, cleaning of oil tankers and shipwrecks may cause various environmental disasters like oil slicks (Bharucha, 2005) in seas. It has serious consequences on living organisms as it can cause damage to marine or terrestrial ecosystems (Chang et al., 2014; ITOPF Ltd, 2011).

### *Natural Gas*

India's natural gas reserves are currently estimated at about 1380.63 billion $m^3$ as on 31.03.2019 (Energy Statistics, 2020). The principal onshore production fields are in Assam, Gujarat, Tamil Nadu and Andhra Pradesh whereas Krishna Godavari Basin on the east coast is one of the most promising areas for offshore production. In recent

**Table 21.1** Fuel-wise installed capacity in MW

| Fuel | MW | Percentage (%) |
|---|---|---|
| Total thermal (coal + gas + oil) | 137,936.18 | 66.63 |
| Hydro (renewable) | 39,291.40 | 18.98 |
| Nuclear | 4780.00 | 2.30 |
| RES | 24,998.46 | 12.09 |
| Total | 207,006.04 | 100.00 |

*Source* Reddy et al. (2013)

years, the lack of market demand in the region and the cost of transporting via pipe line are the main reasons behind decrease in the production of natural gas. Since 2011, India's gas output has decreased as a result of the above-mentioned reasons. In India, GAIL is the state-owned gas company.

India has a significant unconventional energy potential, both from coal bed methane (CBM) and shale gas, in addition to conventional gas resources. Commercial production is still some way off due to limited availability, but CBM activity is picking up steam with a number of private firms, like Reliance and Essar.

According to 2013 data, thermal-based power generation (coal, gas and oil) accounts for 66.63% of the country's total electricity generation (Table 21.1). In this, the installed capacity of coal-fired thermal power stations is 1, 17,833.38 MW, accounting for 56.92% of total electricity generation. Gas-fired combined cycle power plants have the installed capacity of 18,903.05 MW which is 9.13% of total power generation. High power generation cost of diesel oil-based power plants as compared to coal- and gas-fired power plants has reduced its (0.57%) per cent share in total power generation and limited its use only during peak load demand (Reddy et al., 2013). It is not that only non-renewable resources are contributing to total installed capacity but renewable energy sources (RES) also make a substantial contribution. In 2013, renewable energy sources such as industrial and urban waste, small hydropower projects, biomass power, biomass gasifiers and others supplied 12.09% of total electricity.

Having experienced all social and environmental problems related to the electricity generation, governments of all the countries, especially developing countries like India, should formulate energy policies to ensure that its resources are used sustainably. The best alternative to all these problems is the enhancement in utilization of renewable energy sources as they can reduce the current global environmental problems and dependency on fossil fuels for future energy needs.

## Renewable Energy Sources

Renewable energy sources, also known as alternative energy sources, are resources that can be used to generate energy over and over again. They are environment friendly, clean and inexhaustible energy resources. There are a number of renewable

**Table 21.2** Main renewable energy sources and their application

| Energy sources | Applications |
|---|---|
| Hydro power | Electric power generation |
| Biomass | Power generation, gasification, digestion |
| Geothermal | Urban heating, hydro thermal |
| Solar | Solar domestic systems, dryers, solar cookers |
| Direct solar | Photovoltaic, thermal power generation, solar water heaters |
| Wind | Wind mills, water pump |
| Wave and tide | Design, tidal stream |

*Source* Kumar et al. (2019)

energy resources, but the main renewable energy sources along with their applications are enlisted in Table 21.2.

Biomass energy is one of the dominating renewable energy sources in residential sector and being used since ages in India. But in recent years, India has also expanded its use of other renewable energy sources. As far as initiatives of the Government of India is concerned, the Government of India has formed one exclusive ministry for new and renewable energy development, i.e. Commission for additional Sources of Energy in 1981 which became Ministry of Non-Conventional Energy Sources (MNES) in 1992 and later on in 2006 it was renamed as Ministry of New and Renewable Energy (MNRE). For the promotion of renewable energy sources, MNRE has been formulating various policies on banking facilities such as capital and interest at nominal rates, preferential charges, income tax rebates, renewable energy certificates, renewable purchasing commitments and easy loans. As a result, the energy sector has also started adopting the renewable energy sources at a very large scale.

## *Solar Energy*

It is the world's most abundant energy source, found in direct (solar radiation) and indirect (wind, biomass, hydra, ocean, etc.) forms. The sun emits 3.8 × 1023 kW of energy, of which the Earth intercepts roughly 1.8 × 1014 kW (Thirugnanasambandam et al., 2010). The light and heat (thermal) of the sun can be utilized for various purposes such as solar heat (thermal) can be used for heating of water, cooking, drying, power generation, etc., especially in India, as the average temperature is around 28 °C and the estimated potential of solar energy in India is about 20,000 MW (Kumar & Meena, 2017). It makes India a better place for electricity generation from sun rays. Photovoltaic cells transform solar energy's light into electricity, which can then be used for a variety of purposes like lighting, pumping and so on.

**Solar thermal application**

Solar heat has a lot of potential for thermal applications like water heating, crop drying and cooking, etc. From them, solar cooking is one of the most practical uses of solar power. Solar cooking entails the use of solar cookers, and among the various kinds of solar cookers available today, the box-type solar cooker is the most widely used. At 50% capacity utilization, a household solar water heating system with a 100 L per day capacity can reduce $CO_2$ emissions by around 1237 kg per year, and in hot and sunny regions, it can reduce $CO_2$ emissions by around 1410.5 kg per year (Jyotirmay et al., 2002; Kumar & Kandpal, 2007).

**Solar thermal power**

It has a wide range of applications, particularly in remote areas where electricity is scarce and solar radiation is abundant, making the use of solar energy for electricity generation in these areas a viable option (Demirbas, 2007). The term 'solar thermal electricity power system' refers to a device that uses solar radiation to generate electricity through a solar heat to electricity conversion device (Xiao et al., 2004). Solar collectors are an important part of a solar thermal power system. Different types of solar collectors available are compound parabolic, flat-type, parabolic dish, parabolic trough, etc.

**Solar photovoltaic energy**

Photovoltaic (also known as solar or PV) cells that are made of two layers of silicon use the photovoltaic effect (PV effect) to transform solar light into electricity without releasing any harmful gases, noise or other contaminants into the environment. Each silicon layer on PV cell carries an electric charge. When light strikes the cells, the charges begin to move between the two layers, resulting in the production of electricity. A number of PV cells are connected to make a module, and a light bulb can be powered by a module of about 40 cells. For both space and terrestrial applications, silicon solar cells are most commonly used.

Although solar power has not played a minor role in power generation so far, and the majority of which was added in the previous five years but with the launch of the Jawaharlal Nehru National Solar Mission in 2010, India began to place a much greater focus on solar growth, with the goal that was upgraded in 2014, of 100 GW of solar installations by 2022 (India Energy Outlook-Special Report, 2015). According to a recent IEA analysis, India's investment in solar PV in 2018 was bigger than all fossil fuel sources of electricity generation combined, and the country is on track to meet its 2022 target.

## *Wind Energy*

Wind power today accounts for the biggest share of modern renewable energy in power generation in India. It has the fourth-largest wind market after USA, Germany and China in the world (Chaurasiya et al., 2019). By 2022, the wind capacity is expected to add around 38,500 MW to the total installed capacity (Kumar et al., 2019). If fossil fuel sources are not widely available, wind energy can provide a cost-effective solution in many cases. In India, the initiation of wind power programme began near the end of the Sixth Plan, in 1983–84. At 211 wind monitoring stations across 13 states and union territories (including Andaman and Nicobar Islands, Andhra Pradesh, Gujarat, Karnataka, Kerala, Lakshadweep, Madhya Pradesh, Maharashtra, Orissa, Rajasthan, Tamil Nadu, Uttaranchal, and West Bengal), annual mean wind power density greater than 200 W/m$^2$ (watts per square metre) was observed at 50 m height (Kumar et al., 2010).

With the use of wind turbines, wind technology transforms the energy available in the wind into rotational energy and finally converting into electrical or mechanical power. The only area of interest is that wind turbines can cause noise pollution at times. According to a research, the noise produced by a single wind turbine at a distance of 40 m is around 50–60 dB (Tripathi et al., 2016). However, this is insignificant since the CPCB's permissible noise limits are much higher.

## *Hydropower*

In a hydroelectric power plant, the potential or kinetic energy of water is converted into mechanical energy (water mills) or as electrical energy. The height of the built dam on running water body is used to create kinetic energy of water that helps in generating electricity. It is said to be the world's greatest renewable energy resource for electricity generation. There is about 20% contribution of hydropower globally of total power generation capacities and India took the initiative by building its first hydroelectric power plant in Karnataka at Shivana Samudra (Tripathi et al., 2016).

On a global scale, India has enormous hydroelectric potential and ranks fifth in terms of exploitable hydropotential. According to the Central Electricity Authority (CEA), India has 148,700 MW of hydropower potential that can be economically exploited (Kumar et al., 2010). Andhra Pradesh, Bihar, Gujarat, Jammu & Kashmir, Karnataka, Punjab, Sikkim, Uttaranchal and Uttar Pradesh, all of the above-mentioned states have primary hydroelectric power plants. The hydropower system of India is divided into five major provinces which are named as Northern region, Western Region, Southern region, Eastern region and North-Eastern region. The hydropower potential is the highest of North-Eastern region with 98% untapped. The Northern, Eastern, Western and Southern provinces, respectively, have 79, 77, 23 and 33% unexplored hydropower potential (Subrahmanyam, 2013).

National Hydroelectric Power Corporation (NHPC), Northeast Electric Power Company (NEEPCO), Satluj Jal Vidyut Nigam (SJVNL), Tehri Hydro Development Corporation and NTPC-Hydro are some of the public sector hydroelectric power generating companies in India (Tripathi et al., 2016).

Although the generation of electricity by hydroelectric power plants does not create any pollution and helps in reducing the amount of greenhouse gases emitted during the burning of fossil fuels, however the development of these plants stimulates many issues like displacement and resettlement of local people, loss of biodiversity due to land clearance or deforestation, disrupts the aquatic ecosystem, etc. However, when compared with its production capacity, these problems are considered very less.

## *Geothermal Energy*

The heat stored in the Earth's interior is used to generate geothermal energy. India is still lagging behind in making full-fledged use of geothermal energy. As India has a potential for geothermal energy and many geothermal sites are also available for electricity generation, government has tried to utilize this by putting forward various plans to increase the current total installed generating capacity by two times. According to studies conducted by the Indian Geological Survey, the nation has about 340 hot springs associated with mid-continental rifts, subduction, sedimentary basins, and Cretaceous–Tertiary volcanism. They are spread across seven geothermal provinces. (i) Cambay, (ii) Godavari, (iii) Mahanadi, (iv) Sohana, (v) Son Narmada-Tapi rift zone (SONATA), (vi) The Himalayas and (vii) West coast are among these provinces. Recently, one new location discovered as geothermal site is Tattapani in Chhattisgarh. Barren Island has become one of the most important geothermal provinces in the Indian subcontinent as a result of the recent volcanic eruption. The energy released by one-third of the springs is estimated to be $40.9 \times 1018$ cal. This is the same amount of energy produced by 5.7 billion tons of coal or 28 million barrels of oil (Chandrasekharam, 2000; Kumar et al., 2010). In the coming years, India may become a major contributor in electricity generation with the help of geothermal sources in the world.

## *Biomass Energy*

Plants use solar energy and carbon dioxide to produce sugar and oxygen through a process called photosynthesis. This organic matter produced during this process is called as biomass (includes both terrestrial and aquatic). Biomass includes wood, logging waste, animal dung and vegetable matter consisting of leaves and agricultural waste. Biomass combustion provides a significant amount of energy that can be used, making biomass a viable power source. The amount of energy released during combustion is determined by a variety of factors, including weather, chemical and thermal conditions. A number of other products of biomass such as biodiesel and bioethanol can be used to generate energy by various technological ways. In India, biomass energy is being used from ancient times and was the first renewable energy source for energy generation. India's biomass energy potential is believed to be around 19,500 MW (Kumar & Meena, 2017).

**Table 21.3** List of pollutants produced due to burning of biomass causing various health problems on living organisms

| Name of pollutant | Health effects |
|---|---|
| Particulate matter (PM 10 and PM 2.5) | Eye burning sensation, bronchitis, heart-related problems, causes cancer |
| Carbon dioxide ($CO_2$) | Fast breathing rate, sudden rise in heart breathing rate |
| Methane ($CH_4$) | Choking of wind pipe that may lead to suffocation, headaches |
| Carbon monoxide (CO) | Can cause death, heart attacks |
| Oxides of nitrogen ($NO_X$) | Cough, chest pain, difficulty in breathing |
| Sulphur dioxide ($SO_2$) | Eye burning sensation, chest problems, asthma-related issues |

*Source* Kumar et al. (2019)

The traditional use of biomass for cooking or for energy generation may lead to significant health problems (Table 21.3). However, the use of new technologies has overruled these problems and resulted in the use of biomass at a faster rate in power generation.

## Initiatives Taken by Government of India for Promoting Renewable Energy in India

The Government of India has taken various steps to increase the share of renewable energy in providing energy access significantly and to reduce the dependency on fossil fuels for energy generation. Few of them are mentioned below:

- District Advisory Committees were formulated in various districts of the country for renewable energy promotion at grass root level.
- The Jawaharlal Nehru National Solar Mission (JNNSM) was initiated in 2010 as part of the National Action Plan on Climate Change (NAPCC), with the goal of installing 20 GW of grid-connected solar power by 2022 (CII, 2021). Later in 2015, the Government of India has raised the target of the JNNSM to 100 GW by 2022 (IFMR Lead, Centre for Development Finance, 2015).
- Energy Clubs have been established in AICTE-approved or recognized engineering colleges throughout the country to educate young educators on different aspects of renewable energy.
- 'Akshay Urja' a bimonthly newsletter by Ministry of New and Renewable Energy (MNRE) is being published to disseminate information about technological developments, education in the field of renewable energy, etc.
- Akshayaurja (earlier called Aditya Solar shops) shops programme has been promoted since 1995 in major cities to widen the sale and services of all renewable

energy devices. The ministry announced to discontinue this programme in 12th plan period (2017) (Saur News Bureau, 2017).

- In 2020, the Ministry of New and Renewable Energy unveiled a new website to disseminate information about new renewable energy schemes and projects, using cutting-edge technology to make it more informative, interactive and user-friendly (The Economic Times-Rise, 2020).
- Private sector developers were encouraged to make use of advanced technologies in renewable energy projects to draw on maximum energy from these projects.
- In COP 21 (Conference of Parties) in December 2015, India signed Paris Agreement and committed to reduce 30–35% of its GHG emissions below 2005 levels by 2030.
- India has presented its Intended Nationally Determined Contribution (INDC) ahead of COP 21, stating its aim of installing 175 GW of renewable power capacity by 2022, as well as a target of increasing India's share of non-fossil-based capacity to 40% by 2030 (MNRE, Annual Report, 2018–19).
- Pradhan Mantri Kisan Urja Suraksha evam Utthaan Mahabhiyan (PM-KUSUM) scheme was launched for farmers in 2019 by MNRE. With a total central financial support of Rs. 34,422 crore, this scheme seeks to build 25,750 MW of solar pumps and other renewable power plants throughout the country by 2022 (Singh, 2020).
- In 2015, India and France launched the International Solar Alliance (ISA), a treaty-based international intergovernmental organization. Its primary goal was to put solar energy on the global agenda and ensure that everyone have access to renewable energy.
- The MNRE announced a National Lab Policy on Testing, Standardization and Certification to encourage research and development in renewable energy technology. The innovative ideas in the field of renewable energy have also been recognized and encouraged through '*Abhinav Soch-Nayi Sambhawanayen*' award given by the government.

## Achievements in Renewable Energy

- In the past five years, India's use of renewable energy has increased from 20 to 40 GW (from April 2014 to March 2019).
- Out of the target set by Government of India to install 175 GW cumulative capacity by 2022, a total of 78.31 GW of renewable energy capacity has been installed by March 2019 (MNRE, Annual Report, 2018–19).
- India is now ranked fourth and fifth in the world in terms of wind and solar power consumption, respectively, and fifth in terms of overall renewable energy installed capacity (MNRE, 2021).
- Renewable energy installation capacity has grown by 226% in the past five years, and it accounts for 23.30% of the country's total installed electricity capacity (up to 29 February 2020) (MNRE, 2021).

- In the year 2019, a record of more than 120 billion units of renewable electricity was generated (MNRE, Annual Report, 2018–19).
- The wind energy capacity has reached to 5.5 GW in 2016–17, which is the highest level achieved ever (MNRE, 2021).
- In the last five years, solar power capacity has risen by more than 11 times, from 2.6 GW to 28.18 GW (MNRE, Annual Report, 2018–19). As a result, the target under JNNSM has been increased to 100 MW.
- As far as growth in solar power in India is concerned, MNRE has approved the installation of 41 solar parks with an aggregate capacity of more than 26,144 MW in 21 states of India. After this approval, a solar park with 1000 MW capacity has been installed in Kurnool and the work on installing the largest solar park in Pavagada is going on (Kumar & Majid, 2020).
- In 2015, the Ministry of Skill Development and Entrepreneurship established the Skill Council for Green Jobs, a non-profit society that successfully trained over 77,000 Suryamitra trainees (BW Online Bureau, 2021). This society seeks to train undergraduates, ITI/diploma holders and others in installation and maintenance so that National Solar Mission programmes can be implemented throughout the country.

## Conclusion

The paper highlights the current status of energy production, uses and adverse impacts on materials, environment and human beings. India faces a tough challenge in bridging the demand–supply gap. Power sector reforms are high on government agenda and hopefully energy reforms will lead to energy security and efficiency. The country is mainly dependent on fossil fuels especially on coal for power generation, but extensive use and their combustion are causing environmental pollution and deteriorating the health of abiotic and biotic components. Due to the international treaties, accords and conventions, all the countries are committed to reduce the emission of $CO_2$ and other greenhouse gases. This could be possible only by increasing the production and development of energy gradually by using alternative sources such as solar, wind and geothermal. This will help in reducing dependency on imported fossil fuels. In this endeavour, Indian scientists, social scientists, various states and central government are doing their best.

## References


Annual Energy Outlook. (2010). *U.S. energy information administration*. Available at http://www.ascension-publishing.com/BIZ/HD18-2010.pdf. Accessed June 28, 2017.

Beerbaum, S., & Weinrebe, G. (2000). Solar thermal power generation in India: A techno-economic analysis. *Renewable Energy, 21*, 153–174.

Bharucha, E. (2005). *Textbook of environmental studies for undergraduate courses*. University Press, University Grants Commission.
BP Energy Outlook. (2019). *Insights from the evolving transition scenario—India*. Available at https://www.bp.com/content/dam/bp/business-sites/en/global/corporate/pdfs/energy-economics/energy-outlook/bp-energy-outlook-2019-country-insight-india.pdf. Accessed May 24, 2020.
BW Online Bureau. (2021). Fiinovation assists skill council for green jobs to assess impact of Suryamitra programme. Available at http://bweducation.businessworld.in/article/Fiinovation-Assists-Skill-Council-For-Green-Jobs-To-Assess-Impact-Of-Suryamitra-Programme-/15-01-2021-365736/. Accessed February 9, 2021.
Carlson, C. L., & Adriano, D. C. (1993). Environmental impacts of coal combustion residues. *Journal of Environmental Quality, 22*(2), 227–247.
Chandrasekharam, D. (2000). Geothermal energy resources of India: Country update. In *Proceedings World Geothermal Congress 2000* (pp. 133–145), Kyushu, Tohoku, Japan.
Chang, S. E., Stone, J., Demes, K., & Piscitelli, M. (2014). Consequences of oil spills: A review and framework for informing planning. *Ecology and Society, 19*(2), 26.
Chaurasiya, P. K., Warudkar, V., & Ahmed, S. (2019). Wind energy development and policy in India: A review. *Energy Strategy Reviews, 24*, 342–357.
CII (Confederation of Indian Industry). (2021). Renewable energy. Available at https://www.cii.in/sectors.aspx?enc=prvePUj2bdMtgTmvPwvisYH+5EnGjyGXO9hLECvTuNsdd4UMpGp0Zgxc80wpSgZt. Accessed February 9, 2021.
Demirbas, M. F. (2007). Electricity production using solar energy. *Energy Sources, Part A: Recovery, Utilization, and Environmental Effects, 29*(6), 563–569.
Energy Statistics. (2020). National Statistical Office, Ministry of Statistics and Programme Implementation, Government of India. Available at http://www.mospi.gov.in/sites/default/files/publication_reports/ES_2020_240420m.pdf. Accessed May 24, 2020.
GMI. (2015). *Coal mine methane country profiles*. Global Methane Initiative. Available at http://globalmethane.org/documents/Toolsres_coal_overview_fullreport.pdf. Accessed January 31, 2021.
IFMR LEAD, Centre for Development Finance. (2015). Jawaharlal Nehru National Solar Mission, Shakti Sustainable Energy Foundation. Available at http://www.ifmrlead.org/wp-content/uploads/2015/10/NAPCC/1_JNNSM%20Brief_CDF_IFMRLEAD.pdf. Accessed February 9, 2021.
India Energy Outlook. (2015). *World energy outlook special report*. International Energy Agency. Available at http://www.iea.org/publications/freepublications/publication/IndiaEnergyOutlook_WEO2015.pdf. Accessed May 21, 2020.
International Energy Association-IEA. (2002). Coal in the energy supply of India. Available at http://www.iea.org/Textbase/publications/free_new_Desc.asp?PUBS_ID=1101. Accessed May 24, 2020.
International Energy Agency-IEA. (2016). *World energy outlook 2016*. IEA. Available at https://doi.org/10.1787/weo-2016-en. Accessed June 28, 2017.
International Energy Agency-IEA. (2020). *India 2020*. IEA. Available at https://www.iea.org/reports/india-2020. Accessed May 24, 2020.
ITOPF Ltd (International Tanker Owners Pollution Federation Limited). (2011). Effects of oil pollution on the marine environment. Available at https://www.itopf.org/fileadmin/data/Documents/TIPS%20TAPS/TIP_13_Effects_of_Oil_Pollution_on_the_Marine_Environment.pdf. Accessed February 03, 2021.
Jyotirmay, M., Kumar, N. K., & Wagner, H. J. (2002). Energy and environmental correlation for renewable energy systems in India. *Energy Sources, Part A: Recovery, Utilization, and Environmental Effects, 24*(1), 19–26.
Kumar, A., & Kandpal, T. C. (2007). $CO_2$ emissions mitigation potential of some renewable energy technologies in India. *Energy Sources, Part A: Recovery, Utilization, and Environmental Effects, 29*(13), 1203–1214.
Kumar, A., Kumar, K., Kaushik, N., Sharma, S., & Mishra, S. (2010). Renewable energy in India: Current status and future potentials. *Renewable and Sustainable Energy Reviews, 14*, 2434–2442.

Kumar, C. R., & Majid, M. A. (2020). Renewable energy for sustainable development in India: Current status, future prospects, challenges, employment and investment opportunities. *Energy, Sustainability and Society, 10*(2), 1–36.

Kumar, S., Bhattacharyya, B., & Gupta, V. K. (2014). Present and future scenario in India. *Journal of the Institution of Engineers (India): Series B, 95*, 247–254.

Kumar, S., & Meena, R. L. (2017). Renewable energy sources-policies in India. *International Journal of Applied Environmental Sciences, 12*(2), 293–297.

Kumar, S., Rawat, M. K., & Gupta, S. (2019). An evaluation of current status of renewable energy sources in India. *International Journal of Innovative and Exploring Engineering, 8*(10), 1234–1239.

Lighty, J. S., Veranth, J. M., & Sarofim, A. F. (2000). Combustion aerosols: Factors governing their size and composition and implications to human health. *Journal of the Air and Waste Management Association, 50*(9), 1565–1618.

Lindsey, R. (2020). Climate change: Atmospheric carbon dioxide. Available at https://www.climate.gov/news-features/understanding-climate/climate-change-atmospheric-carbon-dioxide#:~:text=The%20global%20average%20atmospheric%20carbon,plus%20or%20minus%200.1%20ppm. Accessed February 9, 2021.

MNRE, Annual Report. (2018–19). Available at https://mnre.gov.in/img/documents/uploads/file_f-1608040317211.pdf. Accessed February 9, 2021.

MNRE-Ministry of New and Renewable Energy. (2021). Available at https://mnre.gov.in/. Retrieved February 9, 2021.

Oliveira, M. L. S., Marostega, F., Taffarel, S. R., Saikia, B. K., Waanders, F. B., DaBoit, K., Baruah, B. P., & Silva, L. F. O. (2014). Nano-mineralogical investigation of coal and fly ashes from coal-based captive power plant (India): An introduction of occupational health hazards. *Science of the Total Environment, 468–469*, 1128–1137.

Poornima, E. H., Rajadurai, M., Rao, T. S., Anupkumar, B., Rajamohan, R., Narasimhan, S. V., Rao, V. N. R., & Venugopalan, V. P. (2005). Impact of thermal discharge from a tropical coastal power plant on phytoplankton. *Journal of Thermal Biology, 30*, 307–316.

Prasad, A. K., Singh, R. P., & Kafatos, M. (2006). Influence of coal based thermal power plants on aerosol optical properties in the Indo-Gangetic basin. *Geophysical Research Letters, 33*, 1–4.

Rapier, R. (2020). Fossil fuels still supply 84 percent of world energy—and other eye openers from BP's annual review. Available at https://www.forbes.com/sites/rrapier/2020/06/20/bp-review-new-highs-in-global-energy-consumption-and-carbon-emissions-in-2019/?sh=6793290466a1. Accessed February 3, 2021.

Reddy, V. S., Kaushik, S. C., & Panwar, N. L. (2013). Review on power generation scenario of India. *Renewable and Sustainable Energy Reviews, 18*, 43–48.

Saur News Bureau. (2017). Government decides to discontinue Akshay Urja shops programme. Available at https://www.saurenergy.com/solar-energy-news/Government-decides-discontinue-akshay-urja-shops-programme. Accessed February 9, 2021.

Seaton, A., MacNee, W., Donaldson, K., & Godden, D. (1995). Particulate air pollution and acute health effects. *Lancet, 345*, 176–178.

Singh, S. (2020). Budget 2020: Major focus on PM KUSUM scheme to help farmers generate income from solar power. Available at https://energy.economictimes.indiatimes.com/news/renewable/budget-2020-major-focus-on-pm-kusum-scheme-to-help-farmers-generate-income-from-solar-power/73832492. Accessed February 9, 2021.

Subrahmanyam, D. S. (2013). Status of electric power generation in India with special emphasis on hydropower expansion. *International Journal of Renewable Energy and Environmental Engineering, 1*(1), 2348–2357.

TERI. (1998). *Teri energy data directory and year book 1998/99*. The Energy and Resource Institute.

The Economic Times-Rise. (2020). India's renewable energy ministry launches its new website to keep pace with time. https://economictimes.indiatimes.com/small-biz/productline/power-generation/indias-renewable-energy-ministry-launches-its-new-website-to-keep-pace-with-time/articleshow/74649515.cms. Accessed February 9, 2021.

Thirugnanasambandam, M., Iniyan, S., & Goic, R. (2010). A review of solar thermal technologies. *Renewable and Sustainable Energy Reviews, 1*, 312–322.
Tripathi, L., Mishra, A. K., Dubey, A. K., Tripathi, C. B., & Baredar, P. (2016). Renewable energy: An overview on its contribution in current energy scenario of India. *Renewable and Sustainable Energy Reviews, 60*, 226–233.
Xiao, C., Luo, H., Tang, R., & Zhong, H. (2004). Solar thermal utilization in China. *Renewable Energy, 29*, 1549–1556.
Zhao, Y., Wang, S., Aunan, K., Seip, H. M., & Hao, J. (2006). Air pollution and lung cancer risks in China—A meta-analysis. *Science of the Total Environment, 366*, 500–513.

The manufacturer's authorised representative in the EU is Springer Nature Customer Service Centre GmbH, Europaplatz 3, 69115 Heidelberg, Germany. If you have any concerns regarding our products, please contact ProductSafety@springernature.com

Printed and bound by CPI Group (UK) Ltd, Croydon, CR0 4YY

30/09/2026

02225241-0009